Metaheuristic Algorithms in Optimal Design of Engineering Problems

Metaheuristic Algorithms in Optimal Design of Engineering Problems

Guest Editors

Łukasz Knypiński
Ramesh Devarapalli
Marcin Kaminski

MDPI

Basel • Beijing • Wuhan • Barcelona • Belgrade • Novi Sad • Cluj • Manchester

Guest Editors

Łukasz Knypiński
Faculty of Control, Robotics and Electrical Engineering
Poznan University of Technology
Poznan
Poland

Ramesh Devarapalli
Institute of Chemical Technology
IndianOil Odisha Campus
Bhubaneswar
India

Marcin Kaminski
Department of Electrical Machines, Drives and Measurements
Wroclaw University of Science and Technology
Wroclaw
Poland

Editorial Office

MDPI AG
Grosspeteranlage 5
4052 Basel, Switzerland

This is a reprint of the Special Issue, published open access by the journal *Algorithms* (ISSN 1999-4893), freely accessible at: https://www.mdpi.com/journal/algorithms/special_issues/J620881IOY.

For citation purposes, cite each article independently as indicated on the article page online and as indicated below:

Lastname, A.A.; Lastname, B.B. Article Title. *Journal Name* **Year**, *Volume Number*, Page Range.

ISBN 978-3-7258-5075-4 (Hbk)
ISBN 978-3-7258-5076-1 (PDF)
https://doi.org/10.3390/books978-3-7258-5076-1

Contents

About the Editors

Łukasz Knypiński

Łukasz Knypinski received his MS degree, PhD, and DSc in Electrical Engineering from Poznan University of Technology, Poland, in 2007, 2016, and 2024, respectively. Currently, he is working at Poznan University of Technology, in the Mechatronics and Electrical Machines Division, Institute of Electrical Engineering and Electronics as an Assistant Professor. His research interests are electrical machines, in particular, permanent magnet brushless DC machines, permanent magnet synchronous motors and metaheuristic optimization algorithms. He has published over 100 conference and journal papers on electrical machines, electromagnetics and optimization.

Ramesh Devarapalli

Ramesh Devarapalli is an Assistant Professor at the Institute of Chemical Technology—IndianOil Odisha Campus (ICT-IOC Bhubaneswar, India), specializing in Electrical, Electronics, and Instrumentation Engineering. He holds a Ph.D. from the Indian Institute of Technology (ISM), Dhanbad (2021), an MBA from Andhra University (2019), and an M.Tech. from the Indian Institute of Technology BHU, Varanasi (2012). His primary research interests include renewable energy sources, artificial intelligence and optimization algorithms, and hydrogen energy, with a focus on fuel cells. Dr. Devarapalli has published over 100 research articles and is recognized for his contributions to the fields of power systems and electrical engineering. He is actively involved in teaching courses related to his areas of expertise and continues to advance research in sustainable energy technologies.

Marcin Kaminski

Marcin Kaminski is a faculty member in the Department of Electrical Machines, Drives and Measurements at the Faculty of Electrical Engineering, Wroclaw University of Science and Technology, Poland. His research interests focus on electrical machines, drives, and measurement systems, with contributions to both academic research and teaching in these areas. Dr. Kaminski is actively involved in advancing the field through research publications and engagement with the academic community.

algorithms

MDPI

Editorial

Metaheuristic Algorithms in Optimal Design of Engineering Problems

Łukasz Knypiński [1,*], Ramesh Devarapalli [2,*] and Marcin Kamiński [3]

1 Faculty of Automatic Control, Robotic and Electrical Engineering, Poznan University of Technology, 60-965 Poznan, Poland
2 Department of Electrical/Electronics and Instrumentation Engineering, Institute of Chemical Technology, Indianoil Odisha Campus, Bhubaneswar 751013, India
3 Department of Electrical Machines, Drives and Measurements, Faculty of Electrical Engineering, Wroclaw University of Science and Technology, 50-372 Wroclaw, Poland; marcin.kaminski@pwr.edu.pl
* Correspondence: lukasz.knypinski@put.poznan.pl (Ł.K.); dr.r.devarapalli@gmail.com (R.D.)

1. Introduction

Metaheuristic optimization algorithms (MOAs) are widely used to optimize the design process of engineering problems [1,2]. MOAs are successfully applied to the optimal design of electromagnetic devices [3], power dispatch problems [4], the search for the tune parameters of controllers [5] for different control systems, and many other applications [6,7]. Among metaheuristic algorithms, the most commonly used [8] are the Particle Swarm Optimization (PSO) algorithm, developed in 1995 [9], and the grey wolf optimization (GWO) algorithm, developed in 2014 [10].

Metaheuristic optimization algorithms are still intensively developed in many research centers all over the world [11]. Nowadays, researchers working in the area of the optimal design of technical objects have a huge number of various optimization algorithms. The algorithms have different properties and are characterized by different convergence. The final solution depends on the random coefficient, and metaheuristic characteristic parameters vary for different optimization algorithms. Due to the above reasons, the optimization process should be repeated several times for different starting populations [12–14].

In order to improve efficiency, modifications of classical methods are often developed [15]. The aim of developing modifications is to reduce the number of objective function calls and to increase the reliability of the optimization algorithm [16]. An increasing number of manuscripts on heuristic optimization methods are concerned with hybrid optimization algorithms [17,18]. Hybrid algorithms can consist of two, three, or even more different algorithms. In the available global literature, there are various algorithm architectures: parallel [19], serial [20], or even mixed [21].

Very often, after the proper selection of an optimization algorithm, it is necessary to execute the adaptation of the algorithm to the optimized system or optimized technical object [22–24]. Proper adaptation of the algorithm by correctly selecting the characteristic parameters of the selected algorithm can also improve the efficiency of the algorithm.

In this Special Issue, entitled "Metaheuristic Algorithms in Optimal Design of Engineering Problems", the Editorial Office received 15 submissions from researchers worldwide. After a rigorous peer-review process, nine manuscripts were accepted for publication.

The Special Issue is concerned with the application of metaheuristic algorithms to solving various optimization problems. The authors publishing their works presented interesting approaches for researchers working with optimization algorithms, especially a hybrid whale optimization algorithm featuring improved genetic characteristics and a multi-objective majority–minority cellular automata algorithm.

2. Special Issue Contribution

Contribution 1: An extension of the Majority–minority Cellular Automata Algorithm, referred to as the Multi-objective Majority–minority Cellular Automata Algorithm, or MOMmCAA, to solve multi-objective optimization problems. By including the repository management and controlling the density of the multi-objective search space, MOMmCAA can optimize multiple objectives without a processing time. The performance of MOMmCAA was compared to established multi-objective algorithms on benchmark test sets and real-world engineering problems; in the results, MOMmCAA was comparable, giving promise for real applications in complex optimization problems.

Contribution 2 develops the work performed in robotics, focusing on humanoid robots that sense and act on a specific task in any industry. The paper discusses a hybrid H∞/sliding mode controller optimized using Particle Swarm Optimization to control a triple inverted pendulum, which has been considered a benchmark for driving joints, stability, and balance. The optimized controller could demonstrate robust performance across different types of perturbations with an average error of 0.053 degrees and a steady torque range of 0.13–0.621 N·m.

The third contribution contrasted heuristic versus metaheuristic algorithms (in particular, Steiglitz–McBride, Jaya, the genetic algorithm, and the Grey Wolf Optimizer) to maximize the fit of the optimal parameters found by a dynamic model to the current and angular velocity responses of DC motors. The study is presented with the most accurate parametric estimation average mean squared error of 0.43% but more computation expense; to the contrary, the Stieglitz–McBride algorithm, with an average MSE of 3.32%, is more computationally efficient. Overall, it appears that if precision is the priority, then GWO is likely a better option. For balance between performance and efficiency, however, the Stieglitz–McBride heuristic may provide an alternative with a more efficient solution, knowing that these algorithms' performance will also depend on the particular error functions.

Contribution 4 proposes a new PSO algorithm that exploits the Sobol and Halton random number samplings with the performance comparison of traditional Monte Carlo-based PSO. For nine benchmark problems plus TSP, in all cases, the variant PSO improves over PSO, especially with better iteration efficiency via Sobol-based PSO and decreased computational times. These results indicate that using Sobol and Halton methods for generating random numbers makes optimization algorithms more efficient.

Contribution 5: Hybrid Improved Whale Optimization Algorithm with Enhanced Genetic Properties for Optimal Application Mapping on a 2D Network on Chip (NoC) Platform. This paper introduces a hybrid of an advanced whale optimization algorithm and an improved genetic algorithm that can perform better in power reduction, energy consumption, and latency than all other state-of-the-art algorithms for various benchmarks and real-time applications. Conclusions derived from the obtained results indicate the efficiency of the proposed hybrid approach. It shows superiority in converging better on both synthetic and real-world task graphs.

The cost-reducing benefit paid by the ridesharing concept encourages more riders to ride in this means of transportation. Contribution 6 is a discount-guaranteed ridesharing concept that assures that the kept discount should always be a minimum to help drivers and riders increase the latter's acceptability. The researchers developed a new metaheuristic algorithm based on a differential evolution incorporating a self-adaptation scheme. They applied this to the DGRP with optimal performance and the fastest convergence compared with existing algorithms.

Contribution 7 recommends a new modified fractional order proportional integral derivative (FOPID) controller aimed at managing frequency stability issues due to the integration of intermittent wind turbines in power systems. A cascaded FOPD presents the proposed method–FOPID controller for coordinated LFC and SMES, optimized with the DOSA for its parameters. The comparative scenario of four scenarios showed how the proposed control technique would outsmart many of the state-of-the-art methods while providing satisfactory responses to such load changes, disturbances posed by wind turbines, or uncertainty due to parameters from the system.

Contribution 8 relates to this issue that partial shading conditions on the photovoltaic panel cause the presence of more than one peak power point at the P-V curve, degrading the efficiency of even the most superior MPPT algorithm adopted for the optimum extraction of power from this array. This study discusses the performance optimization through a single-objective nonlinear optimization problem by using some of the latest metaheuristic algorithms like Cat Swarm Optimization (CSO), grey wolf optimization (GWO), and a newly proposed Chimp Optimization algorithm (ChOA). The MATLAB/SIMULINK results indicate that all metaheuristic methods converge towards the global Maximum Power Point (MPP), and the ChOA shows better performance than the existing algorithms.

Contribution 9 addresses the Transportation Problem (TP) as a unique linear programming problem to minimize the costs of distribution between multiple sources and different destinations through two new adaptations of the Particle Swarm Optimization (PSO) algorithm: Trigonometric Acceleration Coefficients PSO (TrigAc-PSO) and Four Sectors Varying Acceleration Coefficients PSO (FSVAC-PSO). Thirty-two problems of varied sizes have been tested on extended experimental tests that ascertain new PSO variations that outperform classical exact techniques, such as Vogel's Approximation Method and Total Differences Method, as well as already developed PSO variants: Decreasing Weight PSO. The solutions are two efficient and practical approaches to the Transportation Problem: TrigAc-PSO and FSVAC-PSO.

3. Final Remarks

These papers highlight the meaningful progress in the development of metaheuristic optimization algorithms for the optimal design of engineering problems. This Special Issue offers manuscripts presenting interesting algorithms for researchers, students, and practitioners. The Special Issue Editors thank all authors, reviewers, and the editorial team for making this Special Issue possible.

List of Contributions

1. Seck-Tuoh-Mora, J. C.; Hernandez-Hurtado, U.; Medina-Marín, J.; Hernández-Romero, N.; Lizárraga-Mendiola, L. Multi-Objective Majority–Minority Cellular Automata Algorithm for Global and Engineering Design Optimization. *Algorithms* **2024**, *17*, 433. https://doi.org/10.3390/a17100433.
2. Shafeek, Y.A.; Ali, H.I. Application of Particle Swarm Optimization to a Hybrid H_∞/Sliding Mode Controller Design for the Triple Inverted Pendulum System. *Algorithms* **2024**, *17*, 427. https://doi.org/10.3390/a17100427.
3. Munciño, D.M.; Damian-Ramírez, E.A.; Cruz-Fernández, M.; Montoya-Santiyanes, L.A.; Rodríguez-Reséndiz, J. Metaheuristic and Heuristic Algorithms-Based Identification Parameters of a Direct Current Motor. *Algorithms* **2024**, *17*, 209. https://doi.org/10.3390/a17050209.
4. Kannan, S.K.; Diwekar, U. An Enhanced Particle Swarm Optimization (PSO) Algorithm Employing Quasi-Random Numbers. *Algorithms* **2024**, *17*, 195. https://doi.org/10.3390/a17050195.
5. Saleem, S.; Hussain, F.; Baloch, N.K. IWO-IGA—A Hybrid Whale Optimization Algorithm Featuring Improved Genetic Characteristics for Mapping Real-Time Applications onto 2D Network on Chip. *Algorithms* **2024**, *17*, 115. https://doi.org/10.3390/a17030115.
6. Hsieh, F.-S. A Self-Adaptive Meta-Heuristic Algorithm Based on Success Rate and Differential Evolution for Improving the Performance of Ridesharing Systems with a Discount Guarantee. *Algorithms* **2024**, *17*, 9. https://doi.org/10.3390/a17010009.
7. Amiri, F.; Eskandari, M.; Moradi, M.H. Improved Load Frequency Control in Power Systems Hosting Wind Turbines by an Augmented Fractional Order PID Controller

Optimized by the Powerful Owl Search Algorithm. *Algorithms* **2023**, *16*, 539. https://doi.org/10.3390/a16120539.
8. Nagadurga, T.; Devarapalli, R.; Knypiński, Ł. Comparison of Meta-Heuristic Optimization Algorithms for Global Maximum Power Point Tracking of Partially Shaded Solar Photovoltaic Systems. *Algorithms* **2023**, *16*, 376. https://doi.org/10.3390/a16080376.
9. Aroniadi, C.; Beligiannis, G.N. Applying Particle Swarm Optimization Variations to Solve the Transportation Problem Effectively. *Algorithms* **2023**, *16*, 372. https://doi.org/10.3390/a16080372.

Author Contributions: Special Issue Editorial by Ł.K., R.D. and M.K. All authors have read and agreed to the published version of the manuscript.

Funding: This research received no external funding.

Conflicts of Interest: The authors declare no conflicts of interest.

References

1. Tomar, V.; Bansal, M.; Singh, P. Metaheuristic Algorithms for Optimization: A Brief Review. *Eng. Proc.* **2023**, *59*, 238. [CrossRef]
2. Cui, E.H.; Zhang, Z.; Chen, C.J.; Wong, W.K. Applications of nature-inspired metaheuristic algorithms for tackling optimization problems across disciplines. *Sci. Rep.* **2024**, *14*, 9403. [CrossRef] [PubMed]
3. Knypiński, Ł. Performance analysis of selected metaheuristic optimization algorithms applied in the solution of an unconstrained task. *COMPEL* **2022**, *41*, 1272–1284. [CrossRef]
4. Granados, J.; Uturbey, W.; Valadão, R.L.; Vasconcelos, J.A. Many-objective optimization of real and reactive power dispatch problems. *Int. J. Electr. Power Energy Syst.* **2023**, *146*, 108725. [CrossRef]
5. Saha, A.; Chiranjeevi, T.; Devarapalli, R.; Ram Babu, N.; Dash, P.; Garcìa Màrquez, F.P. Analysis of multiple-area renewable integrated hydro-thermal system considering artificial rabbit optimized PI (FOPD) cascade controller and redox flow battery. *Arch. Control. Sci.* **2023**, *33*, 861–884. [CrossRef]
6. Wang, Z.; Yuan, W. Maximum power point tracking controller for photovoltaic system based on chaos quantum particle swarm optimization–moth-flame optimization hybrid model. *Arch. Electr. Eng.* **2024**, *73*, 3–663. [CrossRef]
7. Kommadath, R.; Maharana, D.; Kotecha, P. A metaheuristic-based efficient strategy for multi-unit production planning with unique process constraints. *Appl. Soft Comput.* **2023**, *134*, 109871. [CrossRef]
8. Furio, C.; Lamberti, L.; Pruncu, C.I. An Efficient and Fast Hybrid GWO-JAYA Algorithm for Design Optimization. *Appl. Sci.* **2024**, *14*, 9610. [CrossRef]
9. Kennedy, J.; Eberhart, R. Particle swarm optimization. In Proceedings of the International Conference on Neural Networks, Perth, WA, Australia, 27 November–1 December 1995. [CrossRef]
10. Mirjalili, S.; Mohammad Mirjalili, S.; Lewis, A. Grey Wolf Optimizer. *Adv. Eng. Softw.* **2014**, *69*, 46–61. [CrossRef]
11. Velasco, L.; Guerrero, H.; Hospitaler, A. A Literature Review and Critical Analysis of Metaheuristics Recently Developed. *Arch. Computat. Methods Eng.* **2024**, *31*, 125–146. [CrossRef]
12. Thirunavukkarasu, N.; Lala, H.; Sawle, Y. Reliability index based optimal sizing and statistical performance analysis of stand-alone hybrid renewable energy system using metaheuristic algorithms. *Alex. Eng. J.* **2023**, *74*, 387–413. [CrossRef]
13. Nassef, A.M.; Abdelkareem, M.A.; Maghrabie, H.M.; Baroutaji, A. Review of Metaheuristic Optimization Algorithms for Power Systems Problems. *Sustainability* **2023**, *15*, 9434. [CrossRef]
14. Tang, K.; Meng, C. Particle Swarm Optimization Algorithm Using Velocity Pausing and Adaptive Strategy. *Symmetry* **2024**, *16*, 661. [CrossRef]
15. Knypiński, Ł. Constrained optimization of line-start PM motor based on the gray wolf optimizer. *Eksploat. I Niezawodn.–Maint. Reliab.* **2021**, *23*, 1–10. [CrossRef]
16. Pu, R.; Li, S.; Zhou, P.; Yang, G. Improved Chimp Optimization Algorithm for Matching Combinations of Machine Tool Supply and Demand in Cloud Manufacturing. *Appl. Sci.* **2023**, *13*, 12106. [CrossRef]
17. Shehadeh, H.A. A hybrid sperm swarm optimization and gravitational search algorithm (HSSOGSA) for global optimization. *Neural Comput. Appl.* **2021**, *33*, 11739–11752. [CrossRef]
18. Ezzeldin, R.; Zelenakova, M.; Abd-Elhamid, H.F.; Pietrucha-Urbanik, K.; Elabd, S. Hybrid Optimization Algorithms of Firefly with GA and PSO for the Optimal Design of Water Distribution Networks. *Water* **2023**, *15*, 1906. [CrossRef]
19. Knypiński, Ł. A novel hybrid cuckoo search algorithm for optimization of a line-start PM synchronous motor. *Bull. Pol. Acad. Sci. Tech. Sci.* **2023**, *71*, 1–8. [CrossRef]
20. Geetha, M.; Chandra Guru Sekar, R.; Marichelvam, M.K.; Tosun, Ö. A Sequential Hybrid Optimization Algorithm (SHOA) to Solve the Hybrid Flow Shop Scheduling Problems to Minimize Carbon Footprint. *Processes* **2024**, *12*, 143. [CrossRef]
21. Kelner, V.; Capitanescu, F.; Léonard, O.; Wehenkel, L. A hybrid optimization technique coupling an evolutionary and a local search algorithm. *J. Comput. Appl. Math.* **2008**, *215*, 448–456. [CrossRef]
22. Tekerek, A.; Dortreler, M. The Adaptation Gray Wolf Optimizer to Data Clustering. *J. Polytech.* **2022**, *25*, 1761–1767. [CrossRef]

23. Henrichs, E.; Lesch, V.; Straesser, M.; Kounev, S.; Krupitzer, C. A literature review on optimization techniques for adaptation planning in adaptive systems: State of the art and research directions. *Inf. Softw. Technol.* **2024**, *49*, 106940. [CrossRef]
24. Marchetti, A.G.; François, G.; Faulwasser, T.; Bonvin, D. Modifier Adaptation for Real-Time Optimization—Methods and Applications. *Processes* **2016**, *4*, 55. [CrossRef]

algorithms

MDPI

Article

An Enhanced Particle Swarm Optimization (PSO) Algorithm Employing Quasi-Random Numbers

Shiva Kumar Kannan [1] and Urmila Diwekar [2,*]

[1] Ridge High School, Basking Ridge, Bernards, NJ 07920, USA; shivakannankumar@gmail.com
[2] Vishwamitra Research Institute, Crystal Lake, IL 60012, USA
* Correspondence: urmila@vri-custom.org; Tel.: +1-(630)-886-3047

Abstract: This paper introduces an innovative Particle Swarm Optimization (PSO) Algorithm incorporating Sobol and Halton random number samplings. It evaluates the enhanced PSO's performance against conventional PSO employing Monte Carlo random number samplings. The comparison involves assessing the algorithms across nine benchmark problems and the renowned Travelling Salesman Problem (TSP). The results reveal consistent enhancements achieved by the enhanced PSO utilizing Sobol/Halton samplings across the benchmark problems. Particularly noteworthy are the Sobol-based PSO improvements in iterations and the computational times for the benchmark problems. These findings underscore the efficacy of employing Sobol and Halton random number generation methods to enhance algorithm efficiency.

Keywords: enhanced PSO; SOBOL; Halton; quasi-random numbers

1. Introduction

Particle Swarm Optimization (PSO) is a metaheuristic algorithm for optimization problems. PSO was first introduced in 1995 by James Kennedy and Russell Eberhart [1]. The algorithm is based on the concept of social behavior, where particles (potential solutions) move towards the optimal solution through interactions with other particles in the search space. PSO has been widely used in various fields, including engineering, science, and finance, due to its simplicity, robustness, and efficiency. Despite its success, PSO suffers from several limitations. One of the main limitations is its slow convergence rate, which can be attributed to the premature convergence [2] of the particles towards local optima. This issue can be addressed by introducing efficient improvement techniques in PSO. Several enhancement ideas have been proposed in the past to improve the convergence rate of the PSO algorithm, and they are listed below. Firstly, the inertia weight technique was suggested by Russell Eberhart and Ying Shi [3]. The inertia weight technique is a well-known approach for enhancing the convergence speed of PSO. The inertia weight is used to control the movement of particles in the search space. The idea is to maintain a balance between exploration and exploitation of the search space. The inertia weight is updated at each iteration based on a predefined formula, which controls the speed and direction of particle movement. Various formulas have been proposed for updating the inertia weight, such as linear, nonlinear, and adaptive. The choice of the inertia weight formula depends on the optimization problem and the PSO parameters. Second, the concept of a mutation operator was proposed [4]. A mutation operator is a powerful tool for enhancing the diversity of the PSO population. The mutation operator randomly modifies the position of a particle to generate a new solution in the search space. This operation can prevent premature convergence by introducing new solutions that may lead to better solutions. The mutation operator can be applied at different stages of the PSO algorithm, such as before or after the velocity update. Similar to the mutation operator, another operator known as the crossover operator has also been applied to the PSO algorithm [5–7]. This concept involves the mixture of the attributes of different solutions to gain exploration and to achieve high diversity in potential results to avoid premature convergence. Third, the opposition-based learning technique was suggested [8]. Opposition-based learning (OBL) is a technique that uses the opposite of the current best solution to generate new solutions. The idea behind OBL is that the opposite of the best solution may represent a good direction for exploration in the search space. OBL can improve the diversity and convergence speed of PSO by generating new solutions that are different from those of the current population. Fourth, hybridization with other metaheuristics has been proposed [9]. Hybridization with other metaheuristics is a common approach for improving the efficiency of PSO. The idea is to combine the strengths of different metaheuristics to overcome their weaknesses. For example, PSO can be combined with genetic algorithms (GA), simulated annealing (SA), or ant colony optimization (ACO). The hybridization approach can enhance the exploration and exploitation capabilities of PSO, leading to better solutions in less time. Fifth, dynamic parameter tuning was presented [9]. The PSO parameters, such as the swarm size, maximum velocity, and acceleration coefficients, significantly impact the algorithm's performance. Dynamic parameter tuning is a technique that adjusts the PSO parameters based on the search history during the optimization process. The idea is to adapt the PSO parameters to the problem characteristics and the search progress to improve the convergence speed and solution quality. In conclusion, efficient improvement techniques in PSO can enhance the algorithm's convergence speed and solution quality. The approaches discussed in this paper [9], including the inertia weight technique, mutation operator, opposition-based learning, hybridization with other metaheuristics, and dynamic parameter tuning, can be used individually or in combination to address the limitations of PSO. Tareq M. Shami and a team [10] of researchers conducted a comprehensive survey on PSO. The survey discusses techniques such as varying the inertia weight and hybridizations, which are discussed above. The survey also states that the ability of PSO to be hybridized with other optimization algorithms has contributed to its popularity. Another technique described in the survey is velocity clamping [10], a technique introduced by Eberhart and Kennedy. Velocity clamping involves setting bounds for the values of the velocities of the particles in all the dimensions. Another approach to improving the efficiency of the PSO algorithm discussed in this paper [10] is varying the controlling parameters, such as using the varying inertia weight technique in which inertia weight changes throughout the optimization process, or acceleration coefficient techniques in which the two constant controlling parameters for PSO other than the inertia, are chosen in different ways to yield optimal solutions while evading premature convergence. Many other approaches have been discussed both in the survey and elsewhere. The choice of approach depends on the problem characteristics and the available computational resources. However, most of these approaches can provide problem-dependent solution methods. In this paper, we proposed a new approach to replace the random numbers used for this method with quasi-random numbers [11–13], like Halton and Sobol, by maintaining the k-dimensional uniformity of these quasi-random numbers. This not only provides a generalized approach to any optimization problem, but this method can be used in conjunction with the earlier enhancement techniques like the inertia weight technique, mutation operator, opposition-based learning, hybridization with other metaheuristics, and dynamic parameter tuning. In this research, two enhanced versions of PSO (one using Sobol random numbers and the other using Halton random numbers) were proposed with the intention of speeding up the convergence of the standard PSO algorithm. To test the efficiency improvement of the two proposed enhancements of the standard PSO algorithm, the number of iterations taken to achieve the optimum of the well-known cigar, ellipsoid, and paraboloid functions [14], along with the number of iterations taken to obtain an optimal path for the famous Travelling Salesman Problem (TSP) [15–18], were noted. Following this, improvement in terms of the optimum of the objective function and the number of iterations needed to reach the global optimum were calculated for both the PSO enhanced with Sobol random number samplings and the PSO enhanced with Halton random number samplings, with respect to the standard PSO, which uses Monte Carlo random number samplings. All the results for each benchmark function and TSP unanimously show efficiency improvement due to the use of the Sobol and Halton sequences. Additionally, we noted that the more decision variables an optimization problem has, the improvement due to Sobol and Halton sequences increases. In conclusion, both the enhancements of the standard PSO presented in this research, one utilizing Sobol random numbers and the other utilizing Halton random numbers, consistently show efficiency improvement and a better optimum, meaning that they successfully have increased the speed of convergence of the standard PSO algorithm.

In addition, we have also compared our enhanced PSO with the SALP meta-heuristic variant [19]. This is a recent algorithmic approach developed to improve the convergence rate. Our enhancement is compared with the SALP to see whether the approach of avoiding clusters in random number generation can make the enhanced PSO algorithm perform better than the SALP algorithm. The algorithms are compared for the four benchmark problems.

The time-varying inertia factor was introduced to improve the converge performance of the PSO [20,21]. The authors in [22] introduced a time-varying acceleration coefficient in addition to the time-varying inertia factor. They introduced a PSO concept called a self-organizing hierarchical particle swarm optimizer with a time-varying acceleration coefficient. For the velocity update in PSO, only the social part and cognitive part were considered, and to avoid stagnation in the search space, a time-varying mutation step size was used as well.

An adaptive particle swarm optimization (APSO) that features better search efficiency than the classical particle swarm optimization (PSO) is presented in [23]. It was engineered to perform a global search over the entire search space with faster convergence speed. The APSO algorithm was carried out in two main steps. In the first step, the algorithm evaluated the population distribution and particle fitness based on which a real-time evolutionary state was estimated. This enabled the automatic control of inertia weight, acceleration coefficients, and other algorithmic parameters in real-time to improve the search efficiency and convergence speed in the subsequent step.

The multimodal PSO approach proposed in [24] addressed the issues associated with poor local search ability and the requirement of prior knowledge to initialize the PSO algorithm parameters. The authors in [24] proposed an optimizer based on a distance-based locally informed PSO variant. The algorithm eliminated the need to specify the *niche* parameters in the PSO and enhanced its fine-searching capabilities. To guide the search direction of the particles, each particle used local best information instead of the global best as in the conventional PSO, and the neighborhood of the particle was measured using the Euclidean distance to perform the local best search.

In [25], a hybrid algorithm that combined particle swarm optimization with simulated annealing behavior (SA-PSO) was proposed. The SA-PSO algorithm takes advantage of the good solution quality provided by the simulated annealing and fast searching ability inherent to the particle swarm optimization. It was concluded that the hybrid algorithm could have higher efficiency, better quality and faster convergence speed than conventional PSO variants.

An economic environmental hydrothermal scheduling problem classified as a multi-objective nonlinear constrained optimization was solved using PSO [26]. The algorithm adopted an elite archive set to conserve Pareto optimal solutions and provide multiple evolutionary directions for individuals, while neighborhood searching and chaotic mutation strategies enhance the search capability and diversity of the population. The PSO algorithm also incorporated a constraint handling scheme designed to adjust the constraint violation of hydro and thermal plants.

To avoid parameter selection and overcome the premature convergence problem in the PSO optimization, an adaptive fuzzy particle swarm optimization based on a fuzzy inference system, which incorporated a variable neighborhood search strategy and hybrid evolution, was proposed in [27] and applied to the parameter estimation of nonlinear Hydro Turbine Regulation Systems. The results concluded that the new algorithm's parameter error and the objective function were significantly smaller than the other algorithms. The estimated model could accurately reflect the dynamic characteristics of the real system, proving that the fuzzy PSO variant was an effective and efficient parameter estimation method.

Support vector machine (SVM)-based classifiers need accurate parameter estimation for high-accuracy classification, and in [28], an improved variant of the PSO is used to estimate the SVM parameters. Specifically, dynamic adjustment of inertia is proposed to optimize the parameters of the SVM. The computation of the inertia weight in the PSO proposed in [28] involves the nonlinear reduction in the inertia weight as the number of iterations increases. In particular, the introduction of random function inertia weight in the PSO avoids falling into the local extremum and, at the same time, increases diversity and the global searching ability during the optimization process.

In addition to the above-mentioned PSO algorithm comparisons, we compare the iteration efficiency of the enhanced PSO approach employing the Sobol sequence using the random inertia PSO variant for the benchmark problems.

2. Materials and Methods

Particle Swarm Optimization

There are countless examples of swarms in the real world, ranging from flocking birds to hunting wolves. When searching for nourishment, the individuals of these swarms, called particles to understand PSO, begin random exploration and then start gravitating towards the findings of other swarm individuals. While following signs of nourishment and searching randomly in space, these particles move towards their objective through knowledge of their discoveries and discoveries of the swarm. Similarly, in the PSO algorithm, proposed decision variable sets gravitate towards the optimal findings of each other, themselves, and the whole swarm together to achieve the globally optimal set of decision variables, yielding the optimal function value. The sets of decision variables are called position vectors. They are updated by vectors called velocity vectors (based on the physics principle that change in position is proportional to the velocity), by randomness, and by attraction towards their personal best sets of decision variables and the unanimous global best set of decision variables found by the whole swarm. The global optimum of the function is achieved through gradual movement towards optimal findings of the swarm. The following are the steps of the PSO algorithm.

- Initialize Parameters:
  - Define the population size (number of particles), Np; Define the number of decision variables (dimension), D;
  - Define the maximum number of iterations, T;
  - Define the inertia weight, ω;
  - Define acceleration constants: cognitive and social, c_1, c_2;
  - Initialize the position and velocity of each particle randomly within the search space.

The initially proposed set of solutions is grouped together to form the population matrix, which is written as follows:

$$\begin{bmatrix} x_0^{00} & \cdots & x_0^{0(D-1)} \\ \vdots & \ddots & \vdots \\ x_0^{(Np-1)0} & \cdots & x_0^{(Np-1)(D-1)} \end{bmatrix}$$

Each row in this matrix represents a potential decision variable vector intended to optimize the objective function. Each element in a row is the position with respect to a particular dimension of the particle that corresponds with that row. The subscript for each element is 0, as these values are the initial values in the matrix (0th iteration).

The velocity matrix can be represented as follows:

$$\begin{bmatrix} v_0^{00} & \cdots & v_0^{0(D-1)} \\ \vdots & \ddots & \vdots \\ v_0^{(Np-1)0} & \cdots & v_0^{(Np-1)(D-1)} \end{bmatrix}$$

This represents the velocity of each particle in all the dimensions. The velocity matrix changes the position matrix, and that is inspired by the physics concept that displacement is proportional to velocity. The superscript j is the index of the dimension.

- Set the best-known position for each particle n, P_n^{ij} to its initial position.

$$P^j = x^{0j}$$

- Evaluate Fitness:

  Evaluate the fitness (objective function value) for each particle based on its current position.
- Update the personal best position for each particle G_n^j if its current fitness is better than its previous best fitness.
- Update Global Best:
  - Determine the particle with the best fitness among all particles in the swarm, P_n^*.
  - Update the global best position with the particle's position with the best fitness, G_n^*.
  - Update Velocities and Positions:

  For each particle, update its velocity and position based on the following formulae.

The velocity of each particle at iteration n is updated according to the equation given below.

$$\begin{aligned} v_{n+1}^{ij} = \omega \times v_n^{ij} + c_1 \times r_1 \times \left(P_n^j - X_n^{ij}\right) + \\ c_2 \times r_2 \times \left({G_n}^j - X_n^{ij}\right) \end{aligned} \tag{1}$$

where r_1 and r_2 are random numbers. So, we have two random numbers per variable.

The corresponding position is updated according to the following equation:

$${x_{n+1}}^{ij} = {x_n}^{ij} + v_{n+1}^{ij} \tag{2}$$

- Check Stopping Criteria:
  - If the maximum number of iterations is reached or a satisfactory solution is found, stop the algorithm.
  - Otherwise, go back to step 2 and repeat the process.
- Output:

  Return the global best position as the solution to the optimization problem.

Quasi-random sequence enhancements

In this work, we hypothesize that the success of PSO depends on the choice of appropriate random samples. At each iteration of PSO, two random numbers are generated for each decision variable, as shown in Equation (1). These random numbers are computer-generated random numbers and are called pseudorandom numbers or Monte Carlo random numbers. A sequence of random numbers must have two essential properties: uniformity, i.e., they are equally probable everywhere, and independence, i.e., the current value of a random variable has no relation with the previous values. Figure 1 shows the two random variables generated using a computer (Monte Carlo or pseudorandom numbers) with samples equal to 100. As seen in the figure, for uniformity, the points should be equally distributed; that is not the case here. We need more samples to cover the points equally in that 2-dimensional space. This means more iterations of the algorithm. To circumvent this problem and to increase the efficiency of PSO, we are presenting a construct based on quasi-random numbers. Some well-known quasi-random sequences are Halton, Hammersley, Sobol, Faure, Korobov, and Neiderreiter [11–13]. The choice of an appropriate quasi-Monte Carlo sequence is a function of discrepancy. Discrepancy is a quantitative measure of the deviation of the sequence from the uniform distribution. Therefore, it is desirable to choose a low discrepancy sequence. The Halton, SOBOL, and Hammersley are some examples of low-discrepancy sequences. Here, we are working with the two sequences, Halton and SOBOL, as described below.

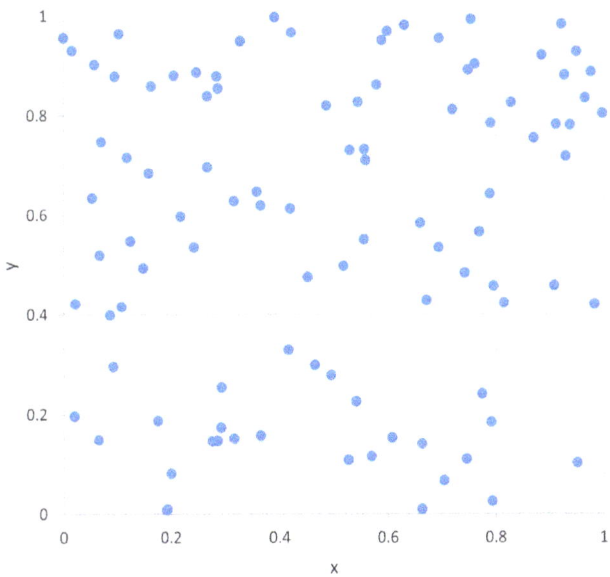

Figure 1. Two-dimensional pseudorandom numbers (100 points).

Halton Sequence Points:

The design of Halton points is given below. Any integer n can be written in radix-R notation (R is an integer) as follows:

$$n \equiv n_m n_{m-1} \ldots n_2 n_1 n_0 \tag{3}$$

$$n = n_0 + n_1 R + n_2 R^2 + \ldots + n_m R^m \tag{4}$$

where $m = [\log_R n] = [\ln n / \ln R]$ (the square brackets denote the integral part). A unique fraction between 0 and 1 called the inverse radix number can be constructed by reversing the order of the digits of n around the decimal point as follows:

$$\varphi_R(n) = n_0 n_1 n_2 \ldots n_m = n_0 R^{-1} + n_1 R^{-2} + \ldots + n_m R^{-m-1} \tag{5}$$

The Halton points on a k-dimensional cube are given by the following sequence:

$$\vec{z_k}(n) = \left(\varphi_{R_1}(n), \varphi_{R_2}(n), \ldots, \varphi_{R_{k-1}}(n)\right),\ n = 1, 2, \ldots, N+1 \tag{6}$$

where $R_1,\ R_2, \ldots R_{k-1}$ are the first $k-1$ prime numbers. The Halton points are $\vec{x_k}(n) = 1 - \vec{z_k}(n)$.

Figure 2 shows 2-dimensional Halton points (100 samples), which shows better uniformity than Figure 1.

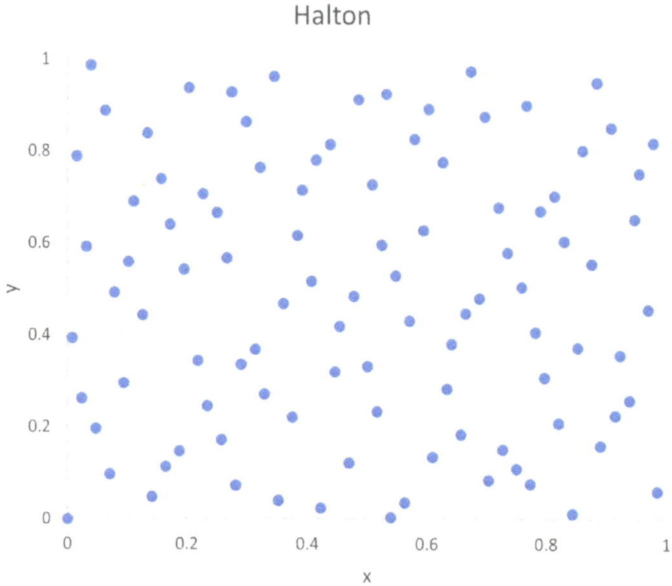

Figure 2. Two-dimensional Halton Points (100).

SOBOL Sequence Points:

Like many other quasi-random sequences, the SOBOL sequence starts from the Van der Corput sequence in base 2. To generate the Vander Corput sequence, consider the k-th point in the Sobol sequence; this integer k can be written as a linear combination of a nonnegative power of base 2 as follows.

$$k = a_0(k) + 2a_1(k) + 2^2a_2(k) + \ldots + 2^r a_r(k) \tag{7}$$

where r is a large number.

Then, the kth element in the Sobol sequence is given by the following equation:

$$x^k = 1/2y_1(k) + 1/2^2y_2(k) + \ldots + 1/2^r y_r(k) \tag{8}$$

where the coefficients $y_i(k)$ can be obtained using the following expression:

$$\begin{Bmatrix} y_1(k) \\ y_2(k) \\ \cdots \\ y_r(k) \end{Bmatrix} = \mathrm{V} \begin{Bmatrix} a_0(k) \\ a_1(k) \\ \cdots \\ a_{r-1}(k) \end{Bmatrix} \mathrm{mod}2 \tag{9}$$

where V is the generation matrix whose elements are 0 or 1. V is an identity matrix for the Vander Corput sequence.

The operation in Equation (9) can be represented as follows:

$$a_0(k)V_1 \oplus a_1(k)\mathbf{V}_2 \oplus a_2(k)v_3 \ldots a_{r-1}(k)V_r$$

where V_i is an element of V and $\oplus$*d*enotes binary addition.

The calculation of generation matrix V involves primitive polynomial A, primitive polynomial of degree d and all the coefficients A1 to Ad−1, which are either 1 or 0 and are given below.

$$P = X^d + A_1X^{d-1} + \ldots + A_{d-1}X + 1 \quad (10)$$

Direction vectors M_i are generated by the recursive equation given below for i > d, and the initial direction vectors, i.e., for i < d, are generated by selecting an odd integer between 0 and 2^d.

$$M_i = 2^1A_1M_{i-1} \oplus 2^2A_2M_{i-2} \oplus \ldots\ldots\ldots \oplus 2^{d-1}A_{d-1}M_{i-d+1} \oplus 2^dM_{i-d} \oplus M_{i-d}. \quad (11)$$

Then, the generation matrix elements can be generated as shown below.

$$V_i^j = \frac{M_i}{2^i} \quad (12)$$

Thus, SOBOL sequence can be generated by generating the V matrix and the Van der Corput sequence in base 2. Figure 3 shows that SOBOL points show better uniformity than pseudo-random numbers from the computer (Figure 1).

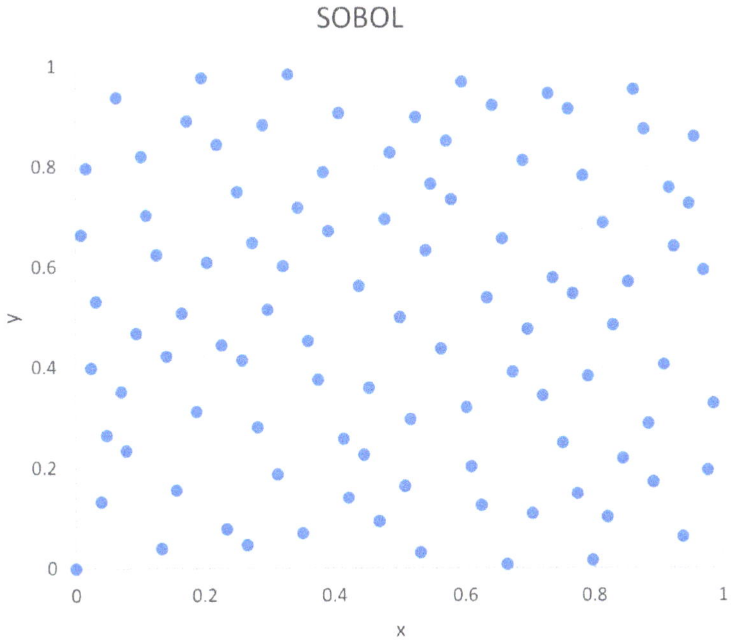

Figure 3. Two-dimensional SOBOL points (100).

We show where the random numbers are used in PSO with traditional PSO in Figure 4 and enhanced PSO with Halton or SOBOL in Figure 5. However, to maintain k-dimensional uniformity, the Halton and SOBOL points cannot be generated one at a time; they must be generated together with the whole sample (for all iterations).

It should be noted that all other efficiency enhancements proposed in the literature can be directly applicable to our new algorithms as we are only changing the random numbers.

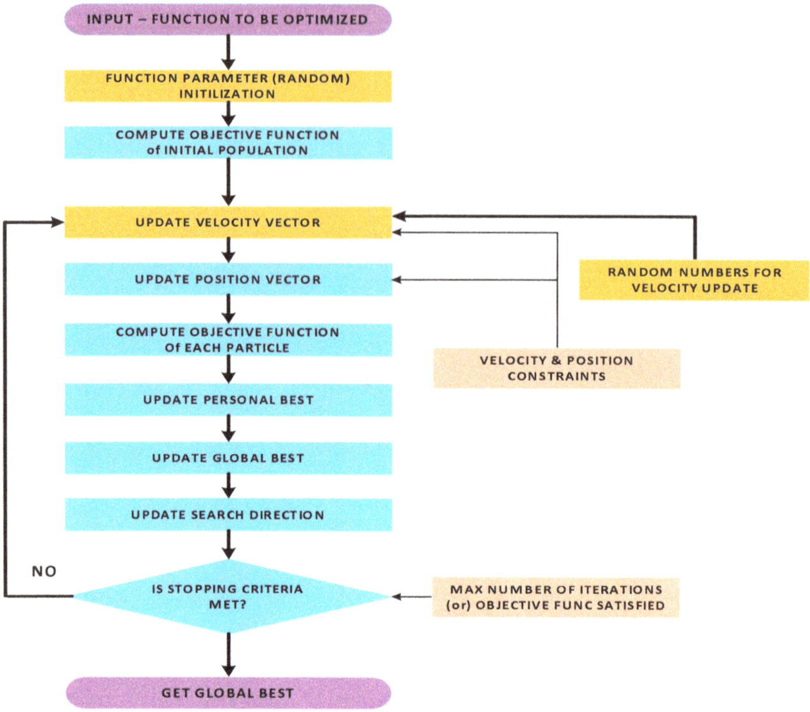

Figure 4. Traditional PSO flowchart.

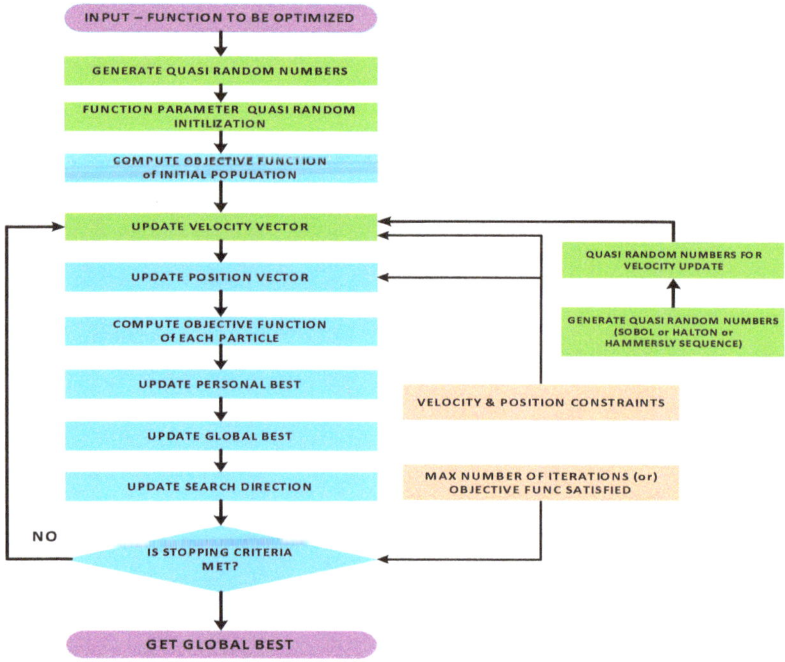

Figure 5. Enhanced PSO flowchart.

3. Results

Test Cases

To perform simulation and to estimate the standard deviations and the computation times, all the following algorithms were on a Windows 10 laptop equipped with an i7 core processor and with inbuilt 16 GB RAM. The simulations were run for 50 iterations and averaged to estimate the elapsed time. The computation elapsed times were estimated using the times associated with the start time and end time of the algorithm execution. These elapsed times were averaged across multiple runs and were used to measure the average elapsed time. The mean values for the number of iterations were estimated and rounded to provide the number of iterations in the following tables. All PSO variants implemented the mutation scheme that used uniformly generated random numbers.

To test the efficiency improvements of the enhanced PSO suggested in this paper, both mixed and continuous versions of three well-known benchmark functions, namely the Cigar, Ellipsoid, and Paraboloid functions, along with the famous Travelling Salesman Problem, were taken as test cases (Table 1). Three algorithms consisting of one PSO code with no enhancement (using Monte Carlo random number samplings), a second PSO code that uses Sobol random number samplings, and a third PSO code that uses Halton random numbers each ran for 5, 10, 15, and 20 decision variables to optimize both the mixed and continuous versions of the three benchmark functions and to solve the TSP problem. The number of iterations taken to achieve an optimum is recorded for all three algorithms to reach the global optimum.

Table 1. Test cases for algorithm testing [11].

	Function	Formula	Range
Continuous Optimization Problems			
1	Cigar	$f(x) = x_1{}^2 + 10^4 \sum_{i=2}^{NC} x_i{}^2$	$[-3,\ 3]^{NC}$
2	Parabolic	$f(x) = \sum_{i-1}^{NC} x_i{}^2$	$[-3,\ 3]^{NC}$
3	Ellipsoid	$f(x) = \sum_{i-1}^{NC} 5^{\frac{i-1}{n-1}} x_i{}^2$	$[-3,\ 3]^{NC}$
Mixed-Variable Combinatorial Optimization Problems			
4	Cigar	$f(x,\ y) = x_i{}^2 + 10^4 \sum_{i=1}^{ND} x_i{}^2 + y_1{}^2 + 10^4 \sum_{i=2}^{ND} y_i{}^2$	$[-3,\ 3]^{NM}$
5	Parabolic	$f(x,\ y) = \sum_{i=1}^{NC} x_i{}^2 + \sum_{i=1}^{ND} y_i{}^2$	$[-3,\ 3]^{NM}$
6	Ellipsoid	$f(x,\ y) = \sum_{i=1}^{NC} 5^{\frac{i-1}{n-1}} x_i{}^2 + \sum_{i=1}^{ND} 5^{\frac{i-1}{n-1}} y_i{}^2$	$[-3,\ 3]^{NM}$

Traveling Salesman Problem Optimization Procedure

The Traveling Salesman Problem is a discrete combinatorial optimization problem [12–15]. The locations of many cities are given, and an optimal order in which the cities are traversed is calculated. A position vector is the suggested order of cities. The velocity vector is a sequence of two-element tuples in which each tuple consists of two indices of elements to be swapped to make the path more optimal. For example, if there are five cities labeled with indices {1, 2, 3, 4, 5}, the population matrix could consist of different suggested orders in which they are to be traversed, such as {2, 4, 3, 5, 1} and {5, 4, 3, 1, 2}. The velocity vector for the first suggested order of cities could be {(1, 2), (3, 2), (4, 5)}, while for the second, it could be {(5, 1)}. In this case, the first element would be swapped with the second, the third with the second after that, and the fifth with the fourth after that, in the first suggested sequence of cities. For the second, the fifth and first elements are to be swapped.

ω, α, and β are pre-determined constants used in this algorithm. A random number is generated for each swap in the previous velocity vector. For each random number that is less than ω, the corresponding swap is included in the new velocity vector. Similarly, this process is carried out for the second term and the third term with α and Beta instead of ω. These three random numbers are generated together to maintain the k-dimensional uniformity of the quasi-random number sequence when Halton or SOBOL is used. Through the usage of this swapping method, the optimal order in which the cities must be visited is attained.

For the ith particle at the iteration index n, to update the velocity, the following equation can be used.

$$th:$$
$$v_{n+1}^{i} = \{\omega * v_n^i\} \oplus \{\alpha * (P_n^i - X_n^i)\} \oplus \left\{\beta * \left(G_n^j - X_n^i\right)\right\} \tag{13}$$

Hence, v_{n+1}^i *is a set of swaps.*

Here, the $\oplus$ is the merging operator, which merges sequences of swaps into a new swap sequence.

$$\alpha < 1, \beta < 1 \; and \; \omega < 1 \tag{14}$$

For the ith particle at the iteration index n, we can apply the new updated velocity v_{n+1}^i to the current position X_n^i, (which is a set of node sequences or a node list in TSP) and obtain the updated position X_{n+1}^i.

$$X_{n+1}^i = X_n^i \xleftarrow{Apply\ Swaps} v_{n+1}^i \tag{15}$$

For $\omega * v_n$, if v_n is a vector of L elements to begin with, then L random numbers are generated and for each random number that is less than ω, the corresponding swap in v_n is used. $\left(P_n^i - X_n^i\right)$ is a swap sequence to move from X_n^i to P_n^i. For example, if P_n^i is {3, 4, 5, 2, 1} and X_n^i *is* $\{5,3,4,1,2\}$, the set of swaps needed to move fromX_n^i to P_n^i is {(1, 3), (3, 2), (4, 5)} (assuming that indices start from 1).

$\alpha * \left(P_n^i - X_n^i\right)$ is a set of swaps that are selected from the swap sequence vector $\left(P_n^i - X_n^i\right)$ of length N based on the N random numbers generated that are less than α.

During initialization, for each particle i, X_0^i is set to a random selection of cities whose IDs are the city node index. If there are N cities to be visited, then X_0^i is a vector of length N.

The following are tables comparing the results produced by Monte Carlo random number sampling with Sobol random number sampling and Halton random number sampling.

Sobol vs. Monte Carlo Random Numbers

We compare the performance of using SOBOL versus conventional random number approaches based on PSO for the three functions (continuous and mixed variable form) in Tables 2–7. In addition, we also compare the SALP algorithm for the continuous ellipse and paraboloid and mixed variable ellipse and paraboloid function optimizations.

Table 2. Comparison of performances of both Monte Carlo and Sobol random number samplings in PSO for the continuous variable Cigar function.

Number of Dimensions	PSO Enhanced (Iterations)	PSO Conventional (Iterations)	PSO Enhanced Standard Deviation	PSO Conventional Standard Deviation	PSO Enhanced Compute Time (secs)	PSO Conventional Compute Time (secs)	Iterations' Improvement Percentage
5	85	107	1.9	5.1	0.31	0.37	21%
10	167	193	2.4	6.3	0.46	0.68	14%
15	183	312	5.2	30.5	0.69	1.1	42%
20	231	447	26.5	30.1	0.87	1.6	49%

Table 3. Comparison of performances of both Monte Carlo and Sobol random number samplings in PSO for the mixed variable Cigar function.

Number of Dimensions	PSO Enhanced (Iterations)	PSO Conventional (Iterations)	PSO Enhanced Standard Deviation	PSO Conventional Standard Deviation	Salp Swarm Standard Deviation	PSO Enhanced Compute Time (secs)	PSO Conventional Compute Time (secs)	Iterations' Improvement Percentage
5	93	97	3.1	6.8	N/A	0.47	0.41	4%
10	119	156	6.1	9.1	N/A	0.54	0.71	24%
15	147	209	11.2	11.8	N/A	0.73	0.87	30%
20	195	264	12.6	15.8	N/A	0.91	1.1	27%

Table 4. Comparison of performances of both Monte Carlo and Sobol random number samplings in PSO optimization for the continuous variable Paraboloid function.

Number of Dimensions	PSO Enhanced (Iterations)	PSO Conventional (Iterations)	Salp Swarm (Iterations)	PSO Enhanced Standard Deviation	PSO Conventional Standard Deviation	Salp Swarm Standard Deviation	PSO Enhanced Compute Time (secs)	PSO Conventional Compute Time (secs)	Iterations' Improvement Percentage
5	42	54	43	0.9	3.2	0.8	0.17	0.21	23%
10	67	91	71	1.2	4.5	0.9	0.26	0.37	26%
15	82	136	88	1.91	7.1	0.89	0.33	0.53	40%
20	102	178	120	1.95	7.9	1.1	0.42	0.59	43%

Table 5. Comparison of performances of both Monte Carlo and Sobol random number samplings in PSO for the mixed variable Paraboloid function.

Number of Dimensions	PSO Enhanced (Iterations)	PSO Conventional (Iterations)	Salp Swarm (Iterations)	PSO Enhanced Standard Deviation	PSO Conventional Standard Deviation	Salp Swarm Standard Deviation	PSO Enhanced Compute Time (secs)	PSO Conventional Compute Time (secs)	Iterations' Improvement Percentage
5	42	47	85	2.2	3.5	1.5	0.19	0.23	11%
10	62	68	98	3.5	4.9	1.69	0.27	0.38	20%
15	78	106	352	2.5	3.5	35.1	0.38	0.53	27%
20	94	151	364	4.5	8.2	30.2	0.46	0.64	38%

Table 6. Comparison of performances of both Monte Carlo and Sobol random number samplings in PSO for the continuous variable Ellipsoid function.

Number of Dimensions	PSO Enhanced (Iterations)	PSO Conventional (Iterations)	Salp Swarm (Iterations)	PSO Enhanced Standard Deviation	PSO Conventional Standard Deviation	Salp Swarm Standard Deviation	PSO Enhanced Compute Time (secs)	PSO Conventional Compute Time (secs)	Iterations' Improvement Percentage
5	48	60	44	2.2	3.7	0.875	0.20	0.24	20%
10	70	96	75	4.1	3.1	1.1	0.31	0.42	27%
15	92	154	96	3.9	5.1	2.8	0.41	0.68	40%
20	105	184	132	4.2	9.2	2.9	0.52	0.88	43%

Table 7. Comparison of performances of both Monte Carlo and Sobol random number samplings in PSO for the mixed variable Ellipsoid function.

Number of Dimensions	PSO Enhanced (Iteration)	PSO Conventional (Iterations)	Salp Swarm (Iterations)	PSO Enhanced Standard Deviation	PSO Conventional Standard Deviation	Salp Swarm Standard Deviation	PSO Enhanced Compute Time (secs)	PSO Conventional Compute Time (secs)	Iterations' Improvement Percentage
5	45	53	90	3.5	4.6	11.9	0.26	0.26	15%
10	62	89	109	4.9	8.3	29.1	0.42	0.43	30%
15	84	128	342	4.4	6.7	54.3	0.52	0.65	35%
20	90	165	350	3.2	7.8	63.6	0.63	0.88	46%

Halton vs. Monte Carlo:

The results of Halton-based enhanced PSO are presented in Tables 8–13.

Table 8. Comparison of performances of both Monte Carlo and (scrambled) Halton random number samplings in PSO for the continuous variable Cigar function.

Number of Dimensions	PSO Enhanced (Iterations)	PSO Conventional (Iterations)	PSO Enhanced Standard Deviation	PSO Conventional Standard Deviation	PSO Enhanced Compute Time (secs)	PSO Conventional Compute Time (secs)	Iterations' Improvement Percentage
5	61	101	3.2	5.1	0.51	0.35	40%
10	111	189	5.1	6.3	0.77	0.64	42%
15	188	298	6.8	30.5	1.1	1.03	37%
20	220	451	10.1	30.1	1.3	1.51	52%

Table 9. Comparison of performances of both Monte Carlo and (scrambled) Halton random number samplings in PSO for the mixed variable Cigar function.

Number of Dimensions	PSO Enhanced (Iterations)	PSO Conventional (Iterations)	PSO Enhanced Standard Deviation	PSO Conventional Standard Deviation	PSO Enhanced Compute Time (secs)	PSO Conventional Compute Time (secs)	Iterations' Improvement Percentage
5	96	95	2.87	4.9	0.67	0.40	1%
10	107	156	4.12	5.8	0.81	0.65	32%
15	157	204	9.75	14.5	1.03	0.88	23%
20	190	265	13.3	8.31	1.30	1.16	29%

Table 10. Comparison of performances of both Monte Carlo and Halton random number samplings in PSO for the continuous variable Paraboloid function.

Number of Dimensions	PSO Enhanced (Iterations)	PSO Conventional (Iterations)	PSO Enhanced Standard Deviation	PSO Conventional Standard Deviation	PSO Enhanced Compute Time (secs)	PSO Conventional Compute Time (secs)	Iterations' Improvement Percentage
5	38	52	0.9	3.1	0.41	0.20	30%
10	64	88	1.1	4.4	0.6	0.33	28%
15	79	142	1.15	6.5	0.76	0.52	45%
20	103	174	1.6	5.5	0.92	0.55	41%

Table 11. Comparison of performances of both Monte Carlo and Halton random number samplings in PSO for the mixed variable Paraboloid function.

Number of Dimensions	PSO Enhanced (Iterations)	PSO Conventional (Iterations)	PSO Enhanced Standard Deviation	PSO Conventional Standard Deviation	PSO Enhanced Compute Time (secs)	PSO Conventional Compute Time (secs)	Iterations' Improvement Percentage
5	38	46	1.37	2.66	0.46	0.21	18%
10	60	77	3.1	4.31	0.62	0.33	23%
15	71	107	3.9	5.78	0.81	0.48	35%
20	92	146	6.99	7.52	0.94	0.67	38%

It can be seen from the above tables (Tables 2–13) that the quasi-random sequence-based enhanced PSO demonstrates superior performance compared to a conventional random number-based PSO for both continuous and mixed variable problems. The enhancement increases generally with a higher number of variables specifically for continuous variable problems. SOBOL works better for mixed variable functions than Halton, but Halton shows better convergence than SOBOL when considering most of the continuous variable benchmark problems.

Table 12. Comparison of performances of both Monte Carlo and Halton random number samplings in PSO for the continuous variable Ellipsoid function.

Number of Dimensions	PSO Enhanced (Iterations)	PSO Conventional (Iterations)	PSO Enhanced Standard Deviation	PSO Conventional Standard Deviation	PSO Enhanced Compute Time (secs)	PSO Conventional Compute Time (secs)	Iterations' Improvement Percentage
5	38	60	2.7	3.7	0.6	0.24	40%
10	75	96	3.08	3.1	1.0	0.42	24%
15	86	154	3.01	5.1	1.27	0.68	45%
20	152	184	3.44	9.2	1.44	0.88	35%

Table 13. Comparison of performances of both Monte Carlo and Halton random number samplings in PSO for the mixed variable Ellipsoid function.

Number of Dimensions	PSO Enhanced (Iterations)	PSO Conventional (Iterations)	PSO Enhanced Standard Deviation	PSO Conventional Standard Deviation	PSO Enhanced Compute Time (secs)	PSO Conventional Compute Time (secs)	Iterations' Improvement Percentage
5	42	53	2.54	4.6	0.50	0.26	21%
10	61	89	2.75	8.3	0.70	0.43	32%
15	110	128	6.19	6.7	0.91	0.65	14%
20	120	165	7.1	7.8	1.2	0.88	27%

Constant Inertia vs. Random Inertia:

As suggested earlier, in new variants proposed in the literature, adaptive parameter enhancements provide improvements. We can also replace the random numbers used in these variants with the quasi-random numbers suggested in our new algorithm as these two improvements are mutually exclusive. We illustrate this by comparing the constant inertia algorithm, which is the traditional approach to PSO, to the random inertia proposed as an enhancement [Tables 14–17].

Table 14. Comparison of random inertial weight-aided Sobol and conventional PSO performances for benchmark functions with 5 dimensions.

Function Type	PSO Sobol Contant Weight Inertia	PSO Conventional Random Weight Inertia	PSO Sobol Random Weight Inertia	Standard Deviation PSO Sobol Contant Weight Inertia	Standard Deviation PSO Conventional Random Weight Inertia	Standard Deviation PSO Sobol Random Weight Inertia	Improvement in Random Inertia (Conventional vs. Sobol)
Cigar—Continuous	85	54	48	1.9	1.88	3.9	11.1%
Ellipse—Continuous	48	35	30	2.2	1.2	1.8	14.3%
Parabola—Continuous	42	31	29	0.9	1.7	1.3	6.45%
Cigar—Mixed	93	108	95	3.1	14.3	3.6	12.03%
Ellipse—Mixed	45	57	47	3.5	8.5	4.6	17.54%
Parabola—Mixed	42	52	43	2.2	15.8	5.5	17.3%

In this section, we compare the performance of the PSO with the constant inertia weight and that of the PSO with the random inertia weight. Random inertia weight has been used in reference [20,21] to improve the iteration efficiency. The traditional constant inertia PSO, our proposed enhanced PSO with quasi-random numbers with constant inertia PSO, is compared with random inertia variants for both algorithms. We use Sobol random numbers for the enhanced PSO and its random inertia variant.

Table 15. Comparison of performances of random inertial weight-aided Sobol and conventional PSO for the benchmark functions with 10 dimensions.

Function Type	PSO Sobol Contant Weight Inertia	PSO Conventional Random Weight Inertia	PSO Sobol Random Weight Inertia	Standard Deviation PSO Sobol Contant Weight Inertia	Standard Deviation PSO Conventional Random Weight Inertia	Standard Deviation PSO Sobol Random Weight Inertia	Improvement in Random Inertia (Conventional vs. Sobol)
Cigar—Continuous	167	76	67	2.4	2.6	1.8	11.8%
Ellipse—Continuous	70	35	30	4.1	1.8	1.7	14.3%
Parabola—Continuous	67	31	29	1.2	2.3	1.6	6.5%
Cigar—Mixed	119	177	136	6.1	9.2	16.1	23.2%
Ellipse—Mixed	62	99	84	4.9	8.0	17.2	15.2%
Parabola—Mixed	62	89	71	3.5	8.2	4.5	20.2%

Table 16. Comparison of performances of random inertial weight-aided Sobol and conventional PSO for the benchmark functions with 15 dimensions.

Function Type	PSO Sobol Contant Weight Inertia	PSO Conventional Random Weight Inertia	PSO Sobol Random Weight Inertia	Standard Deviation PSO Sobol Contant Weight Inertia	Standard Deviation PSO Conventional Random Weight Inertia	Standard Deviation PSO Sobol Random Weight Inertia	Improvement in Random Inertia (Conventional vs. Sobol)
Cigar—Continuous	183	96	85	5.2	2.1	2.2	11.5%
Ellipse—Continuous	92	63	56	3.9	1.2	1.6	11.1%
Parabola—Continuous	82	61	52	1.91	1.3	1.9	14.8%
Cigar—Mixed	147	260	166	11.2	44.8	16.4	36.2%
Ellipse—Mixed	84	140	100	4.4	11.9	6.9	28.6%
Parabola—Mixed	78	133	86	2.5	30.1	5.7	35.3%

Table 17. Comparison of performances of random inertial weight-aided Sobol and conventional PSO for the benchmark functions with 20 dimensions.

Function Type	PSO Sobol Contant Weight Inertia	PSO Conventional Random Weight Inertia	PSO Sobol Random Weight Inertia	Standard Deviation PSO Sobol Contant Weight Inertia	Standard Deviation PSO Conventional Random Weight Inertia	Standard Deviation PSO Sobol Random Weight Inertia	Improvement in Random Inertia (Conventional vs. Sobol)
Cigar—Continuous	231	118	98	26.5	5.4	3.5	16.9%
Ellipse—Continuous	105	77	66	4.2	2.8	2.7	14.3%
Parabola—Continuous	102	74	62	1.95	2.1	1.9	16.2%
Cigar—Mixed	195	330	234	12.6	37.1	21.2	29.1%
Ellipse—Mixed	90	213	120	3.2	28.4	19.2	43.7%
Parabola—Mixed	94	193	105	4.5	23.6	5.1	45.6%

The velocity of each particle in the PSO algorithm at iteration n is updated according to the following equation:

$$v_{n+1}^{ij} = \omega * v_n^{ij} + c_1 * r_1 * \left(P_n^j - X_n^{ij}\right) + c_2 * r_2 * \left(G_n{}^j - X_n^{ij}\right) \tag{16}$$

In the above equation, ω is the inertia weight. For constant inertia, ω was set to 0.75.

For random inertia weight, ω was set to $0.5 + \frac{rand}{2}$ as per [21]. The comparison results are provided in the table below. The improvement percentage in the table below is the comparison between random inertial weight-aided Sobol and conventional PSO variants.

From the above results, we conclude that the Sobol-based PSO variant performs better in terms of iteration efficiency with respect to the conventional PSO as well as random inertia PSO.

TSP Optimization Results

Tables 18 and 19 provide the TSP optimization comparisons using the PSO variants. TSP is a completely discrete problem, and SOBOL performs better, as can be seen from Figures 6–8, as well as Tables 14 and 15. For TSP, the particles were initialized using the standard uniformly distributed random numbers for both the standard and the enhanced PSOs. The velocity and position updates, which are basically a set of swaps, are controlled using the Sobol and Halton-generated random numbers in the case of enhanced PSO.

Table 18. Comparison of performances for Monte Carlo and Sobol random number samplings in PSO for TSP.

Number of Cities	PSO Enhanced (Iterations)	PSO Conventional (Iterations)	PSO Enhanced Standard Deviation	PSO Conventional Standard Deviation	PSO Enhanced Compute Time (secs)	PSO Conventional Compute Time (secs)	Iterations' Improvement Percentage
10	74	103	16.1	28.9	0.98	1.18	28%
15	414	478	88.7	130.2	7.85	8.7	13%
20	1705	1927	210.2	371.4	16.55	17.8	11.5%

Table 19. Comparison of performances for Monte Carlo and Halton random number samplings in PSO for TSP.

Number of Cities	PSO Enhanced (Iterations)	PSO Conventional (Iterations)	PSO Enhanced Standard Deviation	PSO Conventional Standard Deviation	PSO Enhanced Compute Time (secs)	PSO Conventional Compute Time (secs)	Iterations' Improvement Percentage
10	78	103	18.2	28.9	1.26	1.18	24%
15	431	478	76.7	130.2	8.98	8.7	9.8%
20	1760	1927	212.1	371.4	17.39	17.8	8.6%

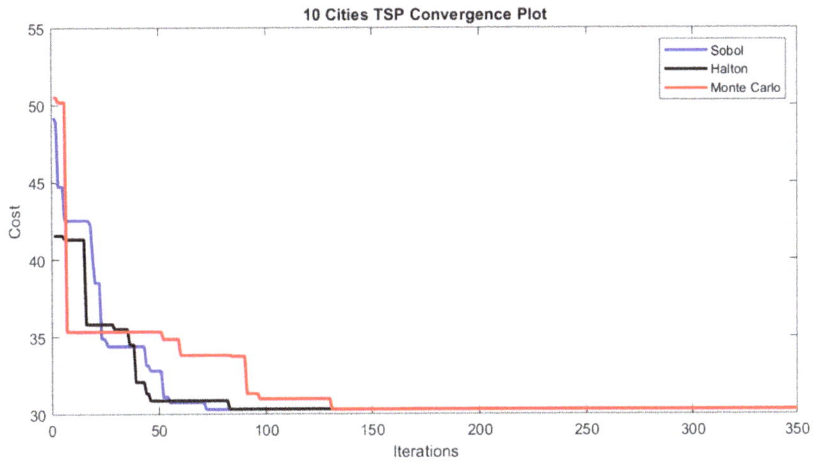

Figure 6. TSP with 10 cities—convergence plot.

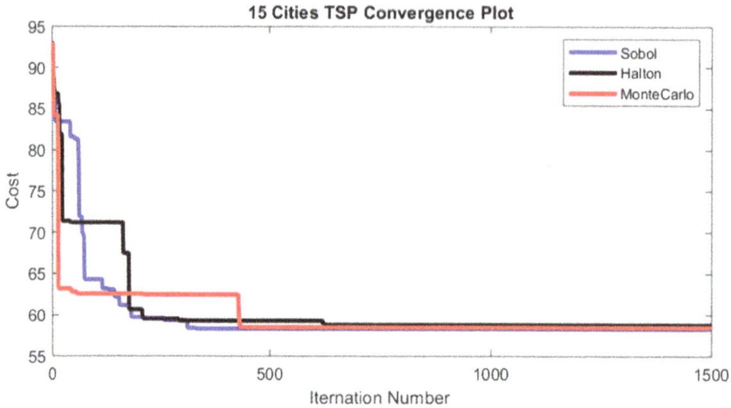

Figure 7. TSP with 15 cities—convergence plot.

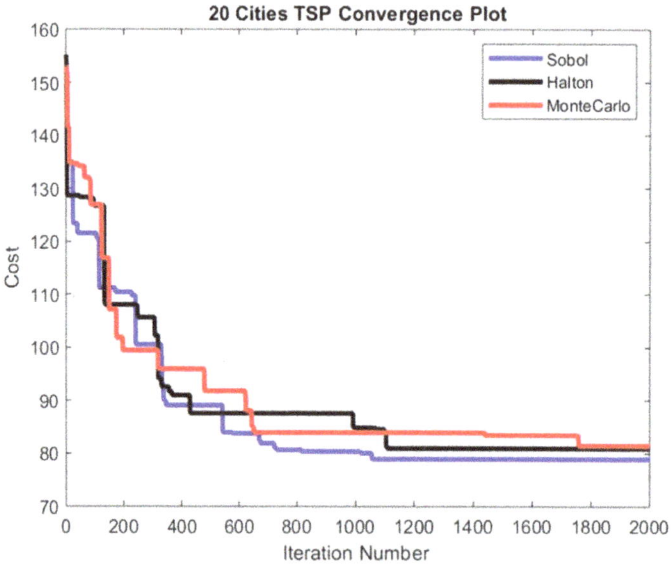

Figure 8. TSP with 20 cities—convergence plot.

4. Discussion

We have augmented the PSO algorithm by integrating Sobol and Halton random number samplings to achieve superior results. Our rationale behind this enhancement is rooted in Sobol and Halton random numbers, which are quasi-random number types. These numbers are generated in such a manner that they are uniformly distributed across multidimensional space, thus mitigating clustering or bias in particle movement. Consequently, this facilitates more extensive exploration, thereby increasing the likelihood of avoiding local optima and accelerating the discovery of the global optimum, as a more significant portion of the space can be explored when biases or clusters are circumvented.

The outcomes indicate that the enhancement applied to the standard PSO through the utilization of quasi-random numbers has consistently improved the number of iterations required for the algorithm to produce the optimal function value across all three benchmark functions. The efficiency gains for continuous functions are more pronounced than those for mixed variable functions.

Additionally, including more decision variables in the problem correlates with more significant improvements in general. As evident from the data, the application of Sobol or Halton sequences to the standard PSO algorithm demonstrates efficiency improvements, suggesting the potential benefits of these sequences to researchers in various optimization fields. Notably, this technique is algorithm-agnostic, as indicated in the introduction, and can thus be employed with other optimization algorithms such as the Genetic Algorithm, Ant Colony Optimization (ACO) algorithm, and other established optimization algorithms.

5. Conclusions

Particle Swarm Optimization harnesses swarm behavior effectively for function optimization but is often hampered by slow convergence. We consistently improved efficiency through two distinct enhancements to the Particle Swarm Optimization algorithm. We proposed the use of quasi-random number sequences to update decision variable values and their rates of change in each iteration to enhance the PSO algorithm. Since quasi-random number sequences do not alter the fundamental algorithm, this concept can be integrated into modified versions of the PSO algorithm suggested previously, as discussed in the introduction. To evaluate whether quasi-random number sequences in PSO yield efficiency improvements, we tested them using continuous and mixed variable Cigar, Ellipsoid, and Paraboloid functions, along with an example of the Traveling Salesman Problem. The results demonstrate that both sequences of quasi-random numbers used to enhance the standard PSO improved efficiency.

Author Contributions: Conceptualization, U.D.; methodology, U.D.; software, S.K.K.; validation, S.K.K.; formal analysis, S.K.K. and U.D.; investigation, S.K.K. and U.D.; resources, U.D.; data curation, S.K.K.; writing—original draft preparation, S.K.K.; writing—review and editing, U.D.; visualization, S.K.K. and U.D.; supervision, U.D.; project administration, U.D. All authors have read and agreed to the published version of the manuscript.

Funding: This research received no external funding.

Data Availability Statement: Data can be obtained from the authors.

Conflicts of Interest: The authors declare no conflicts of interest.

References

1. Shami, T.M.; El-Saleh, A.A.; Alswaitti, M.; Al-Tashi, Q.; Summakieh, M.A.; Mirjalili, S. Particle Swarm Optimization: A Comprehensive Survey. *IEEE Access* **2022**, *10*, 10031–10061. [CrossRef]
2. Kennedy, J.; Eberhart, R. Particle swarm optimization. In Proceedings of the ICNN'95—International Conference on Neural Networks, Perth, WA, Australia, 27 November–1 December 1995; Volume 4, pp. 1942–1948. [CrossRef]
3. Bansal, J.C.; Singh, P.K.; Saraswat, M.; Verma, A.; Jadon, S.S.; Abraham, A. Inertia Weight strategies in Particle Swarm Optimization. In Proceedings of the 2011 Third World Congress on Nature and Biologically Inspired Computing, Salamanca, Spain, 19–21 October 2011; pp. 633–640. [CrossRef]
4. Li, N.; Qin, Y.-Q.; Sun, D.-B.; Zou, T. Particle swarm optimization with mutation operator. In Proceedings of the 2004 International Conference on Machine Learning and Cybernetics (IEEE Cat. No.04EX826), Shanghai, China, 26–29 August 2004; Volume 4, pp. 2251–2256. [CrossRef]
5. Clerc, M.; Kennedy, J. The particle swarm–Explosion, stability, and convergence in a multidimensional complex space. *IEEE Trans. Evol. Comput.* **2002**, *6*, 58–73. [CrossRef]
6. Engelbrecht, A.P. Particle swarm optimization with crossover: A review and empirical analysis. *Artif. Intell. Rev.* **2016**, *45*, 131–165. [CrossRef]
7. Chen, Y.; Li, L.; Xiao, J.; Yang, Y.; Liang, J.; Li, T. Particle swarm optimizer with crossover operation. *Eng. Appl. Artif. Intell.* **2018**, *70*, 159–169. [CrossRef]
8. Zhou, Z.; Li, F.; Abawajy, J.H.; Gao, C. Improved PSO Algorithm Integrated with Opposition-Based Learning and Tentative Perception in Networked Data Centers. *IEEE Access* **2020**, *8*, 55872–55880. [CrossRef]
9. Mandal, B.; Roy, P.K.; Mandal, S. Hybridization of Particle Swarm Optimization with Biogeography-Based Optimization for Reactive Power and Voltage Control. In Proceedings of the 2014 Fourth International Conference of Emerging Applications of Information Technology, Kolkata, India, 19–21 December 2014; pp. 34–39. [CrossRef]

10. Cai, Y.; Yang, S.X. An improved PSO-based approach with dynamic parameter tuning for cooperative target searching of multi-robots. In Proceedings of the 2014 World Automation Congress (WAC), Waikoloa, HI, USA, 3–7 August 2014; pp. 616–621. [CrossRef]
11. Morokoff, W.J.; Caflisch, R.E. Quasi-Random Sequences and Their Discrepancies. *SIAM J. Sci. Comput.* **1994**, *15*, 1251–1279. [CrossRef]
12. Niederreiter, H. *Random Number Generation and Quasi-Monte Carlo Methods*; CBMS-NSF Regional Conference Series in Applied Mathematics 63; Society for Industrial and Applied Mathematics: Philadelphia, PA, USA, 1992.
13. Sobol, I.M. Uniformly distributed sequences with an additional uniform property. *USSR J. Comput. Math. Math. Phys. Engl. Transl.* **1976**, *16*, 1332–1337. [CrossRef]
14. Diwekar, U.M.; Gebreslassie, B.H. Efficient Ant Colony Optimization (EACO) Algorithm for Deterministic Optimization. *Int. J. Swarm Intel. Evol. Comput.* **2016**, *5*, 131. [CrossRef]
15. Hadia, S.K.; Joshi, A.H.; Patel, C.K.; Kosta, Y.P. Solving City Routing Issue with Particle Swarm Optimization. *Int. J. Comput. Appl.* **2012**, *47*, 15.
16. Hadia, S.K.; Joshi, A.H.; Salman, C.A.; Ahmad, I.; Al-Madani, S. Particle swarm optimization for task assignment problem. *Microprocess. Microsyst.* **2002**, *26*, 363–371.
17. Zhong, W.H.; Zhang, J.; Chen, W.N. A novel discrete particle swarm optimization to solve traveling salesman problem. In Proceedings of the IEEE Congress on Evolutionary Computation (CEC), Singapore, 25–28 September 2007; pp. 3283–3287.
18. Wang, K.P.; Huang, L.; Zhou, C.G.; Pang, W. Particle swarm optimization for solving traveling salesman problem. In Proceedings of the 2003 International Conference on Machine Learning and Cybernetic, Xi'an, China, 5 November 2003; Volume 3, pp. 1583–1585.
19. Mirjalili, S.; Gandomi, A.H.; Mirjalili, S.Z.; Saremi, S.; Faris, H.; Mirjalili, S.M. Salp Swarm Algorithm: A bio-inspired optimizer for engineering design problems. *Adv. Eng. Softw.* **2017**, *114*, 163–191. [CrossRef]
20. Umapathy, P.; Venkataseshaiah, C.; Arumugam, M.S. Particle Swarm Optimization with Various Inertia Weight Variants for Optimal Power Flow Solution. *Discret. Dyn. Nat. Soc.* **2010**, *2010*, 462145. [CrossRef]
21. Nikabadi, A.; Ebadzadeh, M. Particle swarm optimization algorithms with adaptive Inertia Weight: A survey of the state of the art and a Novel method. *IEEE J. Evol. Comput.* **2008**, *14*, 305–313.
22. Ratnaweera, A.; Halgamuge, S.K.; Watson, H. Self-Organizing Hierarchical Particle Swarm Optimizer with Time-Varying Acceleration Coefficients. *IEEE Trans. Evol. Comput.* **2004**, *8*, 240–255. [CrossRef]
23. Zhan, Z.H.; Zhang, J.; Li, Y.; Chung, H.-H. Adaptive Particle Swarm Optimization. *IEEE Trans. Syst. Man Cybernetics. Part B Cybern.* **2009**, *39*, 1362–1381. [CrossRef] [PubMed]
24. Qu, B.Y.; Suganthan, P.N.; Das, S. A Distance-Based Locally Informed Particle Swarm Model for Multimodal Optimization. *IEEE Trans. Evol. Comput.* **2013**, *17*, 387–402. [CrossRef]
25. Shieh, H.-L.; Kuo, C.-C.; Chiang, C.-M. Modified particle swarm optimization algorithm with simulated annealing behavior and its numerical verification. *Appl. Math. Comput.* **2011**, *218*, 4365–4383. [CrossRef]
26. Feng, Z.-K.; Niu, W.-J.; Cheng, C.-T. Multi-objective quantum-behaved particle swarm optimization for economic environmental hydrothermal energy system scheduling. *Energy* **2017**, *131*, 165–178. [CrossRef]
27. Liu, D.; Xiao, Z.; Li, H.; Liu, D.; Hu, X.; Malik, O.P. Accurate Parameter Estimation of a Hydro-Turbine Regulation System Using Adaptive Fuzzy Particle Swarm Optimization. *Energies* **2019**, *12*, 3903. [CrossRef]
28. Wang, J.; Wang, X.; Li, X.; Yi, J. A Hybrid Particle Swarm Optimization Algorithm with Dynamic Adjustment of Inertia Weight Based on a New Feature Selection Method to Optimize SVM Parameters. *Entropy* **2023**, *25*, 531. [CrossRef] [PubMed]

algorithms

MDPI

Article

Applying Particle Swarm Optimization Variations to Solve the Transportation Problem Effectively

Chrysanthi Aroniadi and Grigorios N. Beligiannis *

Department of Food Science and Technology, University of Patras, Agrinio Campus, G. Seferi 2, 30100 Agrinio, Greece; c.aroniadi@upatras.gr
* Correspondence: gbeligia@upatras.gr; Tel.: +30-2641074194

Abstract: The Transportation Problem (TP) is a special type of linear programming problem, where the objective is to minimize the cost of distributing a product from a number of sources to a number of destinations. Many methods for solving the TP have been studied over time. However, exact methods do not always succeed in finding the optimal solution or a solution that effectively approximates the optimal one. This paper introduces two new variations of the well-established Particle Swarm Optimization (PSO) algorithm named the Trigonometric Acceleration Coefficients-PSO (TrigAc-PSO) and the Four Sectors Varying Acceleration Coefficients PSO (FSVAC-PSO) and applies them to solve the TP. The performances of the proposed variations are examined and validated by carrying out extensive experimental tests. In order to demonstrate the efficiency of the proposed PSO variations, thirty two problems with different sizes have been solved to evaluate and demonstrate their performance. Moreover, the proposed PSO variations were compared with exact methods such as Vogel's Approximation Method (VAM), the Total Differences Method 1 (TDM1), the Total Opportunity Cost Matrix-Minimal Total (TOCM-MT), the Juman and Hoque Method (JHM) and the Bilqis Chastine Erma method (BCE). Last but not least, the proposed variations were also compared with other PSO variations that are well known for their completeness and efficiency, such as Decreasing Weight Particle Swarm Optimization (DWPSO) and Time Varying Acceleration Coefficients (TVAC). Experimental results show that the proposed variations achieve very satisfactory results in terms of their efficiency and effectiveness compared to existing either exact or heuristic methods.

Keywords: transportation problem; Particle Swarm Optimization; heuristics methods; linear programming

1. Introduction

The Transportation Problem (TP) is one of the most significant types of linear programming problems. The aim of the TP is to minimize the cost of transportation of a given commodity from a number of sources or origins (e.g., factory manufacturing facility) to a number of destinations (e.g., warehouse, store) [1]. Over the years, many classical and stochastic search approaches have been applied for the purpose of solving the TP.

The Northwest Corner method (NWC) is one of the methods that obtains a basic feasible solution to various transportation problems [2]. This process very easily allocates the amounts when few demand and destination stations exist. Moreover, frequently, the exported solution does not approach the optimal. The Minimum Cost Method (MCM) [3] is an alternative method which can yield an initial basic feasible solution. The MCM succeeds in lowering total costs by taking into consideration the lowest available cost values while finding the initial solution. An innovative approach comes from the Vogel Approximation Method (VAM); the VAM is an upgraded version of the MCM which results in a basic feasible solution close to the optimal solution [3]. Both of them take the unit transportation costs into account and obtain satisfactory results; however, VAM is rather slow and computationally intensive for a large range of values. Nevertheless, it has been proven that in problems with a small range of values and a relatively small number of variables, the above exact methods are quite efficient.

In some cases, TP has a complex structure, multifaceted parameters and a huge amount of data to be studied. Therefore, exact methods do not succeed in finding a suitable solution in an acceptable time period; a result, it is unpractical to use them. Taking into consideration the above, apart from conventional solution techniques, various heuristic and metaheuristic methods have been designed to capitalize on their potential capabilities. Specifically, metaheuristic algorithms attempt to find the best feasible solution, surpassing the other technique as much in terms of quality as in computational time [4]. Mitsuo Gen, Fulya Atliparmak and Lin Lin applied a Genetic Algorithm (GA) for a two-stage TP using priority-based encoding, showing that the GA has been receiving great attention and can be successfully applied for combinational optimization problems [5]. Ant Colony Optimization (ACO) algorithms have already proven their efficiency in many complex problems; they constitute a very useful optimization tool for many transportation problems in cases where it is impossible to find an algorithm that finds the optimal solution or in cases where the time interval does not make it possible to approve this solution [6]. The applications of hybrid methods with the combination of two or more heuristic, metaheuristic or even exact methods are also widespread. Interesting research was undertaken in 2019 by Mohammad Bagher Fakhrzad, Fariba Goodarzian and Golmohammadi [7]. In their study, four metaheuristic algorithms, including Red deer Algorithm (RDA), Stochastic Fractal Search (SFS), Genetic Algorithm (GA) and Simulated Annealing (SA), as well as two hybrid algorithms, the RDA and GA (HRDGA) algorithm and the Hybrid SFS and SA (HRDGA) algorithm, were utilized to solve the TP, demonstrating significant effectiveness [7].

Motivated by the above-mentioned applications of metaheuristic algorithms to cope with the TP, this work deals with the application of Particle Swarm Optimization (PSO) to solve the TP effectively. The PSO algorithm was first introduced by Dr. Kennedy and Dr. Eberhart in 1995 and was known as a novel population-based stochastic algorithm, working out complex non-linear optimization problems [8]. The basic idea was originally inspired by simulations of the social behavior of animals such as bird flocking, fish schooling, etc. Possessing their own intelligence, birds of the group connect with each other, sharing their experiences, and follow and trust the mass in order to reach their food or migrate safely without knowing in advance the optimal way to achieve it. The proposed research is expected to enhance the abilities of both the social behavior and personal behavior of the birds. It is observed that the original PSO has deficits in premature convergence, especially for problems with multiple local optimums [9]. The swarm's ability to function with social experience as well as personal experience is determined in the algorithm through two stochastic acceleration components, known as the cognitive and social components [10]. These components have the aptitude to guide the particles in the original PSO method to the optimum point as the correct selection of their values is the key influence on the success and efficiency of the algorithm. Much research has been carried out with a focus on finding out the best combination of these components [10].

First, this paper examines approaches that have already been applied with great success to solve the TP. Adding to the above, two new PSO variations are presented and applied to solve the TP, operating proper transformations of the main PSO parameters. Experimental results show that these new PSO variations have very good performance and efficiency in solving the TP compared to the former methods.

In order to confirm the technical merit and the applied value of our study, 32 instances of the TP with different sizes have been solved to evaluate and demonstrate the performance of the proposed PSO variations. Their experimental results are compared with those of well-known exact methods, proving their superiority over them. One major innovation of the proposed variations is the appropriate combination of acceleration coefficients (parameters c_1, c_2) and inertia weight (parameter w) [11] (see Section 3) in order to come up with better computational results compared to existing approaches. Exhaustive experimental results demonstrate that the performance of the new PSO variations noted significantly higher performance not only compared to the exact methods already applied to solve the TP but also compared to the other PSO variations already introduced in the respective literature. Furthermore, in order to check the stability of the proposed PSO variations, many different combinations of the main PSO parameters were tested and validated.

The contribution of the paper is as follows:

- According to our knowledge, PSO has already been applied for solving the fixed-charged TP, and a heuristic approach was used in order to find the shortest path in a network of routes with a standard number of points connected to each other. For the first time, the PSO-based algorithms are applied to solve the basic TP in a large amount of test instances effectively, not only finding the optimal means of items distribution but also discovering the optimal value.
- Moreover, two new PSO variations are introduced, which sustain balance between exploration and exploitation of the search space. These variations proved to be very efficient in solving the TP, achieving better results compared not only to deterministic but also to other already-known PSO-based methods.
- A thorough experimental analysis has been performed on the PSO variations applied to solve the TP to prove their efficiency and stability.

The remainder of the paper is organized as follows: Section 2 presents the mathematical formulation of the TP. The PSO algorithm is briefly described in Section 3. Section 4 presents the initialization procedure of the basic feasible solutions and the steps of the PSO algorithm for the TP. Both the existing PSO variations as well as the new ones are presented in detail in Section 5. A well-documented case study is conducted in Section 6, in order to compare the performance of five exact methods with the classic PSO and its variations. Lastly, conclusive remarks and future recommendations are presented in Section 7.

2. Transportation Problem (TP)

Many researchers have developed various types of transportation models. The most prevalent was presented by Hitchcock in 1941 [12]. Similar studies were conducted later by Koopmans in 1949 [13] and in 1951 by Dantzig [14]. It is well known that the problem has become quite widespread, so several extensions of transportation model and methods have been subsequently developed. However, how is the Transportation Problem defined?

The TP can be described as a distribution problem, with m suppliers S_i (warehouses or factories) and n destinations D_j (customers or demand points). Each of the m suppliers can be allocated to any of the n destinations at a unit shopping cost c_{ij}, which corresponds to the route from point i to point j. The available quantities of each supplier S_i, $i = 1, 2, \ldots, m$ are denoted as s_i, and those of each destination D_j, $j = 1, 2, \ldots, n$ are denoted as d_j. The objective is to determine how to allocate the available amounts from the supply stations to the destination stations while simultaneously achieving the minimum transport cost and also satisfying demand and supply constraints [12].

The mathematical model of the TP can be formulated as follows:

$$\min Z = \sum_{i=1}^{m} \sum_{j=1}^{n} c_{ij} \cdot x_{ij}; \quad (1)$$

$$\sum_{i=1}^{m} x_{ij} \geq d_j \; for \; j = 1, 2, \ldots, n; \quad (2)$$

$$\sum_{j=1}^{n} x_{ij} \leq s_i \; for \; i = 1, 2, \ldots, m; \quad (3)$$

$$x_{ij} \geq 0 \; for \; i = 1, 2, \ldots, m, \; j = 1, 2, \ldots, n. \quad (4)$$

Equation (1) represents the objective function to be minimized. Equation (2) contains the supply constraints according to which the available number of origin points must be more than or equal to the quantity demanded from the destination points. Respectively, the sum of the amount to be transferred from source S_i to destination D_j must be less than or equal to the available quantity than we possess, as presented in Equation (3). A necessary condition is depicted in Equation (4), as units x_{ij} must take positive and integer values. Without loss of generality, we assume that in this paper, both the supplies and demands are equal following the balanced condition model.

As already mentioned, there are several methods which can lead to finding a basic feasible solution. However, most of the currently used methods for solving transportation problems are considered complex and very expansive in terms of execution time. As a result, it is appealing to seek and discover a metaheuristic approach based on the PSO algorithm to solve the TP efficiently and effectively.

3. Particle Swarm Optimization (PSO)

The Particle Swarm Optimization (PSO) algorithm is considered to be one of the modern innovative heuristic algorithms since its methodology over the years has become extremely prevalent due to its simplicity of implementation; it leads very easily to satisfactory solutions [15]. According to the PSO algorithm, the collective behavior of animals has been analyzed in detail with an eye forward to function as a robust method in order to solve optimization problems in a wide variety of applications [16].

In PSO, each candidate solution can be defined as a particle and the whole swarm can be considered as the population of the algorithm. The particles improve themselves by cooperating and sharing information among the swarm, and they succeed in learning and improving to provide the highest possible efficiency. More precisely, each particle through the search space is intended to find the best value for its individual fitness and, simultaneously, to minimize the objective function by satisfying all the constraints of the problem. Each particle is studied from a perspective that contains three different parameters: position; velocity; and its previous best positions.

Consequently, in n-dimensional search space, each particle of the swarm is represented by $x_{ij} = (x_{i1}, x_{i2}, \ldots, x_{ij})$, and the equation of its position is as follows:

$$x_{ij}(t+1) = x_{ij}(t) + v_{ij}(t+1),\ i = 1,\ 2,\ldots, n\ \kappa\alpha\iota\ j = 1,\ 2,\ldots,\ n, \tag{5}$$

where $x_{ij}(t+1)$ is the current position, $x_{ij}(t)$ is the previous position and $v_{ij}(t+1)$ is the velocity which determines the movement of each particle in the current iteration $(t+1)$.

Respectively, the velocity of the particle is denoted by v_{ij} and is given by the following equation:

$$v_{ij}(t+1) = w\ v_{ij}(t) + c_1 r_1 \left(pbest_{ij}(t) - x_{ij}(t)\right) + c_2 r_2 \left(gbest_{ij}(t) - x_{ij}(t)\right), \quad i = 1,\ 2,\ldots, n\ \kappa\alpha\iota\ j = 1,\ 2,\ldots., n. \tag{6}$$

where

- $v_{ij}(t+1)$ denotes the velocity in the current iteration and $v_{ij}(t)$ is the velocity in the previous iteration.
- w is the inertia weight, used to balance the global exploitation and local exploitation, providing a memory of the previous particle's direction which prevents major changes in the suggested direction of the particles.
- r_1 and r_2 are two variables which are randomly derived from uniform distribution in range $[0, 1]$.
- c_1 and c_1 are defined as "acceleration coefficients" which have a huge effect on the efficiency of the PSO method. The constant c_1 conveys how much confidence a particle has in itself, while c_2 expresses how much confidence a particle has in the swarm.
- The variable $pbest_{ij}(t)$ is the best position of the particle until the iteration t, whereas $gbest_{ij}(t)$ is the finest position of the whole swarm until the same iteration.
- The term $c_1 r_1 \left(pbest_{ij}(t) - x_{ij}(t)\right)$ is known as the cognitive component; it acts as a kind of memory that stores the best previous positions that the particle has achieved.

The cognitive component reflects the tendency of the particles to return to their best positions.

- The term $c_2r_2\left(gbest_{ij}(t) - x_{ij}(t)\right)$ is called the social component. In this particular case, the particles behave according to the knowledge that they have obtained from the swarm, having as a guide the swarm's best position.

The acceleration coefficients c_1 and c_2, together with the random variables r_1 and r_2, affect to a great extent the evolution of cognitive and social component and hence the velocity value, which, as is known, is mainly responsible for the ultimate direction of the particles.

4. The Basic PSO for Solving the TP

The proposed PSO algorithm used to solve the TP is presented in this section. The primary goal is the initialization of the particles according to the problem's instances. This is achieved through a sub-algorithm (an initialization algorithm), as presented below. Initially, the amounts of the supply and demand were defined in tables. Subsequently, through control conditions, the amounts were randomly distributed, satisfying the constraints of the sums of supply and demand.

First, Algorithm 1 creates two vectors, namely, *Supply* and *Demand*, which are its input, as shown in lines 1 and 2. Next, variables m and n are computed. These variables are equal to the values of parameters *Supply* and *Demand*, respectively. Then, a matrix is created consisting of random real numbers (line 7). In line 10, the elements of the candidate solution matrix are rounded to the nearest integer as the amounts of commodities should be non-negative integer values. In the following lines of the algorithm, a process of readjustment and redistribution of matrix L begins so that its values correspond to the given *Supply* and *Demand* amounts. In lines 11 and 12, the sum of all elements of each row of matrix L is stored in vector *Sumrow*, while the sum of all elements of each column of matrix L is stored in vector *Sumcol*. Then, two new vectors, namely, s and d, are created by subtracting *Sumrow* from *Supply* and *Sumcol* from *Demand*, respectively. In the following lines, for each cell of the final matrix, the shortcomings of the matrix are located and assembled appropriately to each cell by zeroing out the excess amount of vectors s and d. The output of Algorithm 1 is a matrix consisting of the initial solutions (Initial Basic Feasible Solutions—IBFS), which comprises the input of Algorithm 2 (see below). All possible Initial Basic Feasible Solutions (IBFS) are non-negative integer values satisfying the supply and demand constraints.

Next, we present the structure of the basic PSO algorithm, which will be applied to solve the TP (Algorithm 2). The process starts with the initialization of the population size *npop*, the maximum number of iterations t_{max}, the personal and social acceleration coefficients c_1 and c_2, the random variables r_1 and r_2 and, finally, the inertia weight w (line 1). Moreover, subsequently, the *Supply*, *Demand* and *Cost* matrixes are defined (line 2).

Line 6 calculates the total transport cost of each particle. Then, in lines 7 and 8, whether the total cost of the current particle is less than the minimum transport cost calculated up to then is checked. If the statement is true, the value of global best cost is upgraded, and this particle is now defined as the best. This process is continued for all candidate particles. In lines 9 to 14, through an iterative loop, the position and velocity of the particles are calculated using Equations (5) and (6). The algorithm exports the particle with the optimal position and its respective optimal transport cost.

Algorithm 1: Initialization algorithm

```
1.  Define Supply = [s_1, s_2, ..., s_m]
2.  Define Demand = [d_1, d_2, ..., d_n]
3.  Define m = length (Supply)
4.  Define n = length (Demand)
5.  Initialize a solution matrix I = zeros (m, n)
6.  Initialize a supporting table B = zeros (m × n)
7.  Generate a new random number x, x = −log (rand(m × n, 1)
8.  Set x = x/Sum(x)
9.  Take a matrix L = reshape (B × x, [m, n])
10. Set solution as a matrix to round each element of L to the nearest
    integer less than or equal to that element of L as Solution = floor [L]
11. Set Sumrow as the sum of the elements of all rows.
12. Set Sumcol as the sum of the elements of all columns.
13. Set s = Supply − Sumrow and d = Demand − Sumcol
14. for i = 1 to m do
15.     for j = 1 to n do
16.         if s(i) smaller or equal to d(j)
17.             ww = min (d(i), d(j))
18.             main (i,1) = main (i, j) + ww
19.             d(j) = d(j) − s(i)
20.         else if d(j) smaller than s(i)
21.             ww = min (s(i), d(j))
22.             main(i, j) = main (i, j) + ww
23.             s(i) = s(i) − d(j)
24.         end
25.     end
26. end
27. Return L = Solution
```

The exported values of the particle's position, although satisfying demand and supply constraints, were observed to be taking occasionally negative and/or fractional values. These values cannot support the aspect of the solution since the values are quoted in quantities (only positive values are allowed); therefore, appropriate modifications have been made for the final form of the particle position.

Two sub-algorithms were designed to repair the algorithm, replacing negative and fractional volumes with natural numbers without breaking the supply and demand conditions.

Algorithm 2: Particle Swarm Optimization algorithm

```
Set the values of t_max, npop, w, c1, c2, r1, r2
Define Supply, Demand and Cost matrices
Define the initial particles
Initialize particle position as
particle(i).Position = initial_particle(i).Position
Initialize velocity as particle(i).Velocity = zeros (m, n)
Calculate the particle(i).Cost = fitnessfun(particle(i).Position)
Update the personal best position and the particle best cost
Update the global best position
    for it = 1: t_max (number of iterations)
        for i = 1: npop (number of particles)
            Calculate particle's velocity: particle(i).Velocity
            Update particle's position: particle(i).Position
            Update personal best position
            Update global best position, Global best
        end
    end
Return GlobalBest.Cost
Return GlobalBest
```

Algorithm 3 takes as input a matrix—*particle*(*i*). *Position*—that has negative values in its cells. The aim is for the negative elements to be eliminated as in [17]. Through an iterative process, which is illustrated in line 3, the algorithm checks each line of the cell of the table and sets as *neg* the value of the cell with the negative value. Subsequently, it searches the maximum element of the column where its negative element was found, as shown in lines 5 and 6. The cardinal value of the negative value is subtracted from the cell with the largest negative value, while the cell with the negative value is set to zero. In line 9, a cell is randomly selected from the row that corresponds to the negative element. If the value is positive, the cardinal of the negative element is subtracted from it. Simultaneously, a cell of this row is counterbalanced by adding to it the cardinal of the negative cell as shown in line 13. Algorithm 3 exports the *particle*(*i*).*Position* with non-negative values, while sustaining the supply and demand conditions.

Applying the above transformation, the result is a matrix with positive but also fractional elements. Algorithm 4 takes as its input the matrix of particles' positions after removing the negative elements. In line 3, a new matrix named *pos* is defined as containing the integer elements of matrix *particle*(*i*).*Position*. In line 4, a vector named *sumrow* is created which contains the sum of each row of the *pos* matrix; while in line 5, a vector named *sumcol* is created, containing the sum of each column. In lines 6 and 7, the differences between the quantities of the *Supply* and *sumrow* and *Demand* and *sumcol* matrices are noted, respectively, in order to record the quantities missing from the *pos* matrix. Then, through an iterative loop, the u cell of $s(u)$ is compared with the v cell of $d(v)$. The minimum quantity of these two is selected and entered into the *pos* matrix, reallocating the integer amounts in an appropriate manner to satisfy the available supply and demand items. The algorithm terminates when vectors s and v are zeroed and the integer quantities are inserted into *pos* matrix, which is the output of Algorithm 4. The final solution is a non-negative integer solution matrix satisfying the requested constraints.

Algorithm 3: Negative values repair algorithm

```
for k = 1:m
    for l = 1:n
        if particle(i).Position(k, l) < 0
            Set neg = particle(i).Position(k, l)
            Find the maximum element of the i-th column,
             as max= (particle(i).Position(:, l))
            Find the exact position of the maximum element
            Change the value of maximum
             as particle(i).Position(a, b) = max_element − |neg|
            Set the negative element particle(i).Position(k, l) = 0
            Select a random number of the k-th row
            l_count = 1
            while ((j_count <= n)
                if particle(i).Position(k, l_count) > 0
                    Set particle(i).Position(k, l_count) = particle(i).Position(k, l_count) − |neg|

                    Balance one element in row k as:
                  particle(i).Position(a, l_count) = particle(i).Position(a, l_count) + |neg|
                end
            l_count = l_count + 1
            end
        end
    end
end
```

Algorithm 4: Amend fractions

```
Set as input particle(i).Position
Enter Supply, Demand
pos = floor(particle(i).Position)
Set as sumrow the sum of the elements of the row.
Set as sumcol the sum of the column's elements
Take s = Supply − sumrow
Take d = Demand − sumcol
for u = 1:m
        for v = 1:n
                If s(u) ≤ d(v)
                        ww = min(s(u), d(v))
                        pos(u, v) = pos(u, v) + ww
                        d(v) = d(v) − s(u)
                        s(u) = 0
                else if d(v) < s(u)
                        ww = min(s(u), d(v))
                        pos(u, v) = pos(u, v) + ww
                        s(u) = s(u) − d(v)
                        d(v) = 0
                end
        end
end
Return particle(i).position = pos
```

5. Variations of PSO

This section presents two already-known and two new variations of the classical implementation of the PSO, which are presented and used in this contribution to solve the TP. These variations are investigated in order to improve the performance of the classical PSO algorithm.

5.1. Decreasing Weight Particle Swarm Optimization (DWPSO)

The inertia weight w is the most influential parameter with respect to both the success rate and the function evaluation [18]. In DWPSO, the inertia factor is linearly decreasing. The decision to use this variation was not arbitrary; DWPSO is one of the classic and very effective PSO variations since its superiority remains imperishable over years. Through DWPSO, the algorithm focuses on diversity at former iterations and on convergence at latter ones [18]. The right and proper selection of the inertia weight provides a balance among global and local exploitation and results in fewer iterations, on average, to find a sufficiently optimal solution [19]. Exploitation is the capacity of particles to converge to the same peak of the objective function and remain there without wanting to obtain better solutions in their wider field. On the contrary, in the exploration condition, the particles are in constant search, discovering beneficial solutions. After constant research regarding the figurative of inertia weight, Shi and Eberhart concluded that values in the interval [0.9, 1.2] have a positive effect on the improvement of the solution [20]. A linearly decreasing inertia weight with $c_1 = 2, c_2 = 2$ and w between 0.4 and 0.9 was used by Shi and Eberhart, too. According to their claim, w_{new} is the new inertia weight, which linearly decreases from 0.9 to 0.4.

Equation (7) for DWPSO is given as

$$w_{new}(t) = w_{max} - \frac{(w_{max} - w_{min}) \cdot t}{t_{max}}, \tag{7}$$

where w_{max} is set as 0.9, performing extensive exploration, and w_{min} is equal to 0.4, performing more exploitation. Moreover, t is the current iteration of the algorithm and t_{max} is the maximum number of iterations. A large portion of researchers' results illustrate that linearly decreasing in the inertia weight can greatly improve the performance of PSO, having better results than the classic implementation of the algorithm.

5.2. Time-Varying Acceleration Coefficients (TVAC)

In population-based optimization methods, proper control of global and local exploration is essential for the efficient identification of the optimum solution.

Rathweera et al. introduced the TVAC in PSO [11]. According to their research, the cognitive parameter c_1 starts with a high value and linearly decreases to a low value, whereas the social parameter c_2 starts with a low value and linearly increases to a high value [21]. On the one hand, with a large value for the cognitive parameter and small value for the social parameter at the beginning, particles are moving by their own experience according to their own best positions, being able to move freely without following the mass. On the other hand, a small value for the cognitive parameter and a large value for the social parameter help the particles to escape from the area around their personal best positions and allow them to enhance the global search in the latter stages of the optimization procedure, converging toward the global optima. This concept can be mathematically represented as

$$c_1 = c_{1i} - \left(c_{1i} - c_{1f}\right) \cdot \frac{t}{t_{max}}; \tag{8}$$

$$c_2 = c_{2i} - \left(c_{2i} - c_{2f}\right) \cdot \frac{t}{t_{max}}, \tag{9}$$

where c_{1i} defines the value of c_1 in the first iteration equal to 2.5 and c_{1f} defines the value of c_1 in the last iteration equal to 0.5. Respectively, the value of c_2 in the first iteration is c_{2i} and is set to 0.5, while the value of c_2 in the first iteration is c_{2f} and is set to 2.5 [21].

5.3. Trigonometric Acceleration Coefficients-PSO (TrigAC-PSO)

In this subsection, a new variation is introduced. According to this variation, the impact of parameters c_1 and c_2 is extensively studied. First, each particle is guided by the knowledge and experience gained by the swarm (the value of c_2 is considerably bigger than the value of c_1). Next, relying on the learning mechanism, each particle builds its own strategy and acquires its own experience (the value of c_2 is becoming smaller while the value of c_1 is becoming bigger (see Equations (10) and (11)). This decrement of c_2 and increment of c_1 take place until both parameters are equalized to 2 in the last generation of the algorithm.

The following equations are used to calculate the cognitive and social acceleration parameters:

$$c_1 = \frac{c_{1f}}{2} + \sin \frac{2 \cdot c_{1i} \cdot t}{t_{max}} \cdot \frac{\pi}{2}; \tag{10}$$

$$c_2 = c_{2i} + \cos \frac{c_{2f} \cdot \pi \cdot t}{2 \cdot t_{max}} - \frac{1}{2}. \tag{11}$$

Here, in the first iteration, c_{1i}, which is the personal acceleration value, is equal to 0.5, while c_{2i}, which is the social acceleration value, is equal to 3.5. In the last iteration of the algorithm, both personal c_{1f} and social c_{2f} are equal to 2.

The value of inertia weight w varies according to the number of the current iteration t and the number of maximum iterations t_{max}.

It is described as follows in Equation (12):

$$w = \frac{t_{max} - t}{t_{max}}. \tag{12}$$

5.4. Four Sectors Varying Acceleration Coefficients PSO (FSVAC-PSO)

In the following section, a new variation is developed. This variation is novel and comprises the major technical merit of this contribution. The major role in this variation is the multiple changes of the coefficient parameters c_1 and c_2. In this case, the solution is approached both from the knowledge of the particle and from the experience of the whole swarm. The number of iterations is divided into four sectors. Starting from the first iteration, the social and cognitive acceleration coefficient is initialized to 2. In the first sector of iterations, the value of c_1 is increasing while the value of c_2 is decreasing. As a result, the particle is mostly influenced by its own knowledge, while the influence of the swarm on it is limited; in the second sector, the value of c_1 is decreasing while the value of c_2 is increasing to an equilibrium between the knowledge of the particle gained at the previous sector and the experience of the swarm; in the third sector, the value of c_1 is decreasing while the value of c_2 is increasing—explicitly, the particles are allowed to move towards the global best position, following the swarm's movements; as a result, information about the global best is reallocated to all the particles for more exploration before the swarm finally converges [11]; in the fourth sector, the particles head toward both their own personal best and global best observed by the whole swarm—the concept of this variation is based on the combination of all types of different searching behaviors, as they arise for different values of the coefficient acceleration parameters, culminating in equilibrium between exploitation and exploration of the search space; finally, in the last iteration, the two coefficient parameters are equated.

The formulation is represented in detail below:

- In the first Iteration, as already mentioned:

$$c_1 = 2$$

$$c_2 = 2;$$

- In the first sector:

$$\left.\begin{array}{c} c_1 = \frac{(2\cdot c_{1f} - c_{1i})\cdot t}{t_{max}} - 1 \\ c_2 = \frac{c_{2i}}{2\cdot c_{2f}} \cdot \frac{t}{t_{max}} \end{array}\right\} \tag{13}$$

where $c_{1i} = 2$, $c_{1f} = 3$,$c_{2i} = 2$ and $c_{2f} = 1$;

- In the second sector:

$$\left.\begin{array}{c} c_1 = \frac{c_{1f}\cdot c_{1i}\cdot t}{2\cdot t_{max}} - 1 \\ c_2 = 0.5 + \frac{c_{2f}+c_{2i}}{2} \cdot \frac{t}{t_{max}} \end{array}\right\} \tag{14}$$

where $c_{1i} = 3$, $c_{1f} = 2$, $c_{2i} = 1$ and $c_{2f} = 2$;

- In the third sector:

$$\left.\begin{array}{c} c_1 = \frac{3}{2} - \frac{(c_{1i} - c_{1f})\cdot t}{t_{max}} \\ c_2 = 0.5 + \frac{2\cdot c_{2i}+1}{c_{2f}} \cdot \frac{t}{t_{max}} \end{array}\right\}, \tag{15}$$

where $c_{1i} = 2$, $c_{1f} = 0.5$, $c_{2i} = 2$ and $c_{2f} = 2.5$;

- In the fourth sector:

$$\left.\begin{array}{c} c_1 = \frac{(4c_{1i} + c_{1f})\cdot t}{t_{max}} \\ c_2 = \frac{3}{2} + (c_{2f} - c_{2i}) \cdot \frac{t}{t_{max}} \end{array}\right\}, \tag{16}$$

where $c_{1i} = 0.5$, $c_{1f} = 2$, $c_{2i} = 2.5$ and $c_{2f} = 2$;

- In the last iteration:

$$c_1 = 2$$

$$c_2 = 2.$$

In the above formulations, c_{1i}, c_{1f}, c_{2i} and c_{2f} are initial and final values of cognitive and social components acceleration factors, respectively. To improve the solution quality, these coefficients are updated in such a way that the values increase and decrease at a steady pace. According to this approach, the solution avoids being trapped into a local optimum, as shown by the experimental results presented in Section 6.

As for the inertia weight w, Equation (12) is used to provide the necessary momentum for particles to roam across the search space.

6. Case Studies and Experimental Results

In this section, the proposed variations of the PSO algorithm are applied in thirty two well-known numerical examples of the TP, as shown in Table 1. The numerical examples of this study come from the research of B. Amaliah, who compared five different methods, which will be presented briefly below, regarding their performance in solving the TP [22].

Vogel's Approximation Method (VAM) is an iterative procedure such that in each step, proper penalties for each available row and column are taken into account through the least cost and the second-least cost of the transportation matrix [22]. The Total Differences Method 1 (TDM1) was introduced by Hosseini in 2017. The method is based on VAM's innovation to use penalties for all rows and columns of the transportation matrix. The TDM1 was developed by calculating penalty values only for rows of the transportation matrix [23]. Amaliah et al., in 2019, represented their new method, known as the Total Opportunity Cost Matrix Minimal Cost (TOCM-MT). This method has a mechanism with which to check the value of the least-cost cell before allocating the maximum units x_{ij}; this is in contradiction to the TDM1, which directly allocates the maximum units x_{ij} to the least cost [24]. Juman and Hoque, in 2015, developed a formulation method called the Juman and Hoque method (JHM). Their study is based on the distribution of supply and demand quantities, taking into account the two minimum-cost cells and their redistribution through penalties [25]. Finally, the last method presented is known as the Bilqis Chastine Erma Method (BCE), which constitutes an enhanced version of the JHM [26].

Table 1. Detail of 32 numerical examples of the TP.

No	From Journal	Name	Problem Size	Optimal Solution
1	Srinivasan and Thompson (1977)	Pr.1	3·4	880
2	Deshmukh (2012)	Pr.2	3·4	743
3	Ramadan and Ramadan (2012)	Pr.3	3·3	5600
4	Schrenk et al. (2011)	Pr.4	3·4	59
5	Samuel (2012)	Pr.5	3·4	28
6	Imam et al. (2009)	Pr.6	3·4	435
7	Adlakha and Kowalski (2009)	Pr.7	4·5	390
8	Kaur et al. (2018)	Pr.8	3·5	1580
9	G. Patel et al. (2017)	Pr.9	4·4	49
10	Ahmed et al. (2016b)	Pr.10	4·4	410
11	Ahmed et al. (2016b)	Pr.11	3·4	2850
12	Ahmed et al. (2016a)	Pr.12	3·5	183
13	Uddin and Khan (2016)	Pr.13	3·4	799
14	Uddin and Khan (2016)	Pr.14	3·5	273
15	Das et al. (2014a)	Pr.15	3·4	1160
16	Khan et al. (2015a)	Pr.16	3·4	200
17	Azad and Hossain (2017)	Pr.17	3·4	240
18	Morade (2017)	Pr.18	3·3	820
19	Jude (2016)	Pr.19	3·4	190
20	Jude (2016)	Pr.20	4·4	83
21	Hosseini (2017)	Pr.21	3·4	3460
22	Amaliah et al. (2019)	Pr.22	3·4	910
23	Amaliah et al. (2019)	Pr.23	4·4	1670
24	Amaliah et al. (2019)	Pr.24	4·4	2280
25	Amaliah et al. (2019)	Pr.25	3·4	2460
26	Amaliah et al. (2019)	Pr.26	3·3	291
27	Juman and Hoque (2015)	Pr.27	3·3	4525
28	Juman and Hoque (2015)	Pr.28	3·4	920
29	Juman and Hoque (2015)	Pr.29	3·4	809
30	Juman and Hoque (2015)	Pr.30	3·4	417
31	Juman and Hoque (2015)	Pr.31	4·5	3458
32	Juman and Hoque (2015)	Pr.32	4·6	109

The whole algorithmic approach was implemented using MATLAB R2021b. The algorithm was tested on a set of different dimensional problems. All parameters of the proposed algorithm were selected after exhaustive experimental testing. Each of the four variations was tested using different parameter values, and those values whose computational results were superior to other values were selected. The number of iterations is set to 100. The parameter r_1 is set as a random number derived from the uniform distribution in range [0, 1], and r_2 is set as the complement of r_1; that is, $r_2 = 1 - r_1$. This modification plays a significant part as it is different from the customary application where both r_1 and r_2 are randomly derived uniformly from range [0, 1]. Using the former relationship between r_1 and r_2, we manage to achieve stronger control over these parameters' values.

In the following table (Table 2) and Figure 1, the performance of both the exact methods and the PSO-based ones are presented for 30 Monte Carlo runs; more precisely, the best value achieved by each method is depicted. The last column presents the optimal solution of each numerical problem. As shown, the Vogel method manages to find 9 out of the 32 test instances (28.13%); the Total Differences Method 1 (TDM1) succeeds in finding more optimal solutions than the Vogel method by finding 13 out of 32 optimal solutions (40.63%); using the TOCM-MT method, the results show that the method's performance is better still, finding the optimum in 23 out of 32 test instances (71.9%); the JHM method, which accumulated 21 optimal solutions, was less effective than TOCM-MT (65.62%); the BCE method, which achieved 27 out of 32 test instances (84.4%), proved to be the most efficient compared to all previously mentioned methods; the classic PSO, the TVAC, the TrigAC-PSO and the FSVAG-PSO achieve the optimum in 31 out of 32 test instances (96.88%), while the PWPSO achieves the optimum in 30 out of the 32 (93.76%).

Table 2. The optimal solution of each method for the 32 test instances.

No.	Name	VAM	TDM1	TOCM-MT	JHM	BCE	PSO	DWPSO	TVAC	TrigAC-PSO	FSVAC	Optimal (Op)
1	Pr.1	955	880	880	880	880	880	880	880	880	880	880
2	Pr.2	779	779	743	743	743	743	743	743	743	743	743
3	Pr.3	5600	5600	5600	5600	5600	5600	5600	5600	5600	5600	5600
4	Pr.4	59	59	61	59	59	59	59	59	59	59	59
5	Pr.5	28	28	28	28	28	28	28	28	28	28	28
6	Pr.6	475	475	435	460	435	435	435	435	435	435	435
7	Pr.7	390	400	390	390	390	390	390	390	390	390	390
8	Pr.8	1600	1595	1580	1580	1580	1580	1580	1580	1580	1580	1580
9	Pr.9	49	53	53	49	49	49	49	49	49	49	49
10	Pr.10	470	435	435	420	410	410	411	410	410	410	410
11	Pr.11	2850	2850	2850	2850	2850	2850	2850	2850	2850	2850	2850
12	Pr.12	187	186	187	183	187	183	183	183	183	183	183
13	Pr.13	859	859	799	799	799	799	799	799	799	799	799
14	Pr.14	273	273	273	273	273	290	273	273	273	273	273
15	Pr.15	1220	1160	1160	1170	1170	1160	1160	1160	1160	1160	1160
16	Pr.16	204	200	200	218	204	200	200	200	200	200	200
17	Pr.17	248	248	240	240	240	240	240	240	240	240	240
18	Pr.18	820	820	820	820	820	820	820	820	820	820	820
19	Pr.19	190	190	190	190	192	190	190	190	190	190	190
20	Pr.20	92	83	83	83	83	83	83	83	83	83	83
21	Pr.21	3520	3570	3460	3460	3460	3460	3460	3460	3460	3460	3460
22	Pr.22	990	990	910	960	910	910	910	910	910	910	910
23	Pr.23	1680	1670	1670	1690	1670	1670	1670	1670	1670	1670	1670
24	Pr.24	2400	2400	2400	2310	2280	2280	2286	2281	2284	2288	2280
25	Pr.25	2980	2980	2500	2500	2460	2460	2460	2460	2460	2460	2460
26	Pr.26	327	291	291	327	291	291	291	291	291	291	291
27	Pr.27	5125	4550	5225	4525	4525	4525	4525	4525	4525	4525	4525
28	Pr.28	960	960	930	920	920	920	920	920	920	920	920
29	Pr.29	859	849	809	809	809	809	809	809	809	809	809
30	Pr.30	476	465	417	417	417	417	417	417	417	417	417
31	Pr.31	3778	3572	3513	3487	3487	3458	3458	3458	3458	3458	3458
32	Pr.32	112	117	109	112	109	109	109	109	109	109	109

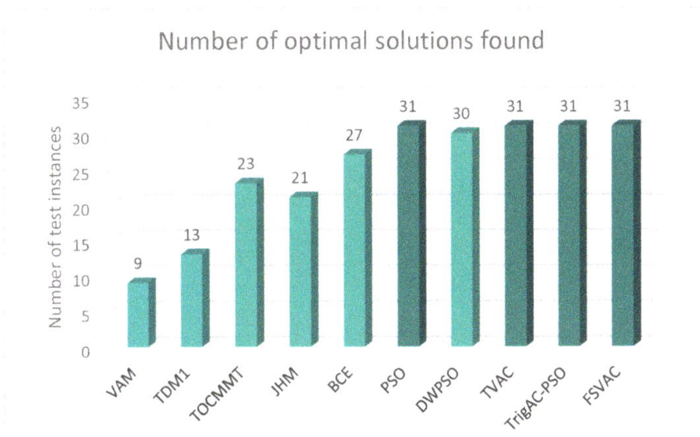

Figure 1. The number of optimal solutions that every method achieved.

One significant finding of our research is that the new PSO variation, TrigAC-PSO, which is first presented in this study, achieved very good results. The following table

(Table 3) examines the deviation of VAM, TDM1, TOCM-MT, JHM, BCE, PSO, DuPSO, TVAC, TrigAC-PSO and FSVAC-PSO. The measurement of deviation shows the difference between the observed value and the expected value of a variable, and it is given by the following formula:

$$Dev = \frac{x_{ij} - optimal}{optimal}, \tag{17}$$

where x_{ij} is the current solution.

Table 3. The deviation (dev) of the methods for 32 numerical examples.

	VAM	TDM1	TOCMMT	JHM	BCE	PSO	DWPSO	TVAC	TrigAC-PSO	FSVAC
Pr.01	0.085227	0	0	0	0	0	0	0	0	0
Pr.02	0.048452	0.048452	0	0	0	0	0	0	0	0
Pr.03	0	0	0	0	0	0	0	0	0	0
Pr.04	0	0	0.033898	0	0	0	0	0	0	0
Pr.05	0	0	0	0	0	0	0	0	0	0
Pr.06	0.091954	0.091954	0	0.057471	0	0	0	0	0	0
Pr.07	0	0.025641	0	0	0	0	0	0	0	0
Pr.08	0.012658	0.009494	0	0	0	0	0	0	0	0
Pr.09	0	0.081633	0.081633	0	0	0	0	0	0	0
Pr.10	0.146341	0.060976	0.060976	0.02439	0	0	0.002439	0	0	0
Pr.11	0	0	0	0	0	0	0	0	0	0
Pr.12	0.021858	0.016393	0.021858	0	0.021858	0	0	0	0	0
Pr.13	0.075094	0.075094	0	0	0	0	0	0	0	0
Pr.14	0	0	0	0	0	0.062271	0	0	0	0
Pr.15	0.051724	0	0	0.008621	0.008621	0	0	0	0	0
Pr.16	0.02	0	0	0.09	0.02	0	0	0	0	0
Pr.17	0.0333	0.0333	0	0	0	0	0	0	0	0
Pr.18	0	0	0	0	0	0	0	0	0	0
Pr.19	0	0	0	0	0.010526	0	0	0	0	0
Pr.20	0.108434	0	0	0	0	0	0	0	0	0
Pr.21	0.017341	0.031792	0	0	0	0	0	0	0	0
Pr.22	0.087912	0.087912	0	0.054945	0	0	0	0	0	0
Pr.23	0.005988	0	0	0.011976	0	0	0	0	0	0
Pr.24	0.052632	0.052632	0.052632	0.026316	0	0	0.002632	0.000439	0.001754	0.003509
Pr.25	0.211382	0.211382	0.01626	0.01626	0	0	0	0	0	0
Pr.26	0.123711	0	0	0.123711	0	0	0	0	0	0
Pr.27	0.132596	0.005525	0.154696	0	0	0	0	0	0	0
Pr.28	0.043478	0.043478	0.01087	0	0	0	0	0	0	0
Pr.29	0.061805	0.049444	0	0	0	0	0	0	0	0
Pr.30	0.141487	0.115108	0	0	0	0	0	0	0	0
Pr.31	0.092539	0.032967	0.015905	0.008386	0.008386	0	0	0	0	0
Pr.32	0.027523	0.073394	0	0.027523	0	0	0	0	0	0
Average	0.05292	0.03583	0.014023	0.01405	0.002168	0.001946	0.000158	0.000013	0.000054	0.00011

Considering Table 3 and Figure 2, it is evident that method VAM, TDM1, TOCM-MT and 1HM appear to be more inefficient, deviating from the optimal solution at a significant scale. More precisely, the results of Table 3 show that the solutions achieved by VAM differ from the optimal solution by 5.29%, the results of TDM1 by 3.58%, the results of TOCM-MT by 1.4% and the results of JHM by 1.4%. BCE method presented higher levels of efficiency since the values of deviation were negligible. Analysis of the data of Table 3 reveals that the percentage of the deviation in classic PSO, as well as in its variations, was almost zero. Furthermore, the TVAC method was nearest to the optimal solution, followed by TrigAC-PSO, FSVAC-PSO and finally by DWPSO.

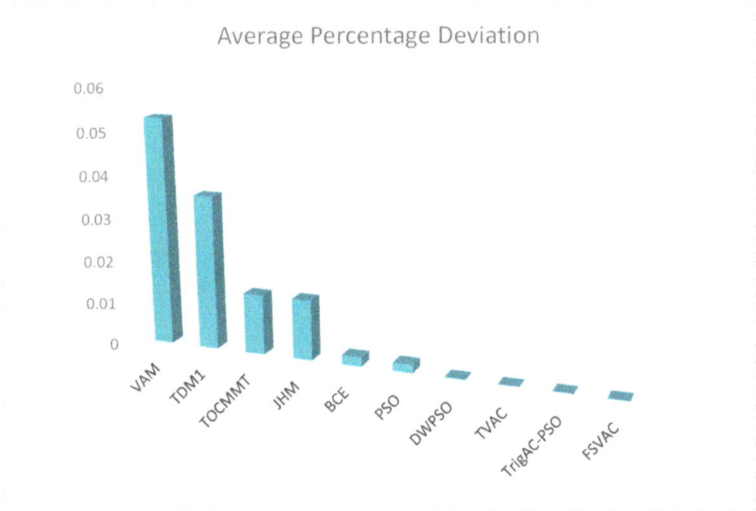

Figure 2. Average percentage deviation for each method.

The findings from the current study provide us with the basic information for an extensive meta-analysis, allowing us to investigate which of the presented PSO variations has better performance in solving the TP. To serve this cause, many experiments were carried out which investigated different values of PSO population size (number of particles). The classic PSO, as well as each one of its variations (DWPSO, TVAC, TrigAC-PSO, FSVAC-PSO), were tested for 10, 15 and 20 particles for all 32 numerical examples. The results presented in Tables 4–6 show the performance of the classic PSO as well of its variations for 30 independent runs. The number of generations was stable and equal to 100 for all runs.

Evidence from this study, presented in Table 4, expounds the accuracy rate of each algorithm for 10 particles. The accuracy rate is given by the following formulation:

$$Accuracy = \frac{TOR}{TR}, \tag{18}$$

where TOR is the total number of runs where optimal solution was found and TR is the number of runs.

Table 4 shows that the classic PSO obtained 38.33% accuracy rate. A significant increase in accuracy rate, using 10 particles, was evident in DWPSO, which achieved 59.58% accuracy, almost twice as much as the percentage of the classic PSO. Moreover, TVAC obtain a 61.45% accuracy rate. The best results came from TrigAC-PSO, since this PSO variation achieved a 62.81% accuracy rate. Last but not least, FSVAC achieved a 59.5% accuracy rate.

Table 4. Accuracy of PSO, DWPSO, TVAC, TrigAC-PSO and FSVAC for 10 particles.

	PSO	DWPSO	TVAC	TrigAC-PSO	FSVAC
Pr.01	0.0333	0.2	0.4666	0.5667	0.2333
Pr.02	0.7667	1	1	1	1
Pr.03	1	1	1	1	1
Pr.04	1	1	1	1	1
Pr.05	0.2333	0.6667	0.8333	0.8667	0.8333
Pr.06	0.3667	1	0.9667	1	1
Pr.07	0.7667	0.9	1	1	0.9
Pr.08	0	0	0.2667	0.1667	0.1334
Pr.09	0.0333	0.3	0.2	0.2667	0.1
Pr.10	0	0	0.0333	0	0.0333

Table 4. *Cont.*

	PSO	DWPSO	TVAC	TrigAC-PSO	FSVAC
Pr.11	0.1	0.1	0.1333	0.1	0.0667
Pr.12	0	0.3	0.4	0.4667	0.3667
Pr.13	0.5334	1	1	0.8667	1
Pr.14	0	0	0	0.3333	0
Pr.15	0.7	1	1	1	1
Pr.16	0	0.4667	0.4667	0.6667	0.5667
Pr.17	0.4333	0.9	0.9333	0.9	0.8333
Pr.18	1	1	1	1	1
Pr.19	0.4667	0.7	0.7667	0.5	0.7333
Pr.20	0.4667	0.8333	0.6667	0.8667	0.7667
Pr.21	0.5667	0.5333	0.3333	0.5333	0.3939
Pr.22	0.3667	1	0.9667	0.8333	1
Pr.23	0.0333	0.0333	0.1667	0.1667	0.1667
Pr.24	0	0	0	0	0
Pr.25	0.5333	0.9667	0.9333	0.9667	1
Pr.26	0.4667	0.6	0.7	0.7	0.7333
Pr.27	0.0333	0.4667	0.5667	0.4	0.4667
Pr.28	0.2	0.2667	0.0667	0.2	0
Pr.29	0.8333	1	1	1	1
Pr.30	0.8667	1	1	1	1
Pr.31	0.4	0.8333	0.7	0.7	0.7
Pr.32	0.0667	0	0	0.0333	0
Average	0.383338	0.595834	0.611459	0.6281313	0.594603

The accuracy rate results for 15 particles are presented in Table 5. DWPSO achieved 65.31% accuracy, whereas TVAC reached 66.99%. It is of particular interest that TrigAC-PSO achieved the highest accuracy rate once again by reaching 69.8%. Finally, FSVAC obtained an accuracy rate equal to 66.56%.

Table 5. Accuracy of PSO, DWPSO, TVAC, TrigAC-PSO and FSVAC for 15 particles.

	PSO	DWPSO	TVAC	TrigAC-PSO	FSVAC
Pr.01	0.0667	0.6	0.6333	0.6333	0.4333
Pr.02	0.9667	1	1	1	1
Pr.03	1	1	1	1	1
Pr.04	1	1	1	1	1
Pr.05	0.7	0.9	1	1	1
Pr.06	0.7	1	0.9667	1	1
Pr.07	0.7667	1	1	0.9667	0.9667
Pr.08	0	0.0667	0.3333	0.1667	0.0333
Pr.09	0.0667	0.2333	0.2333	0.2	0.1333
Pr.10	0.2667	0	0	0.0667	0
Pr.11	0.2333	0.0667	0.3	0.2	0.0333
Pr.12	0.1	0.4	0.4333	0.3333	0.2
Pr.13	0.5667	1	1	0.9667	1
Pr.14	0	0.0333	0	0.0667	0.1
Pr.15	0.3	0.9333	0.9667	1	1
Pr.16	0	0.6	0.5	0.9333	0.5333
Pr.17	0.5667	0.9	1	0.9667	1
Pr.18	1	1	1	1	1
Pr.19	0.5333	0.9	0.9	0.6667	1
Pr.20	0.6334	0.9667	0.8667	0.9333	1
Pr.21	0.6667	0.4333	0.6667	1	0.4667
Pr.22	0.6	1	1	1	1
Pr.23	0.0667	0.2	0.3667	0.2333	0.2
Pr.24	0.6666	0	0	0	0
Pr.25	0.6333	1	0.8333	1	1
Pr.26	0.3333	1	0.8	0.6667	1
Pr.27	0.3667	0.4	0.4667	0.8333	1
Pr.28	0.3333	0.1333	0.1	0.3333	0.1667
Pr.29	1	1	1	1	1
Pr.30	0.9333	1	1	1	1
Pr.31	0.3667	1	0.9333	1	1
Pr.32	0.1333	0.1333	0.1	0.1667	0.0333
Average	0.486463	0.653122	0.669894	0.69791875	0.665622

The accuracy rate results for 20 particles are presented in Table 6 and Figure 3. A high percentage of 53.33% was obtained by the classic PSO. Between DWPSO and TVAC, it is evident that both rates were sufficiently close, with accuracy rates ascending up to 66.78% and 66.56%, respectively. FSVAC, the variation which has been proposed and presented in this research, achieved accuracy rate equal to 66%. This new method evinced positive effects in terms of its validity and effectiveness. Last but not least, TrigAC-PSO demonstrated the best performance compared to all other variations, achieving 74.3%. Running the algorithm using 20 particles, TrigAC-PSO found the optimal in 31 out of 32 test instances, reaching 96.88%. Moreover, in 20 out of 32 numerical examples, this variation managed to reach the optimum in all 30 runs, with a success rate of 62.5%. The punctuality of this method rises to 75%; hence, this variation is established, compared to other variations, as the ideal option for the solution of the TP.

Table 6. Accuracy of PSO, DWPSO, TVAC, TrigAC-PSO and FSVAC for 20 particles.

	PSO	DWPSO	TVAC	TrigAC-PSO	FSVAC
Pr.01	0.1	0.6667	0.7333	0.7	0.6
Pr.02	1	1	1	1	1
Pr.03	1	1	1	1	1
Pr.04	1	1	1	1	1
Pr.05	0.7333	1	1	1	1
Pr.06	0.5667	1	1	0.9667	1
Pr.07	0.9	0.9667	0.9667	1	1
Pr.08	0.0667	0.2333	0.3	0.0667	0.1
Pr.09	0.0333	0.4	0.2	0.3333	0.4
Pr.10	0.2	0	0	0.1333	0.0667
Pr.11	0.3333	0.3	0.2	0.4	0.1
Pr.12	0.2333	0.5	0.4667	0.5333	0.7
Pr.13	0.8333	1	1	1	1
Pr.14	0	0.0333	0.0333	0.3	0.1
Pr.15	0.6667	1	0.9667	1	1
Pr.16	0.0333	0.6667	0.7	1	0.6
Pr.17	0.5333	0.9	1	1	1
Pr.18	1	1	1	1	1
Pr.19	0.5333	0.6333	0.5	0.6333	1
Pr.20	1	1	0.9333	0.9333	0.9667
Pr.21	0.7333	0.4	0.9	1	0.3
Pr.22	0.5333	1	0.9	1	0.3
Pr.23	0.1	0.3333	0.3333	0.7333	0.4667
Pr.24	0.0667	0	0	0	0
Pr.25	0.8333	1	1	1	0.9667
Pr.26	0.6667	1	1	1	1
Pr.27	0.4667	0.1333	0	0.6	0.2667
Pr.28	0.4667	0.1	0.1	0.2	0.1
Pr.29	1	1	1	1	1
Pr.30	0.8667	1	1	1	1
Pr.31	0.3667	1	1	1	1
Pr.32	0.2	0.1	0.0667	0.2333	0.0667
Average	0.533331	0.667706	0.665625	0.7427031	0.659381

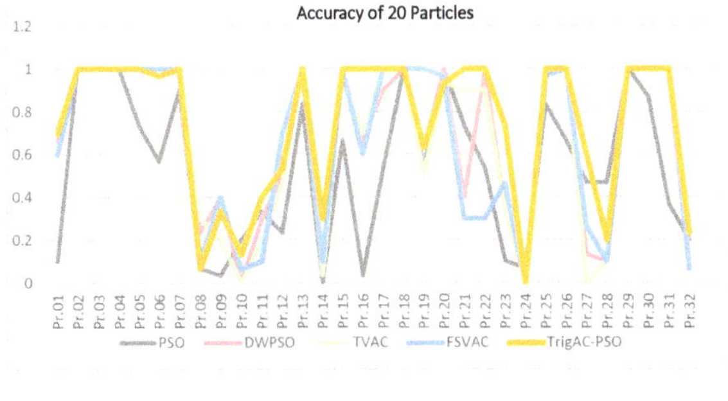

Figure 3. Accuracy for 20 particles.

In summary, the proposed method FSVAC-PSO, although it did not demonstrate the highest average success rate, was very accurate in calculating the optimal solution in cases where the aforementioned variations were unable to approach the optimal solution. In more detail, this research experimented on population sizes of 10, 15, 20 particles over 32 well-known test instances used in the respective literature. For each problem, as already mentioned, 30 independent experimental runs were conducted. In the case of 10 particles, the classical PSO found the optimal solution in only in 3 out of 32 test instances in all 30 runs (9.4%); DWPSO found the optimal solution in 10 out of 32 test instances in all 30 runs (31.25%); while TVAC and Trig-PSO managed to find the optimal solution in 9 out of 32 test instances in all 30 runs (28.13%); finally, FSVAC was shown to be the best PSO variation, finding the optimal solution in 11 out of 32 test instances in all 30 runs (34.4%).

In the case of 15 particles, FSVAC also showed the best performance by finding in the optimal solution in 18 out of 32 test instances in all 30 runs (56.25%); the classic PSO found the optimal solution in 4 out of 32 (12.5%) test instances, and TVAC in 11 out of 32, in all 30 runs; last but not least, both DWPSO and TrigAC-PSO found the optimal value in 13 out of 32 test instances in all 30 runs (40.63%).

In the case of 20 particles, the variations TrigAC-PSO and FSVAC are still more accurate than the other PSO variations since they succeeded in finding the optimal solution in 18 out of 32 and in 17 out of 32 test instances in all 30 runs, respectively. The other PSO variations attained relatively lower success rates in finding the optimal solution in all of their runs.

In the following table (Table 7), the most important statistical measures in the cases of 20 particles for 30 independent runs are represented for all PSO variations. These experimental results demonstrate the very good performance and stability of the proposed PSO variations in solving the TP. As presented, in all cases, the mean value is very close to the best one, showing that all these variations are not only efficient but also quite stable. The value of the Coefficient of Variation (CV), which is the basic measure for proving stability of stochastic algorithms, is, for all PSO variations, quite small; more precisely, the mean CV value is for each PSO variation is as follows: Classic PSO, 2.12%; DWPSO, 1.32%; TVAC, 0.87%; TrigAC-PSO, 0.66%; and FSVAC, 1.26%. These values show that TrigAC-PSO, which is one of the new PSO variations presented in this work, is the most stable one.

Table 7. Statistical measures for 20 particles.

		PSO	DWPSO	TVAC	TrigAC-PSO	FSVAC
Pr.01	Mean	894.7666	884.3	883.6	882.86667	884.4
	St.Dev	22.51847	10.61278	11.2544	7.41263	8.76356
	Min	880	880	880	880	880
	Max	965	928	940	910	917
	cv%	2.51668658	1.200133	1.273701	0.8396096	0.990905
Pr.02	Mean	743	743	743	743	743
	St.Dev	0	0	0	0	0
	Min	743	743	743	743	743
	Max	743	743	743	743	743
	cv%	0	0	0	0	0
Pr.03	Mean	5600	5600	5600	5600	5600
	St.Dev	0	0	0	0	0
	Min	5600	5600	5600	5600	5600
	Max	5600	5600	5600	5600	5600
	cv%	0	0	0	0	0
Pr.04	Mean	59	59	59	59	59
	St.Dev	0	0	0	0	0
	Min	59	59	59	59	59
	Max	59	59	59	59	59
	cv%	0	0	0	0	0
Pr.05	Mean	28.266666	28	28	28	28
	St.Dev	0.4497764	0	0	0	0
	Min	28	28	28	28	28
	Max	29	28	28	28	28
	cv%	1.591190254	0	0	0	0

Table 7. *Cont.*

		PSO	DWPSO	TVAC	TrigAC-PSO	FSVAC
Pr.06	Mean	441.1333333	435	435	435	435
	St.Dev	7.619092775	0	0	0	0
	Min	435	435	435	435	435
	Max	463	435	435	435	435
	cv%	1.72716324	0	0	0	0
Pr.07	Mean	391.5	390.0333	390.4667	390	390
	St.Dev	5.015493237	0.182574	2.55603	0	0
	Min	390	390	390	390	390
	Max	410	391	404	390	390
	cv%	1.281096612	3.290831	0.654611	0	0
Pr.08	Mean	1650.933333	1593.033	1598.7	1592.1333	1629.533
	St.Dev	44.7667775	17.95873	23.31367	12.23824	40.14347
	Min	1580	1580	1580	1580	1580
	Max	1790	1642	1661	1623	1712
	cv%	2.711604194	1.127329	1.45829	0.7686698	2.463495
Pr.09	Mean	52.6	51.3	52.06667	51.6	51.16667
	St.Dev	2.40114915	2.768667	1.79910	1.90462	1.89524
	Min	49	109	49	49	49
	Max	63	122	55	53	53
	cv%	4.564922339	5.397012	3.455389	3.6911285	3.704062
Pr.10	Mean	427.4333333	427.7667	428.1	428.3	425.8
	St.Dev	8.935336025	4.38401	5.16853	6.25410	5.71386
	Min	410	411	411	410	410
	Max	434	431	431	432	430
	cv%	2.09046308	1.02486	1.20732	1.4602153	1.341913
Pr.11	Mean	2913.833333	2934.8	2864.633	2857.1	3036.767
	St.Dev	186.8538214	220.8248	226.892	11.66885	380.4261
	Min	2850	2850	2850	2850	2850
	Max	3850	3945	2977	2891	4554
	cv%	6.412646162	7.524358	1.117014	0.4084159	12.52734
Pr.12	Mean	190.5666667	184.4333	186.0333	184.33333	183.9
	St.Dev	7.623391106	1.50134	4.60496	1.49327	1.39827
	Min	183	183	183	183	183
	Max	206	186	200	186	186
	cv%	4.00038015	0.814029	2.475347	0.8100976	0.760345
Pr.13	Mean	805.3666667	799	799	799	799
	St.Dev	16.77535823	0	0	0	0
	Min	799	799	799	799	799
	Max	878	799	799	799	799
	cv%	2.082946678	0	0	0	0
Pr.14	Mean	319.9333333	302.2	290.4	290.1	292.2667
	St.Dev	21.78251962	17.70525	6.69328	16.159442	8.10250
	Min	292	273	273	273	273
	Max	378	335	317	327	306
	cv%	6.808455809	5.858785	2.304849	5.5703008	2.772298
Pr.15	Mean	1186.1	1160	1160.067	1160	1160
	St.Dev	73.12122669	0	0.36514	0	0
	Min	1160	1160	1160	1160	1160
	Max	1401	1160	1162	1160	1160
	cv%	6.164845012	0	0.031476	0	0
Pr.16	Mean	217.1333333	202.9667	202.7333	200	203.2333
	St.Dev	7.946950546	5.979755	6.53338	0	6.76034
	Min	200	200	200	200	109
	Max	237	218	220	200	119
	cv%	3.659940381	2.946176	3.222647	0	3.326397
Pr.17	Mean	244.6333333	241.6667	240	240	240
	St.Dev	6.025711195	5.195046	0	0	0
	Min	240	240	240	240	240
	Max	256	259	240	240	240
	cv%	2.463160319	2.149674	0	0	0

Table 7. *Cont.*

		PSO	DWPSO	TVAC	TrigAC-PSO	FSVAC
	Mean	820	820	820	820	820
	St.Dev	0	0	0	0	0
Pr.18	Min	820	820	820	820	820
	Max	820	820	820	820	820
	cv%	0	0	0	0	0
	Mean	190.8	190.7	190.9333	190.56667	190
	St.Dev	0.924755326	0.952311	0.98026	0.81720	0
Pr.19	Min	190	190	190	190	190
	Max	192	192	192	192	190
	cv%	0.484672603	0.499377	0.513407	0.4288264	0
	Mean	83	83	83.3	83.2	83.1
	St.Dev	0	0	0.91538	0.761124	0.54772
Pr.20	Min	83	83	83	83	83
	Max	83	83	86	86	86
	cv%	0	0	1.098902	0.914813	0.659113
	Mean	3484.5	3536.467	3468.2	3460	3536.1
	St.Dev	58.99371679	71.42864	34.7804	0	63.5839
Pr.21	Min	3460	3460	3460	3460	3460
	Max	3745	3646	3645	3460	3644
	cv%	1.693032481	2.019774	1.00284	0	1.798138
	Mean	928.3	910	913.6	910	910
	St.Dev	28.61534433	0	14.8686	0	0
Pr.22	Min	910	910	910	910	910
	Max	990	910	990	910	910
	cv%	3.08255352	0	1.627476	0	0
	Mean	1679.1	1671.133	1670.733	1671	1671.367
	St.Dev	14.23291474	1.136642	0.58329	2.34888	2.02541
Pr.23	Min	1670	1670	1670	1670	1670
	Max	1724	1675	1672	1679	1678
	cv%	0.847651405	0.068016	0.034912	0.1405674	0.121183
	Mean	2403.966667	2366.4	2372.8	2361.5667	2333.833
	St.Dev	44.98005944	29.09627	31.0332	18.34287	117.1320
Pr.24	Min	2280	2317	2320	2322	2292
	Max	2495	2430	2424	2390	2420
	cv%	1.871076669	1.229559	1.307874	0.7767247	1.47626
	Mean	2468.766667	2460	2460	2460	2460
	St.Dev	24.21695521	0	0	0	0
Pr.25	Min	2460	2460	2460	2460	2460
	Max	2563	2460	2460	2460	2460
	cv%	0.980933335	0	0	0	0
	Mean	292.6	291	291	291	291
	St.Dev	2.485821865	0	0	0	0
Pr.26	Min	291	291	291	291	291
	Max	299	291	291	291	291
	cv%	0.84956318	0	0	0	0
	Mean	4574.233333	4639.967	4666.433	4535	4634.9
	St.Dev	73.42821187	60.22915	29.96973	18.34910	67.4073
Pr.27	Min	4525	4525	4529	4525	4525
	Max	4753	4675	4677	4585	4675
	cv%	1.605257243	1.298051	0.642241	0.4046109	1.454343
	Mean	941.7	953.3667	953.0667	947.26667	947.9667
	St.Dev	22.49314072	19.67404	13.35957	17.26454	13.60396
Pr.28	Min	920	920	920	920	920
	Max	974	992	960	960	968
	cv%	2.38856756	2.063638	1.401746	1.8225639	1.435060
	Mean	809	809	809	809	809
	St.Dev	0	0	0	0	0
Pr.29	Min	809	809	809	809	809
	Max	809	809	809	809	809
	cv%	0	0	0	0	0

Table 7. *Cont.*

		PSO	DWPSO	TVAC	TrigAC-PSO	FSVAC
Pr.30	Mean	419.5333333	417	417	417	417
	St.Dev	7.103827688	0	0	0	0
	Min	417	417	417	417	417
	Max	445	417	417	417	417
	cv%	1.693268955	0	0	0	0
Pr.31	Mean	3480.066667	3458	3458	3458	3458
	St.Dev	33.27620392	0	0	0	0
	Min	3458	3458	3458	3458	3458
	Max	3587	3458	3458	3458	3458
	cv%	0.956194438	0	0	0	0
Pr.32	Mean	114.3	116.9333	114.4333	114.7	118.4333
	St.Dev	3.761419395	4.532894	3.549485	3.77057	4.44648
	Min	109	109	109	109	109
	Max	122	125	127	119	125
	cv%	3.290830616	3.876477	3.101794	3.2873372	3.754424

The above results urged us to continue the research for an even greater number of particles, in order to study the behavior of new variations in a multi-solution environment.

More specifically, the aforementioned variations were also tested on the set of 40 and 50 particles. In this case, 10 independent runs were carried out for each test instance, reducing the chances of finding the optimal solution from the 30 independent runs that we have already performed. Selecting more particles revealed significant results.

The results showed, once again, the consistent superiority of the proposed variations. Table 8 and Figure 4 shows the accuracy achieved by each variation for 40 particles. These results provide further support for the hypothesis that TrigAC-PSO and FSVAC are still more accurate than the other PSO variations, since they attained accuracy rates 88.31% and 77.5%, respectively; the DWPSO method follows with 75.94%, and TVAC with 74.38%; last but not least is the classic PSO with 51.56%, attaining a spectacular 13% increase over the 10-particle accuracy rates, but maintaining a steady performance for 15 and 20 particles.

Table 8. Accuracy of PSO, DWPSO, TVAC, TrigAC-PSO and FSVAC for 40 particles.

	PSO	DWPSO	TVAC	TrigAC-PSO	FSVAC
Pr.01	0.2	0.8	0.8	1	0.8
Pr.02	1	1	1	1	1
Pr.03	1	1	1	1	1
Pr.04	1	1	1	1	1
Pr.05	1	1	1	1	1
Pr.06	0.4	1	1	1	1
Pr.07	0.8	1	1	1	1
Pr.08	0.2	0.2	0.3	0.7	0.1
Pr.09	0	0.5	0.3	0.6	0.3
Pr.10	0.2	0.5	0.2	0.5	0.6
Pr.11	0	0.5	0.4	0.8	0.4
Pr.12	0.2	0.7	0.7	0.7	1
Pr.13	0.7	1	1	1	1
Pr.14	0	0.3	0.5	0.7	0.5
Pr.15	0.8	1	1	1	1
Pr.16	0.2	1	0.8	1	1
Pr.17	0.3	1	1	1	1
Pr.18	1	1	1	1	1
Pr.19	0.2	0.9	0.6	0.9	1
Pr.20	1	0.9	1	1	1
Pr.21	0.6	0.5	0.6	0.7	0.4
Pr.22	0.7	1	1	1	1
Pr.23	0.1	0.7	0.9	1	0.8
Pr.24	0	0	0	0	0
Pr.25	0.7	1	1	1	1
Pr.26	0.8	1	1	1	1
Pr.27	0.8	0.5	0.2	0.9	0.6
Pr.28	0	0.1	0.4	0.6	0.3
Pr.29	1	1	1	1	1
Pr.30	1	1	1	1	1
Pr.31	0.5	1	1	1	1
Pr.32	0.1	0.2	0.1	0.2	0
Average	0.515625	0.759375	0.74375	0.853125	0.775

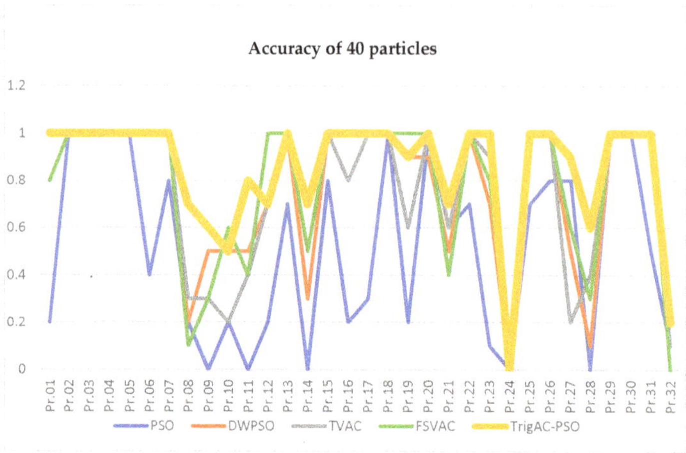

Figure 4. Accuracy for 40 particles.

In the following table (Table 9), the most important statistical measures in the case of 40 particles for 10 independent runs are represented for all PSO variations. According to the particularly low values of the Coefficient of Variation (CV), we can infer that the PSO variations are extremely stable; more precisely, the mean CV value for each PSO variation is as follows: Classic PSO, 2.14%; DWPSO, 0.93%; TVAC, 0.86%; TrigAC-PSO, 0.47%; and FSVAC, 0.81%. These values show that TrigAC-PSO, which is one of the new PSO variations presented in this work, is once again the most stable method.

Table 9. Statistical measures for 40 particles.

		PSO	DWPSO	TVAC	TrigAC-PSO	FSVAC
Pr.01	Mean	900.6	882.6	879.6	880	886.5
	Var	32.69455	7.229569	2.065591	0	13.94633
	Min	880	880	874	880	880
	Max	975	903	882	880	918
	cv%	3.630307	0.819122	0.234833	0	1.57319
Pr.02	Mean	743	743	743	743	743
	Var	0	0	0	0	0
	Min	743	743	743	743	743
	Max	743	743	743	743	743
	cv%	0	0	0	0	0
Pr.03	Mean	5600	5600	5600	5600	5600
	Var	0	0	0	0	0
	Min	5600	5600	5600	5600	5600
	Max	5600	5600	5600	5600	5600
	cv%	0	0	0	0	0
Pr.04	Mean	59	59	59	59	59
	Var	0	0	0	0	0
	Min	59	59	59	59	59
	Max	59	59	59	59	59
	cv%	0	0	0	0	0
Pr.05	Mean	28	28	28	28	28
	Var	0	0	0	0	0
	Min	28	28	28	28	28
	Max	28	28	28	28	28
	cv%	0	0	0	0	0
Pr.06	Mean	443.2	435	435	435	435
	Var	11.51617	0	0	0	0
	Min	435	435	435	435	435
	Max	472	435	435	435	435
	cv%	2.598414	0	0	0	0

Table 9. *Cont.*

		PSO	DWPSO	TVAC	TrigAC-PSO	FSVAC
Pr.07	Mean	393.9	390	390	390	390
	Var	8.2253	0	0	0	0
	Min	390	390	390	390	390
	Max	410	390	390	390	390
	cv%	2.08817	0	0	0	0
Pr.08	Mean	1659.4	1595.4	1611.5	1586.3	1623.8
	Var	74.99363	23.41035	53.60193	18.53555	32.25179
	Min	1580	1580	1571	1580	1580
	Max	1802	1650	1717	1639	1671
	cv%	4.519322	1.467365	3.326213	1.168477	1.986192
Pr.09	Mean	52.4	50.2	50.5	49.7	50.4
	Var	1.577621	1.619328	1.509231	1.251666	1.074968
	Min	50	49	49	49	49
	Max	55	53	53	53	52
	cv%	3.010728	3.225752	2.988576	2.518442	2.132872
Pr.10	Mean	421.7	417	417.2	414.3	414.7
	Var	11.72888	9.092121	6.924995	5.945119	6.848357
	Min	410	410	410	410	410
	Max	434	430	430	425	430
	cv%	2.781333	2.180365	1.659874	1.434979	1.6514
Pr.11	Mean	3273.9	2894	2852.8	2850.2	3016.8
	Var	557.6408	121.5237	4.391912	0.421637	255.7915
	Min	2851	2850	2850	2850	2850
	Max	4360	3237	2864	2851	3513
	cv%	17.03292	4.199159	0.153951	0.014793	8.478901
Pr.12	Mean	190.9	185.2	183.7	183.9	183
	Var	7.218033	4.289522	1.251666	1.449138	0
	Min	183	183	183	183	183
	Max	203	196	186	186	183
	cv%	3.781054	2.316157	0.681364	0.788003	0
Pr.13	Mean	807.4	799	799	799	799
	Var	13.52528	0	0	0	0
	Min	799	799	799	799	799
	Max	827	799	799	799	799
	cv%	1.675165	0	0	0	0
Pr.14	Mean	300.9	284.6	280	276.9	281.8
	Var	15.16905	10.25454	8.628119	6.707376	9.29516
	Min	290	273	273	273	273
	Max	330	302	290	291	292
	cv%	5.041225	3.603141	3.081471	2.42231	3.298495
Pr.15	Mean	1160.4	1160	1160	1160	1160
	Var	0.843274	0	0	0	0
	Min	1160	1160	1160	1160	1160
	Max	1162	1160	1160	1160	1160
	cv%	0.072671	0	0	0	0
Pr.16	Mean	215.2	200	201.6	200	200
	Var	8.243516	0	4.718757	0	0
	Min	200	200	200	200	200
	Max	221	200	215	200	200
	cv%	3.83063	0	2.340653	0	0
Pr.17	Mean	244.7	240	240	240	240
	Var	6.236986	0	0	0	0
	Min	240	240	240	240	240
	Max	256	240	240	240	240
	cv%	2.54883	0	0	0	0
Pr.18	Mean	820	820	820	820	820
	Var	0	0	0	0	0
	Min	820	820	820	820	820
	Max	820	820	820	820	820
	cv%	0	0	0	0	0
Pr.19	Mean	191.4	190.2	190.7	190.1	190
	Var	0.843274	0.632456	0.948683	0.316228	0
	Min	190	190	190	190	190
	Max	192	192	192	191	190
	cv%	0.440582	0.332521	0.497474	0.166348	0
Pr.20	Mean	83.3	83.3	83	83	83
	Var	0.948683	0.948683	0	0	0
	Min	83	83	83	83	83
	Max	86	86	83	83	83
	cv%	1.138876	1.138876	0	0	0

Table 9. *Cont.*

		PSO	DWPSO	TVAC	TrigAC-PSO	FSVAC
Pr.21	Mean	3529.4	3482.4	3479.3	3469.3	3482
	Var	136.2777	28.33608	40.89567	16.54657	28.15828
	Min	3460	3460	3460	3460	3460
	Max	3805	3545	3590	3502	3544
	cv%	3.861213	0.813694	1.175399	0.476943	0.808681
Pr.22	Mean	919.6	910	910	910	910
	Var	16.46005	0	0	0	0
	Min	910	910	910	910	910
	Max	954	910	910	910	910
	cv%	1.789914	0	0	0	0
Pr.23	Mean	1674.4	1670.3	1670.1	1670	1671
	Var	4.718757	0.483046	0.316228	0	2.538591
	Min	1670	1670	1670	1670	1670
	Max	1686	1671	1671	1670	1678
	cv%	0.281818	0.02892	0.018935	0	0.15192
Pr.24	Mean	2412.6	2370	2352.5	2349	2351.1
	Var	20.74823	19.47648	34.42302	21.34895	29.51252
	Min	2371	2341	2287	2315	2296
	Max	2441	2397	2419	2393	2394
	cv%	0.859994	0.821792	1.463253	0.908853	1.255264
Pr.25	Mean	2479	2460	2460	2460	2460
	Var	31.34042	0	0	0	0
	Min	2460	2460	2460	2460	2460
	Max	2540	2460	2460	2460	2460
	cv%	1.264237	0	0	0	0
Pr.26	Mean	292.2	291	291	291	291
	Var	2.699794	0	0	0	0
	Min	291	291	291	291	291
	Max	299	291	291	291	291
	cv%	0.923954	0	0	0	0
Pr.27	Mean	4550.9	4538.4	4578.5	4525.6	4556.4
	Var	54.61471	20.28245	59.81871	1.897367	62.54634
	Min	4525	4525	4525	4525	4525
	Max	4657	4585	4675	4531	4675
	cv%	1.200086	0.446908	1.306513	0.041925	1.372714
Pr.28	Mean	971.9	931	934.1	924.7	930.3
	Var	11.97637	14.96663	15.05139	8.525126	12.5614
	Min	960	920	920	920	920
	Max	993	969	955	947	960
	cv%	1.232263	1.607586	1.611326	0.921934	1.350253
Pr.29	Mean	809	809	809	809	809
	Var	0	0	0	0	0
	Min	809	809	809	809	809
	Max	809	809	809	809	809
	cv%	0	0	0	0	0
Pr.30	Mean	417	417	417	417	417
	Var	0	0	0	0	0
	Min	417	417	417	417	417
	Max	417	417	417	417	417
	cv%	0	0	0	0	0
Pr.31	Mean	3469.5	3458	3458	3458	3458
	Var	12.40296	0	0	0	0
	Min	3458	3458	3458	3458	3458
	Max	3483	3458	3458	3458	3458
	cv%	0.357485	0	0	0	0
Pr.32	Mean	115	118.6	121.4	115	117.1
	Var	2.94392	8.126773	8.448537	4.944132	2.330951
	Min	109	109	109	109	112
	Max	119	129	129	123	120
	cv%	2.559931	6.852254	6.959256	4.299245	1.990565

The following table (Table 10) and Figure 5 present the accuracy for the 50 particles. The accuracy for each PSO variation is as follows: Classic PSO, 52.5%; DWPSO, 71.3%; TVAC, 76.56%; TrigAC-PSO, 86.88%; and FSVAC, 82.19%. The two new variations range at the highest levels. These are particularly promising results, demonstrating that the increase in the particle's number leads to an increase in the PSO variation's accuracy, especially in the case of TrigAC-PSO and FSVAC. The results of 50 particles are equal to or better than the results that are currently presented. Overall, TrigAC-PSO was the one that obtained the most robust results.

Table 10. Accuracy of PSO, DWPSO, TVAC, TrigAC-PSO and FSVAC for 50 particles.

	PSO	DWPSO	TVAC	TrigAC-PSO	FSVAC
Pr.01	0.2	1	1	1	1
Pr.02	1	1	1	1	1
Pr.03	1	1	1	1	1
Pr.04	1	1	1	1	1
Pr.05	1	1	1	1	1
Pr.06	0.7	1	1	1	1
Pr.07	0.9	1	1	1	1
Pr.08	0.2	0.3	0.3	0.4	0.1
Pr.09	0	0.6	0.5	0.8	0.5
Pr.10	0.1	0.1	0.4	0.7	0.5
Pr.11	0.3	0.4	0.6	1	0.5
Pr.12	0.2	0.7	0.5	0.8	1
Pr.13	0.8	1	1	1	1
Pr.14	0	0.3	0.2	0.7	1
Pr.15	0.8	1	1	1	1
Pr.16	0.1	1	1	1	1
Pr.17	0.3	0.7	1	1	1
Pr.18	1	1	1	1	1
Pr.19	0.2	0.9	0.8	0.9	1
Pr.20	0.7	1	1	1	1
Pr.21	0.5	0.3	0.6	1	0.4
Pr.22	0.8	1	1	1	1
Pr.23	0.2	0.8	0.6	1	0.8
Pr.24	0	0	0.1	0	0
Pr.25	1	1	1	1	1
Pr.26	0.7	1	1	1	1
Pr.27	0.5	0.3	0.1	0.6	0.5
Pr.28	0.1	0.4	0.7	0.7	0.7
Pr.29	1	1	1	1	1
Pr.30	1	1	1	1	1
Pr.31	0.5	1	1	1	1
Pr.32	0	0	0.1	0.2	0.3
Average	0.525	0.74375	0.765625	0.86875	0.821875

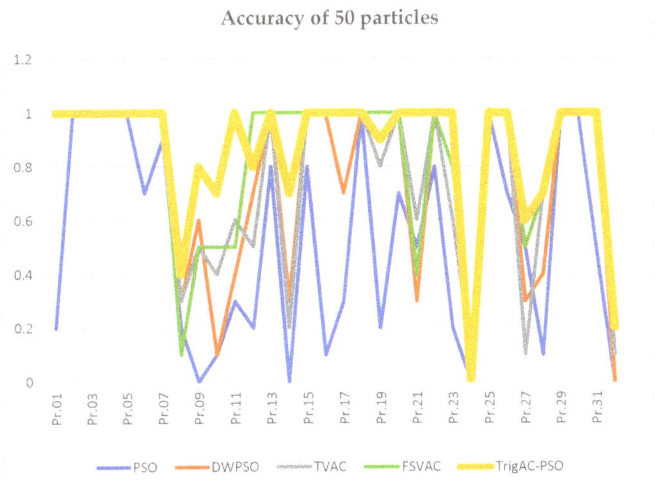

Figure 5. Accuracy for 50 particles.

The results of Table 11 lead to similar conclusions. In order to examine the stability for the 50 particles, it is worth comparing the CV values of the proposed variations with those of the traditional variations. Superior results are seen from TrigAC-PSO, as the CV value is equal to 0.4%, followed by the FSVAC with 0.59%. The other values of variations ranged as follows: Classic PSO, 2.19%; DWPSO, 0.77%; and TVAC, 0.83%.

Table 11. Statistical measures for 50 particles.

		PSO	DWPSO	TVAC	TrigAC-PSO	FSVAC
	Mean	896.7	880	880	880	880
	Var	17.79544	0	0	0	0
Pr.01	Min	880	880	880	880	880
	Max	929	880	880	880	880
	cv%	1.984548	0	0	0	0
	Mean	743	743	743	743	743
	Var	0	0	0	0	0
Pr.02	Min	743	743	743	743	743
	Max	743	743	743	743	743
	cv%	0	0	0	0	0
	Mean	5600	5600	5600	5600	5600
	Var	0	0	0	0	0
Pr.03	Min	5600	5600	5600	5600	5600
	Max	5600	5600	5600	5600	5600
	cv%	0	0	0	0	0
	Mean	59	59	59	59	59
	Var	0	0	0	0	0
Pr.04	Min	59	59	59	59	59
	Max	59	59	59	59	59
	cv%	0	0	0	0	0
	Mean	28	28	28	28	28
	Var	0	0	0	0	0
Pr.05	Min	28	28	28	28	28
	Max	28	28	28	28	28
	cv%	0	0	0	0	0
	Mean	438.9	435	435	435	435
	Var	6.279597	0	0	0	0
Pr.06	Min	435	435	435	435	435
	Max	448	435	435	435	435
	cv%	1.430758	0	0	0	0
	Mean	392	390	390	390	390
	Var	6.324555	0	0	0	0
Pr.07	Min	390	390	390	390	390
	Max	410	390	390	390	390
	cv%	1.613407	0	0	0	0
	Mean	1677.3	1611.7	1601.7	1585.9	1634.5
	Var	73.73986	47.30058	31.18778	7.125073	41.31518
Pr.08	Min	1580	1580	1580	1580	1580
	Max	1794	1713	1672	1595	1705
	cv%	4.396343	2.934825	1.947168	0.449276	2.527696
	Mean	52.4	50.3	50.2	49.4	50.8
	Var	0.966092	1.888562	1.549193	0.843274	2.043961
Pr.09	Min	51	49	49	49	49
	Max	53	53	53	51	54
	cv%	1.843687	3.754597	3.086043	1.707032	4.023546
	Mean	424.1	421	418.6	412.6	416.1
	Var	9.362455	8.589399	9.057839	4.299871	8.292567
Pr.10	Min	410	410	410	410	410
	Max	432	430	430	421	430
	cv%	2.207605	2.040237	2.163841	1.04214	1.992926
	Mean	3149.1	2870.4	2856.9	2850	2898.3
	Var	611.9779	44.53014	14.0115	0	95.44405
Pr.11	Min	2850	2850	2850	2850	2850
	Max	4429	2990	2894	2850	3091
	cv%	19.43342	1.551357	0.490444	0	3.293105

Table 11. *Cont.*

		PSO	DWPSO	TVAC	TrigAC-PSO	FSVAC
Pr.12	Mean	187.4	183.9	184.5	183.6	183
	Var	3.687818	1.449138	1.581139	1.264911	0
	Min	183	183	183	183	183
	Max	193	186	186	186	183
	cv%	1.967886	0.788003	0.856986	0.688949	0
Pr.13	Mean	804.6	799	799	799	799
	Var	11.80584	0	0	0	0
	Min	799	799	799	799	799
	Max	827	799	799	799	799
	cv%	1.467293	0	0	0	0
Pr.14	Mean	302.8	284.2	289.2	278.1	273
	Var	15.38975	7.871185	15.59772	8.491172	0
	Min	290	273	273	273	273
	Max	328	291	328	295	273
	cv%	5.082481	2.769594	5.393403	3.05328	0
Pr.15	Mean	1184.3	1160	1160	1160	1160
	Var	76.14321	0	0	0	0
	Min	1160	1160	1160	1160	1160
	Max	1401	1160	1160	1160	1160
	cv%	6.429386	0	0	0	0
Pr.16	Mean	212.9	200	200	200	200
	Var	8.69802	0	0	0	0
	Min	200	200	200	200	200
	Max	221	200	200	200	200
	cv%	4.085496	0	0	0	0
Pr.17	Mean	246.9	241.6	240	240	240
	Var	6.773314	2.796824	0	0	0
	Min	240	240	240	240	240
	Max	256	248	240	240	240
	cv%	2.743343	1.157626	0	0	0
Pr.18	Mean	820	820	820	820	820
	Var	0	0	0	0	0
	Min	820	820	820	820	820
	Max	820	820	820	820	820
	cv%	0	0	0	0	0
Pr.19	Mean	191.6	190.1	190.3	190.1	190
	Var	0.843274	0.316228	0.674949	0.316228	0
	Min	190	190	190	190	190
	Max	192	191	192	191	190
	cv%	0.440122	0.166348	0.354676	0.166348	0
Pr.20	Mean	83.9	83	83	83	83
	Var	1.449138	0	0	0	0
	Min	83	83	83	83	83
	Max	86	83	83	83	83
	cv%	1.72722	0	0	0	0
Pr.21	Mean	3493.6	3468.4	3477.6	3460	3470.8
	Var	57.87765	13.1673	24.84262	0	12.76105
	Min	3460	3456	3460	3460	3460
	Max	3625	3492	3519	3460	3495
	cv%	1.656676	0.379636	0.714361	0	0.367669
Pr.22	Mean	914.7	910	910	910	910
	Var	10.133	0	0	0	0
	Min	910	910	910	910	910
	Max	938	910	910	910	910
	cv%	1.107795	0	0	0	0
Pr.23	Mean	1675.2	1670.2	1672	1670	1670.6
	Var	7.743097	0.421637	3.018462	0	1.577621
	Min	1670	1670	1670	1670	1670
	Max	1694	1671	1679	1670	1675
	cv%	0.462219	0.025245	0.18053	0	0.094434

Table 11. *Cont.*

		PSO	DWPSO	TVAC	TrigAC-PSO	FSVAC
Pr.24	Mean	2423.4	2354.5	2336.8	2347	2355
	Var	13.35165	31.64824	37.90573	26.48689	28.5151
	Min	2401	2288	2280	2307	2309
	Max	2446	2390	2404	2388	2395
	cv%	0.550947	1.34416	1.622121	1.128543	1.210832
Pr.25	Mean	2460	2460	2460	2460	2460
	Var	0	0	0	0	0
	Min	2460	2460	2460	2460	2460
	Max	2460	2460	2460	2460	2460
	cv%	0	0	0	0	0
Pr.26	Mean	292.2	291	291	291	291
	Var	1.932184	0	0	0	0
	Min	291	291	291	291	291
	Max	295	291	291	291	291
	cv%	0.661254	0	0	0	0
Pr.27	Mean	4585.8	4549.3	4552.3	4532	4595.2
	Var	69.97904	45.93486	42.05829	16.57307	75.36843
	Min	4525	4525	4525	4525	4525
	Max	4675	4675	4668	4577	4675
	cv%	1.525994	1.009713	0.923891	0.36569	1.640156
Pr.28	Mean	957.3	928	929.9	923.3	923.6
	Var	21.12424	12.26558	17.85404	8.420214	9.070097
	Min	920	920	920	920	920
	Max	990	960	967	947	949
	cv%	2.206647	1.321722	1.919995	0.911969	0.982037
Pr.29	Mean	809	809	809	809	809
	Var	0	0	0	0	0
	Min	809	809	809	809	809
	Max	809	809	809	809	809
	cv%	0	0	0	0	0
Pr.30	Mean	417	417	417	417	417
	Var	0	0	0	0	0
	Min	417	417	417	417	417
	Max	417	417	417	417	417
	cv%	0	0	0	0	0
Pr.31	Mean	3470.4	3458	3458	3458	3458
	Var	18.42221	0	0	0	0
	Min	3458	3458	3458	3458	3458
	Max	3508	3458	3458	3458	3458
	cv%	0.530838	0	0	0	0
Pr.32	Mean	118.1	122.3	123.7	114.1	112.6
	Var	5.384133	6.498718	8.590046	3.784471	3.306559
	Min	112	11a2	109	109	109
	Max	127	129	129	120	119
	cv%	4.558961	5.313751	6.944257	3.316802	2.936553

The overall results demonstrate two inferences of decisive importance: first, the PSO algorithm and its variations have successfully solved the TP with maximum accuracy and efficiency; second, TrigAC-PSO, beyond any doubt, is the leading option for solving the TP in terms of both stability and the solution's quality.

7. Conclusions

As technology is developing, the need for product improvement and trading is of high priority in obtaining a more economical solution. The PSO algorithm was applied with success in order for the TP to be solved. Furthermore, two new variations were introduced and compared to already-known variations. These variations induced exceptional results and indicated their superiority against the existing variations and the well-known exact methods in the literature. The proposed PSO variations have been tested in a variety of test instances with different combined values of inertia weight as well as social and personal acceleration parameters. It was evidently proven that the solution quality is inseparably linked with the selection of proper values for controlling the algorithm parameters. In order to see the effectiveness and stability of the proposed variations, we compared their results with those of other PSO variations for the same instances. Remarkably, the punctuality of one of our variations rose to 88%, and it was finally established as the ideal option compared to all other variations for the solution of TP.

It can be easily observed that this PSO variationis simple compared to other variations with complex structures. It was a challenge to achieve better results by creating and running simple computational algorithms, proving that keeping a balance between human and artificial intelligence is the key to the success of computational intelligence.

A more comprehensive analysis may be needed in order to examine the TP to a greater extent. Moreover, the proposed PSO variations could be applied to more complex networks such as the Sioux Fall network [27] in order to demonstrate the algorithm's good performance and independence of the network's size. Except for this, some other real constraints can be proposed in order to find the optimal solution for the TP with PSO algorithm variations not only in balanced instances but also in more realistic unbalanced instances in the future. Moreover, combining the proposed PSO variations with other meta-heuristic methods to solve the TP will be an interesting challenge.

Author Contributions: Conceptualization, C.A. and G.N.B.; methodology, C.A. and G.N.B.; software, C.A.; validation, C.A. and G.N.B.; formal analysis, C.A. and G.N.B.; investigation, C.A. and G.N.B.; resources, C.A.; data curation, C.A.; writing—original draft preparation, C.A.; writing—review and editing, C.A. and G.N.B.; visualization, C.A.; supervision, G.N.B.; project administration, G.N.B. All authors have read and agreed to the published version of the manuscript.

Funding: This research received no external funding.

Data Availability Statement: Data and the programming code used in this paper can be sent, upon request, to the interested reader. Please contact: gbeligia@upatras.gr.

Conflicts of Interest: The authors declare no conflict of interest.

References

1. Karagul, K.; Sahin, Y. A novel approximation method to obtain initial basic feasible solution of transportation problem. *J. King Saud Univ. Eng. Sci.* **2019**, *32*, 211–218. [CrossRef]
2. Deshpande, V.A. An optimal method for obtaining initial basic feasible solution of the transportation problem. In Proceedings of the National Conference on Emerging Trends in Mechanical Engineering Patel College of Engineering & Technology (GCET), Vallabh Vidyanagar (ETME-2009), Vallabh Vidyanagar, India, 5–6 March 2010; Volume 20, p. 21.
3. Taha, H.A. *Operations Research: An Introduction*, 8th ed.; Pearson Prentice Hall: Hoboken, NJ, USA, 2007.
4. Mostafa, R.R.; El-Attar, N.E.; Sabbeh, S.F.; Vidyarthi, A.; Hashim, F.A. ST-AL: A hybridized search based metaheuristic computational algorithm towards optimization of high dimensional industrial datasets. *Soft Comput.* **2022**, *27*, 13553–13581. [CrossRef] [PubMed]
5. Gen, M.; Altiparmak, F.; Lin, L. A genetic algorithm for two-stage transportation problem using priority-based encoding. *OR Spectr.* **2006**, *28*, 337–354. [CrossRef]
6. Swiatnicki, Z. Application of ant colony optimization algorithms for transportation problems using the example of the travelling salesman problem. In Proceedings of the 2015 4th International Conference on Advanced Logistics and Transport (ICALT), Valenciennes, France, 20 May 2015. [CrossRef]
7. Fakhrzadi, M.; Goodarziani, F.; Golmohammadi, A.M. Addressing a fixed charge transportation problem with multiroute and different capacities by novel hybrid meta-heuristics. *J. Ind. Syst. Eng.* **2019**, *12*, 167–184.
8. Eberhart, R.; Kennedy, J. A New Optimizer Using Particle Swarm Theory. In Proceedings of the Sixth International Symposium on Micro Machine and Human Science, Nagoya, Japan, 4 October 1995.
9. Salehizadeh, S.M.A.; Yadmellat, P.; Menhaj, M.B. Local Optima Avoidable Particle Swarm Optimization. In Proceedings of the IEEE Swarm Intelligence Symposium, Nashville, TN, USA,, 15 May 2009. [CrossRef]
10. Lin, S.-W.; Ying, K.-C.; Chen, S.-C.; Lee, Z.-J. Particle swarm optimization for parameter determination and feature selection of support vector machines. *Expert Syst. Appl.* **2008**, *35*, 1817–1824. [CrossRef]
11. Ratnweera, S.K.; Halgamuge, H.C. Watson Self-organizing hierarchical particle swarm optimizer with time-varying acceleration coefficients. *IEEE Trans. Evol. Comput.* **2004**, *8*, 240–255. [CrossRef]
12. Hitchcock, F.L. The Distribution of a Product from Several Sources to Numerous Localities. *J. Math. Phys.* **1941**, *20*, 224–230. [CrossRef]

13. Koopmans, T. Optimum Utilization of the Transportation System. *Econometrica* **1949**, *17*, 136–146. [CrossRef]
14. Dantzig, G.B. Application of the simplex method to a traznsportation problem. In *Activity Analysis of Production and Allocation*; Koopmans, T.C., Ed.; John Wiley and Sons: New York, NY, USA, 1951; pp. 359–373.
15. Rosendo, M.; Pozo, A. A hybrid particle swarm optimization algorithm for combinatorial optimization problems. In Proceedings of the IEEE Congress on Evolutionary Computation (CEC), Barcelona, Spain, 18–23 July 2010; pp. 1–8.
16. Wang, D.; Tan, D.; Liu, L. Particle swarm optimization algorithm: An overview. *Soft Comput.* **2017**, *22*, 387–408. [CrossRef]
17. Huang, H.; Zhifang, H. Particle swarm optimization algorithm for transportation problems. In *Particle Swarm Optimization*; Intech: Shanghai, China, 2009; pp. 275–290.
18. Wang, J.; Wang, X.; Li, X.; Yi, J. A Hybrid Particle Swarm Optimization Algorithm with Dynamic Adjustment of Inertia Weight Based on a New Feature Selection Method to Optimize SVM Parameters. *Entropy* **2023**, *25*, 531. [CrossRef] [PubMed]
19. Jain, M.; Saihjpal, V.; Singh, N.; Singh, S.B. An Overview of Variants and Advancements of PSO Algorithm. *Appl. Sci.* **2022**, *12*, 8392. [CrossRef]
20. Shi, Y.; Eberhart, R. A modified particle swarm optimizer. In Proceedings of the IEEE International Conference on Evolutionary Computation, Anchorage, AK, USA, 4–9 May 1998; pp. 69–73.
21. Sengupta, S.; Basak, S.; Peters, R.A., II. Particle Swarm Optimization: A Survey of Historical and Recent Developments with Hybridization Perspectives. *Mach. Learn. Knowl. Extr.* **2019**, *1*, 157–191. [CrossRef]
22. Amaliah, B.; Fatichah, C.; Suryani, E. A new heuristic method of finding the initial basic feasible solution to solve the transportation problem. *J. King Saud Univ.-Comput. Inf. Sci.* **2022**, *34*, 2298–2307. [CrossRef]
23. Hosseini, E. Three new methods to find initial basic feasible solution of transportation problems. *Appl. Math. Sci.* **2017**, *11*, 1803–1814. [CrossRef]
24. Amaliah, B.; Fatichah, C.; Suryani, E. Total opportunity cost matrix—Minimal total: A new approach to determine initial basic feasible solution of a transportation problem. *J. Egypt. Inform.* **2019**, *20*, 131–141. [CrossRef]
25. Juman, Z.A.M.S.; Hoque, M.A. An efficient heuristic to obtain a better initial feasible solution to the transportation problem. *Appl. Soft Comput.* **2015**, *34*, 813–826. [CrossRef]
26. Amaliah, B.; Fatichah, C.; Suryani, E. A Supply Selection Method for better Feasible Solution of balanced transportation problem. *Expert Syst. Appl.* **2022**, *203*, 117399. [CrossRef]
27. Sun, D.; Chang, Y.; Zhang, L. An ant colony optimisation model for traffic counting location problem. *Transport* **2012**, *165*, 175–185. [CrossRef]

algorithms

MDPI

Article

Application of Particle Swarm Optimization to a Hybrid H_∞/Sliding Mode Controller Design for the Triple Inverted Pendulum System

Yamama A. Shafeek * and Hazem I. Ali

Control and Systems Engineering Department, University of Technology-Iraq, Baghdad 10001, Iraq; hazem.i.ali@uotechnology.edu.iq
* Correspondence: cse.21.20@grad.uotechnology.edu.iq

Abstract: The robotics field of engineering has been witnessing rapid advancements and becoming widely engaged in our lives recently. Its application has pervaded various areas that range from household services to agriculture, industry, military, and health care. The humanoid robots are electro–mechanical devices that are constructed in the semblance of humans and have the ability to sense their environment and take actions accordingly. The control of humanoids is broken down to the following: sensing and perception, path planning, decision making, joint driving, stability and balance. In order to establish and develop control strategies for joint driving, stability and balance, the triple inverted pendulum is used as a benchmark. As the presence of uncertainty is inevitable in this system, the need to develop a robust controller arises. The robustness is often achieved at the expense of performance. Hence, the controller design has to be optimized based on the resultant control system's performance and the required torque. Particle Swarm Optimization (PSO) is an excellent algorithm in finding global optima, and it can be of great help in automatic tuning of the controller design. This paper presents a hybrid H_∞/sliding mode controller optimized by the PSO algorithm to control the triple inverted pendulum system. The developed control system is tested by applying it to the nominal, perturbed by parameter variation, perturbed by external disturbance, and perturbed by measurement noise system. The average error in all cases is 0.053 deg and the steady controller effort range is from 0.13 to 0.621 N.m with respect to amplitude. The system's robustness is provided by the hybrid H_∞/sliding mode controller and the system's performance and efficiency enhancement are provided by optimization.

Keywords: metaheuristic optimization algorithm; particle swarm optimization; engineering design; H_∞; sliding mode control; triple inverted pendulum; robust control; uncertainty; torque

1. Introduction

Humanoids are robots designed in the shape of humans to carry out a variety of tasks. They are built to receive data from their environment through suitable sensors and to take actions by moving their joints in a certain way to accomplish the required task (Figure 1a). Their sophisticated control system is partitioned into smaller dedicated-purpose controllers. The most crucial part of the overall control system is the one that is responsible for generating the motors' commands to maintain balance and stability. For this control goal, a triple inverted pendulum system is used to develop, test, and enhance the control design. Mechanically, the torso, thighs, and shanks of the humanoid are represented by three links connected by joints that are driven by electric direct current (DC) motors, as shown in Figure 1b. Hence, the triple inverted pendulum system describes the basic dynamics of a humanoid. The angles that are made by the three links with the vertical axis are to be controlled by the torques produced by two DC motors. The torques are provided by the drivers to the upper two joints through two belt pulleys. The angles are measured by three potentiometers. The horizontal bars are added to ease the balance by increasing the moment of inertia [1].

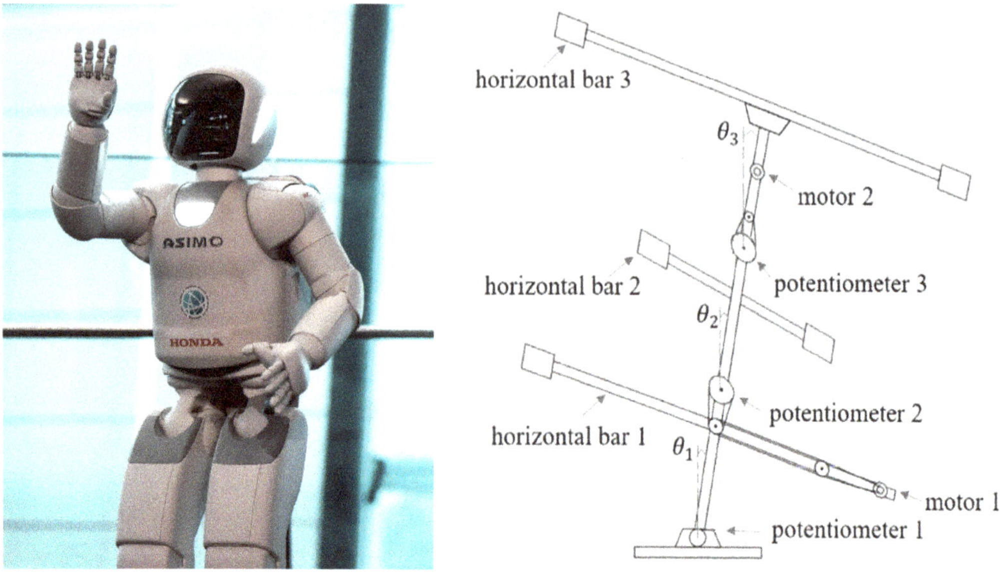

(**a**) (**b**)

Figure 1. Resemblance of humanoid structure to triple inverted pendulum: (**a**) humanoid structure [2]; (**b**) triple inverted pendulum structure [1].

The triple inverted pendulum is an unstable system which is difficult to control. Besides being an unstable system, it is subject to perturbations, due to uncertainties in the system's friction parameters and moments of inertia. In addition, external disturbances like wind gusts affect the system response. Noise in measurement readings is another source of uncertainty that affects the system response. In such systems, where perturbations occur, the controller used must take the uncertainties into account during its design to provide robustness against them. The other challenge of controlling multiple degrees of freedom (DOFs) by fewer control signals is encountered practically. This may occur by the failure of one actuator during operation or may even be a goal in itself for the sake of minimizing the consumed energy, cost, or size of the system. The underactuation makes disturbance rejection more difficult, because the applied disturbance to the unactuated joint would not fulfill the matching condition.

In the last few years, both optimal and non-optimal control methods have been suggested and developed to control the inverted pendulums and robotic systems. Among the researchers who utilized non-optimal control methods are Sharma et al. [3]; they presented a decoupling sliding mode algorithm to control a single inverted pendulum on a cart. The poles of the reduced order system have been placed once near the imaginary axis and in another case away from the imaginary axis, and in the dominant region. It has been shown that locating the poles in the dominant region reduces steady state error and produces better disturbance rejection. Bonifacio et al. [4] used the Attractive Ellipsoid Method (AEM) to stabilize the triple inverted pendulum and compared it to the sliding mode control (SMC) system. It was found that AEM-based control rejects external perturbations more smoothly, while the SMC consumes less energy. Nguyen [5] developed an SMC system for the two-link fully actuated robot arm. In system response, the angles of the arms reach the desired value within 1.5 s. Kharabian et al. [6] proposed a hybrid sliding mode/H-infinity control approach using a fuzzy neural network weighting method to reduce nonlinearity, provide precise trajectory tracking, and to the enhance noise rejection capability of a single-link flexible manipulator system. The proposed hybrid controller reduces the total system nonlinearity, leading to higher performance with respect to noise cancellation, compared to sliding mode controller alone. Yet, the proposed control approach was not applied on manipulators with two or three links, whereas the system's complexity rises and its control becomes harder.

Saif et al. [7] utilized synergetic control theory and fractional calculus to develop a fractional synergetic control (FSC) strategy for a four-DOF robot manipulator; the synergetic control approach was implemented to obtain fast convergence to the equilibrium point. The results of combining synergetic control and fractional calculus in control system design showcased good tracking performance in both joint space and workspace trajectories. However, the method's robustness had not been investigated. Ahmed et al. [8] utilized Time Delay Estimation (TDE)-based model-free control to estimate the external disturbance and the unknown friction parameters of a Puma 560 rigid manipulator, and terminal SMC to obtain system robustness. The system responses show that the control method can suppress the effect of uncertainty and produce effective tracking. However, the effect of measurement noise that can affect the overall control performance has not been considered. Likewise, the effect of measurement noise was not determined by Anjum et al. [9] when they incorporated a fixed-time adaptive sliding mode observer into the fixed-time non-singular terminal SMC design for the Puma 560 manipulator.

For the single-link manipulator, Liu et al. [10] proposed an adaptive tracking controller based on the mismatched disturbance observer. The adaptive approach and disturbance-observer design contribute to effective disturbance rejection and smoother control of the robotic manipulator link. The closed-loop signals are found to be globally uniformly bounded with asymptotically stable tracking error, though the proposed control method is not generalized to more complex multi-link systems. In the same context, Qiu et al. [11] developed a disturbance observer-based adaptive fuzzy control method for the single-link manipulator; the proposed control system acquires finite-time prescribed performance, by which the tracking error enters a prescribed bounded set within finite time, even though the prescribed performance control strategy which is debated may not fully account for all practical prospects.

Flatness-based control has been employed by Rigatos et al. [12] in successive loops for both the three-DOF rigid-link robotic manipulator and three-DOF autonomous underwater vessel. The proposed method separates the controlled system into two differentially flat subsystems, connected in cascade. The response of the two case studies exhibits fast and precise tracking of the desired set points; still, the robustness to uncertainty has not been inspected.

Jabbar et al. [13] proposed a modified backstepping control method to stabilize the rotary double-inverted pendulum, where the control law is a combination of the backstepping control action and uncertainty compensation control action. The simulation results show the control system's ability for exponential stabilization in the presence of uncertainties and disturbances. Yet, the proposed method involves complex design steps and lacks tunability to become consistent within practical applications. Siradjuddin et al. [14] used pole placement to obtain the feedback gain that stabilizes the single-link inverted pendulum at the desired position of the cart. The results show that placing the poles at more negative locations yields faster response at the expense of higher overshoot and greater control action, yet does not afford robustness. Pristovani et al. [15] employed a Multi Input/Multi Output (MIMO) decoupled control system method to implement the push-recovery strategy for the triple inverted pendulum by simplifying the control design into three serial Single Input Single Output (SISO) systems with known and uncertain disturbance models in each inverted pendulum, where the PID controller was used in each link to damp the external force applied on it. The proposed control system achieved 85.71% success in withstanding external forces, but did not consider its suitability for under-actuated systems. Although Masrom et al. [16] implemented Interval Type-2 Fuzzy Logic Control (IT2FLC) to a triple inverted pendulum on two wheels and the developed controller managed to withstand 16% greater disturbance than the type-1 fuzzy logic controller, the researchers considered disturbance applied on the third link only. The control system's robustness against possible disturbances on other links is missing in the study.

Since optimization provides enhancements in control system design, it has been employed in different ways to provide optimal solutions for engineering problems. Soltanpour et al. [17], for instance, presented optimal fuzzy SMC for a two-DOF robotic manipulator. In order to compensate for the information shortages with respect to the system's uncertainties, PSO was used to adjust the parameters of fuzzy membership functions. The proposed control method outperformed the classic SMC in terms of control input smoothness, though it had slightly higher tracking errors. The two-DOF robotic manipulator has also been controlled by an optimal integral SMC based on the pseudo spectral method. Liu et al. [18] applied integral SMC to restrain disturbance and adopted the pseudo spectral method to deal with the constraints. The proposed controller demonstrated the ability to track the desired reference signals accurately within 2 s. The results illustrated that the robotic manipulator system exhibited good robustness and anti-disturbance capabilities when subjected to external disturbances. However, the measurement noise for the system has not been considered.

Oliveira et al. [19] used Grey Wolf Optimization (GWO) with chaotic basis to tune the parameters of a higher-order SMC for the position control of a two-DOF rigid robot manipulator. Tent and Singer maps were applied to the optimization method to balance the exploration/exploitation phases of the algorithm, based on the algorithm itself and the chosen cost function. It was shown that the general repeatability of the algorithm was improved using chaotic maps in the higher-order SMC optimization. However, the suitability of the proposed method has not been tested for three-DOF manipulators. For the electro–hydraulic actuator system, Soon et al. [20] applied PSO to tune the Proportional–Integral–Derivative (PID) sliding surface of the SMC. The use of optimization in controller design has improved the system's performance by 0.6407%. However, uncertainty was not taken into consideration in the study. On the other hand, Jibril et al. [21] used a Linear–Quadratic Regulator (LQR), where the cost is defined through a quadratic function, and pole placement, where the closed-loop poles of the plant are positioned in desired locations in the s-plane. Both control methods were applied for the stabilization of a triple inverted pendulum system and compared; the comparison shows that the pole-placement controller improves the stability of the system more, but with no indications of its robustness against perturbations. As for the triple-link rotary inverted pendulum, Hazem et al. [22] made a comparative study of the Neuro-Fuzzy Friction Estimation Model (NFFEM) and Adaptive Friction Estimation Model (AFEM) methods used to estimate the friction coefficients of the plant. The NFFEMs are trained by a radial-basis function artificial neural network. It has been deduced, based on the root mean square error of the joints' position, that NFFEMs produce much better estimation results than AFEMs, although the researchers had not address the robustness issue.

Singh et al. [23] applied H_∞ and μ-synthesis control for the double inverted pendulum on a cart to achieve disturbance rejection and robust stability. The H_∞ controller seeks to minimize the mixed-sensitivity cost function, while the μ-synthesis controller implements the D-K iteration method to find the stabilizing gain that minimizes the upper structured singular value of the system. The simulation results show that the μ-synthesis control system has a more robust performance. In [24,25], Shafeek et al. proposed a method for enhancing the H_∞ and μ-synthesis control systems' robustness and performance for the triple inverted pendulum, by incorporating PSO and the Gazelle Optimization Algorithm (GOA), respectively, into the controllers' design. It has been shown that utilizing the optimization in control system design allows the possibility of balancing robustness and performance aspects within the system.

Meta-heuristic optimization algorithms are known to be well-suited for a wide range of complex optimization problems. Their key power lies in their ability to handle large search spaces, constraints, and multi-objective cost functions. The following studies are recommended to researchers interested in the field of meta-heuristic optimization applications. In fact, various meta-heuristic optimization algorithms (such as PSO [26–28], Ant Colony Optimization [29], the Bees Algorithm [30], Grey Wolf Optimization [31], the Whale Optimization Algorithm [32], Sunflower Optimization [33], the Gorilla Troops Algorithm [34], and the Chimp Optimization algorithm [35]) have been applied successfully in many engineering problems. In addition, it is also noteworthy that Rubio et al. [36] tuned the high-gain observer and controller gains to enhance the position and velocity perturbation attenuation for inverted pendulums using a genetic optimizer. The suggested method resulted in better perturbation attenuation compared to simplex and Bat optimizers. As for robots, Rubio [37] applied the Bat Algorithm and the Modified Bat Algorithm to minimize both tracking error and control energy consumption by optimizing the control gain. Sorcia-Vázquez et al. [38] also managed the minimization of the tracking error and the control effort for the experimental two-tank system through applying the genetic algorithm for the tuning of the PID and Fractional Order PID (FOPID) controller gains. The tuned FOPID control system performed better than the tuned PID control system in terms of overshoot, settling time, and control signal smoothness.

The literature review shows a gap in a fully robust analysis of the triple inverted pendulum which considers all the possible sources of uncertainties in the system. To fill the gap, this paper contributes from three important perspectives:

- Formulating a hybrid H_∞/SMC design for a triple inverted pendulum system that is robust to parameter variations, external disturbances, and measurement noise;
- Utilizing PSO to tune parameters in the control action based on the Integral Time Absolute Error (ITAE) performance index of the three controlled variable errors and the Integral Square Control Signal (ISCS) performance index of the two actuators' torques;
- Evaluating the robustness, performance, and efficiency of the proposed optimized control system in different cases.

The rest of the paper introduces the mathematical model of the triple inverted pendulum system in Section 2, presents the PSO algorithm in Section 3, develops the hybrid H_∞/SMC design and optimization in Section 4, and tests the control system response to different cases in Section 5. Finally, it discusses the outcomes of the developed control system and the most noticeable findings, and suggests directions for future work in Section 6.

2. Mathematical Model of the Plant

The Mathematical model of the triple link inverted pendulum shown in Figure 1b is derived based on Lagrangian mechanics established by the scientist Joseph-Louis Lagrange. The Lagrangian (L) of many mechanical systems represents the difference between their kinetic (T) and potential (V) energies

$$L = T - V, \quad (1)$$

For the triple inverted pendulum model of Figure 1b, the kinetic and potential energies are defined as:

$$\begin{aligned} T = {} & \tfrac{1}{2} m_1 \left[\left(\tfrac{d}{dt}(h_1 \, sin\theta_1) \right)^2 + \left(\tfrac{d}{dt}(h_1 \, cos\theta_1) \right)^2 \right] + \tfrac{1}{2} I_1 \dot{\theta}_1^2 + \tfrac{1}{2} m_2 \left[\left(\tfrac{d}{dt}(l_1 \, sin\theta_1 + h_2 \, sin\theta_2) \right)^2 + \left(\tfrac{d}{dt}(l_1 \, cos\theta_1 + h_2 \, cos\theta_2) \right)^2 \right] + \\ & \tfrac{1}{2} I_2 \dot{\theta}_2^2 + \tfrac{1}{2} m_3 \left[\left(\tfrac{d}{dt}(l_1 \, sin\theta_1 + l_2 \, sin\theta_2 + h_3 \, sin\theta_3) \right)^2 + \left(\tfrac{d}{dt}(l_1 \, cos\theta_1 + l_2 \, cos\theta_2 + h_3 \, cos\theta_3) \right)^2 \right] + \tfrac{1}{2} I_3 \dot{\theta}_3^2, \end{aligned} \quad (2)$$

$$V = M_3 \, g \, cos\theta_1 + M_2 \, g \, cos\theta_2 + M_3 \, g \, cos\theta_3, \quad (3)$$

respectively, where g represents the acceleration of gravity, and

$$M_1 = m_1 \, h_1 + m_2 \, l_1 + m_3 \, l_1, \quad (4)$$

$$M_2 = m_2\,h_2 + m_3\,l_2, \tag{5}$$

$$M_3 = m_3\,h_3, \tag{6}$$

Then, by applying the stationary action principle,

$$\frac{d}{dt}\left(\frac{\partial L}{\partial \dot{\rho}_k}\right) = \frac{\partial L}{\partial \rho_k}, \tag{7}$$

where ρ_j represents the kth position vector in the generalized coordinates, the equations of motion of the system are obtained [1]:

$$\begin{aligned}
&\begin{bmatrix} J_1 + I_{p1} & l_1\,M_2cos(\theta_1-\theta_2) - I_{p1} & l_1\,M_3cos(\theta_1-\theta_3) \\ l_1\,M_2cos(\theta_1-\theta_2) - I_{p1} & J_2 + I_{p1} + I_{p2} & l_2\,M_3cos(\theta_2-\theta_3) - I_{p2} \\ l_1\,M_3cos(\theta_1-\theta_3) & l_2\,M_3cos(\theta_2-\theta_3) - I_{p2} & J_3 + I_{p2} \end{bmatrix}\begin{bmatrix}\ddot{\theta}_1\\ \ddot{\theta}_2\\ \ddot{\theta}_3\end{bmatrix} \\
&+\begin{bmatrix} C_1 + C_2 + C_{p1} & -C_2 - C_{p1} & 0 \\ -C_2 - C_{p1} & C_{p1} + C_{p2} + C_2 + C_3 & -C_3 - C_{p2} \\ 0 & -C_3 - C_{p2} & C_3 + C_{p2} \end{bmatrix}\begin{bmatrix}\dot{\theta}_1\\ \dot{\theta}_2\\ \dot{\theta}_3\end{bmatrix} \\
&+\begin{bmatrix} l_1M_2\sin(\theta_1-\theta_2)\dot{\theta}_2^2 + l_1M_3\sin(\theta_1-\theta_3)\dot{\theta}_3^2 - M_1g\,\sin(\theta_1) \\ l_1M_2\sin(\theta_1-\theta_2)\dot{\theta}_1^2 + l_2M_3\sin(\theta_2-\theta_3)\dot{\theta}_3^2 - M_2g\,\sin(\theta_2) \\ l_1M_3\sin(\theta_1-\theta_3)\left(\dot{\theta}_1^2 - 2\,\dot{\theta}_1\dot{\theta}_3\right) + l_2M_3\sin(\theta_2-\theta_3)\left(\dot{\theta}_2^2 - 2\,\dot{\theta}_2\dot{\theta}_3\right) - M_3g\,\sin(\theta_3) \end{bmatrix} \\
&+\begin{bmatrix} K_1 & 0 \\ -K_1 & K_2 \\ 0 & -K_2 \end{bmatrix}\begin{bmatrix} t_{m1} \\ t_{m2} \end{bmatrix} = \begin{bmatrix} 1 & -1 & 0 \\ 0 & 1 & -1 \\ 0 & 0 & 1 \end{bmatrix}\begin{bmatrix} d_1 \\ d_2 \\ d_3 \end{bmatrix}
\end{aligned} \tag{8}$$

where $\boldsymbol{\theta} = \begin{bmatrix}\theta_1 & \theta_2 & \theta_3\end{bmatrix}^T$ represents the vector of angles of each link from the vertical line, as depicted in Figure 1b.

$$J_1 = I_1 + m_1\,h_1^2 + m_2\,l_1^2 + m_3\,l_1^2, \tag{9}$$

$$J_2 = I_2 + m_2\,h_2^2 + m_3\,l_2^2, \tag{10}$$

$$J_3 = I_3 + m_3\,h_3^2, \tag{11}$$

I_i, m_i, h_i, and l_i represent the ith link's moment of inertia around its center of gravity, mass, distance from its bottom to the its center of gravity, and length, respectively.

$$I_{pi} = I_{p_i'} + K_i^2\,I_{mi}, \tag{12}$$

$I_{p_i'}$, and K_i represent the ith hinge's belt–pulley system's moment of inertia, and ratio of teeth, respectively, I_{mi} represents the ith motor's moment of inertia, and C_i represents the viscous friction coefficients of the ith hinge,

$$C_{pi} = C_{p_i'} + K_i^2\,C_{mi}, \tag{13}$$

C_{mi}, $C_{p_i'}$ represent the viscous friction coefficients of the ith motor and the ith hinge's belt–pulley system, respectively, t_{mi} represents the control torque of the ith motor, and d_i represents the disturbance torque to the ith link.

Linearizing the model described in Equation (3) [1] around the operating point $\theta_1 = \theta_2 = \theta_3 = 0$, results in the following:

$$\begin{aligned}
&\begin{bmatrix} J_1 + I_{p1} & l_1 M_2 - I_{p1} & l_1 M_3 \\ l_1 M_2 - I_{p1} & J_2 + I_{p1} + I_{p2} & l_2 M_3 - I_{p2} \\ l_1 M_3 & l_2 M_3 - I_{p2} & J_3 + I_{p2} \end{bmatrix} \begin{bmatrix} \ddot{\theta}_1 \\ \ddot{\theta}_2 \\ \ddot{\theta}_3 \end{bmatrix} \\
&+ \begin{bmatrix} C_1 + C_2 + C_{p1} & -C_2 - C_{p1} & 0 \\ -C_2 - C_{p1} & C_{p1} + C_{p2} + C_2 + C_3 & -C_3 - C_{p2} \\ 0 & -C_3 - C_{p2} & C_3 + C_{p2} \end{bmatrix} \begin{bmatrix} \dot{\theta}_1 \\ \dot{\theta}_2 \\ \dot{\theta}_3 \end{bmatrix} \\
&+ \begin{bmatrix} -M_1 g & 0 & 0 \\ 0 & -M_2 g & 0 \\ 0 & 0 & -M_3 g \end{bmatrix} \begin{bmatrix} \theta_1 \\ \theta_2 \\ \theta_3 \end{bmatrix} + \begin{bmatrix} K_1 & 0 \\ -K_1 & K_2 \\ 0 & -K_2 \end{bmatrix} \begin{bmatrix} t_{m1} \\ t_{m2} \end{bmatrix} \\
&= \begin{bmatrix} 1 & -1 & 0 \\ 0 & 1 & -1 \\ 0 & 0 & 1 \end{bmatrix} \begin{bmatrix} d_1 \\ d_2 \\ d_3 \end{bmatrix}
\end{aligned} \tag{14}$$

Arranging the model in state-space representation, given that the system has two inputs (u_1, u_2) and three outputs (θ_1, θ_2, θ_3), and substituting the nominal values of the parameters given in Table 1 [1], yields

$$\begin{aligned}
\dot{\mathbf{x}}(t) &= \begin{bmatrix} 0 & 0 & 0 & 1 & 0 & 0 \\ 0 & 0 & 0 & 0 & 1 & 0 \\ 0 & 0 & 0 & 0 & 0 & 1 \\ 12.54 & -8.26 & -0.39 & -0.043 & 2.75 & -0.36 \\ -4.38 & 36.95 & -3 & 0.086 & -9.57 & 2.29 \\ -6.82 & -22.94 & 11.93 & -0.034 & 6.82 & -2.86 \end{bmatrix} \mathbf{x}(t) + \begin{bmatrix} 0 & 0 \\ 0 & 0 \\ -50 & 6.12 \\ 174.41 & -38.93 \\ -124.2 & 48.62 \end{bmatrix} \mathbf{u}(t) \\
\mathbf{y}(t) &= \begin{bmatrix} 1 & 0 & 0 & 0 & 0 & 0 & 0 \\ 0 & 1 & 0 & 0 & 0 & 0 & 0 \\ 0 & 0 & 1 & 0 & 0 & 0 & 0 \end{bmatrix} \mathbf{x}(t) + \begin{bmatrix} 0 & 0 \\ 0 & 0 \\ 0 & 0 \end{bmatrix} \mathbf{u}(t)
\end{aligned} \tag{15}$$

where the state vector $\mathbf{x}$ is $\begin{bmatrix} \theta_1\ \theta_2\ \theta_3\ \dot{\theta}_1\ \dot{\theta}_2\ \dot{\theta}_3 \end{bmatrix}^{\mathrm{T}}$, the control input vector $\mathbf{u}$ is $\begin{bmatrix} u_1 & u_2 \end{bmatrix}^{\mathrm{T}}$, and the output vector $\mathbf{y}$ is $[\theta_1\ \theta_2\ \theta_3]^{\mathrm{T}}$. Then, for this MIMO system, $n = 6$ states and $m = 2$ inputs.

Table 1. Nominal values of model's parameters [1].

Parameter	Value	Unit
J_1	0.654	kg.m^2
J_2	0.117	kg.m^2
J_3	0.535	kg.m^2
m_1	3.25	kg
m_2	1.9	kg
m_3	2.23	kg
h_1	0.35	m
h_2	0.181	m
h_3	0.245	m
l_1	0.5	m
l_2	0.4	m
$I_{p1'}$	7.95×10^{-3}	kg.m^2
$I_{p2'}$	3.97×10^{-3}	kg.m^2
K_1	30.72	dimensionless
K_2	27	dimensionless
I_{m1}	2.4×10^{-5}	kg.m^2
I_{m2}	4.90×10^{-6}	kg.m^2
C_1	6.54×10^{-2}	N.m.s
C_2	2.32×10^{-2}	N.m.s
C_3	8.80×10^{-3}	N.m.s
C_{m1}	2.19×10^{-3}	N.m.s
C_{m2}	7.17×10^{-4}	N.m.s
$C_{p1'}$	0	N.m.s
$C_{p2'}$	0	N.m.s
g	9.81	m.s^{-2}

3. PSO Algorithm

The PSO algorithm is a population-based metaheuristic search method that is valuable in engineering design optimization (and many other applications, as stated in [35]). The algorithm was developed by James Kennedy and Russell C. Eberhart, inspired by the organized trajectory adjustment used in bird swarms and fish schools, based on their own experiences and the experiences of others in the group (Figure 2).

(a) (b)

Figure 2. Social behavior of (**a**) bird swarm [39] and (**b**) fish school [40].

The algorithm (Algorithm 1) starts by spreading its agents randomly in the predefined search space, to look for the best solution. Then, in each iteration, the agents update their velocities and positions based on mathematical formulas that incorporate their current positions, velocities, and historical information. These update formulas for the agents' velocity (V_{PSO}) and position (X_{PSO}) are the following:

Algorithm 1: PSO algorithm	
Step 1:	Initialize the algorithm's parameters (w_{PSO}, c_{PSO1}, r_{PSO1}, c_{PSO2}, r_{PSO2}, dimension, bounds, number of search agents, iteration counter, and maximum number of iterations)
Step 2:	Set the cost function according to the required application of optimization
Step 3:	Initialize the population
Step 4:	Evaluate the cost function for all candidate solutions
Step 5:	Specify each candidate solution as its best personal solution primarily (p_{best})
Step 6:	Specify the candidate solution that produces the minimum cost function among all agents as the best global solution (g_{best})
Step 7-a:	While (iteration number < maximum number of iterations), do:
Step 7-b-1:	For (all agents of all dimensions), do the following:
Step 7-b-2:	Update agents' velocity (Equation (16))
Step 7-b-3:	Update agents' position (Equation (17))
Step 7-b-4:	Evaluate the cost function for all candidate solutions
Step 7-b-5:	Specify the candidate solution that produces the minimum cost function in its history as the best personal solution obtained so far (p_{best})
Step 7-b-6:	End for
Step 7-c:	Specify the candidate solution that produces the minimum cost function in all agents' history as the best global solution obtained so far (g_{best})
Step 7-d:	Increase the iteration counter by one
Step 7-e:	End while
Step 8:	Return the best solution (g_{best})

$$V_{PSOi}(t+1) = w_{PSO} \times V_{PSOi}(t) + c_{PSO1} \times r_{PSO1} \times (p_{besti}\,(t) - X_{PSOi}(t)) + c_{PSO2} \times r_{PSO2} \times (g_{best}\,(t) - X_{PSOi}(t)), \quad (16)$$

and

$$X_{PSOi}\,(t+1) = X_{PSOi}\,(t) + V_{PSOi}\,(t+1), \quad (17)$$

respectively, where w_{PSO} represents the inertia weight, which is, instead of being a constant value as in [41], chosen to change its value through iterations from $\overline{w}_{PSO}$ to $\underline{w}_{PSO}$ according to the formula

$$w_{PSO} = \overline{w}_{PSO} - (\overline{w}_{PSO} - \underline{w}_{PSO}) \times \frac{t}{maximum\ number\ of\ iterations}, \quad (18)$$

to provide a thorough exploration of the search space in the early stages of optimization. Then its value keeps decreasing, to reduce the particles' tendency to move far away from their current positions, such that the optimization method converges to a solution more efficiently; r_{PSO1} and r_{PSO2} are random numbers between 0 and 1, c_{PSO1} and c_{PSO2} are cognitive and social coefficients, respectively, and g_{best} and p_{besti} represent the best global and personal solutions, respectively. Subsequently, the continued updating guides the agents towards better solutions till a specified stopping criterion is met (usually, the maximum number of iterations). The PSO agents' updating is visualized in Figure 3.

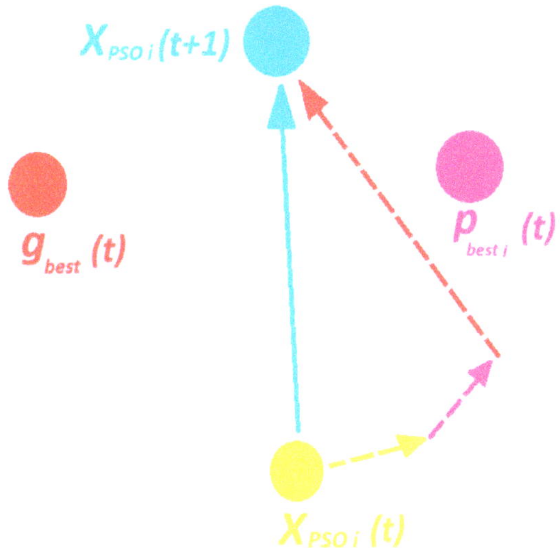

Figure 3. PSO agent update.

4. Hybrid H∞/SMC Controller Design

The control of uncertain systems is required to provide robustness against parameter variations, external disturbances, and measurement noise. Nevertheless, seeking extreme robustness is not advised, because it leads to unnecessary conservativeness that affects the system's performance characteristics. Optimization can be a great aid in enhancing the controller design to calibrate the system's robustness/performance outcome. H_∞ and SMC are well known robust controllers that have shown their effectiveness in many uncertain control systems, so the controller design process is composed of the steps shown in Figure 4 and detailed below:

First, the H_∞ control is used to design the sliding manifold of the SMC to provide a more powerful hybrid controller. To account for parameter variations, the system in Equation (15) is represented as

$$\dot{\mathbf{x}}(t) = (A + \Delta A)\,\mathbf{x}(t) + B\,\mathbf{u}(\mathrm{t}) \tag{19}$$

where ΔA represents the uncertainty matrix bounded by the Q matrix:

$$\mathrm{Q} = \begin{bmatrix} 0 & 0 & 0 & 0 & 0 & 0 \\ 0 & 0 & 0 & 0 & 0 & 0 \\ 0 & 0 & 0 & 0 & 0 & 0 \\ 0.2 & 0.2 & 0.2 & 0.2 & 0.2 & 0.2 \\ 0.2 & 0.2 & 0.2 & 0.2 & 0.2 & 0.2 \\ 0.2 & 0.2 & 0.2 & 0.2 & 0.2 & 0.2 \end{bmatrix} \tag{20}$$

In order to apply decoupling for MIMO systems, a similarity transformation is required to transform the state variable $\mathbf{x}(t)$ to a new state variable $\mathbf{q}(t)$, defined as:

$$\mathbf{q}(t) = H\,\mathbf{x}(t) \tag{21}$$

where H is the $n \times n$ similarity transformation matrix, obtained as

$$H = \begin{bmatrix} N & B \end{bmatrix}^T \tag{22}$$

where the $n \times (n-m)$ matrix N is the null space of the transpose of B matrix. Thus:

$$H = \begin{bmatrix} 0 & 0 & 1 & 0 & 0 & 0 \\ 0.227 & 0.399 & 0 & 0.788 & 0.361 & 0.19 \\ -0.793 & -0.452 & 0 & 0.361 & 0.165 & 0.087 \\ 0.564 & -0.796 & 0 & 0.19 & 0.087 & 0.045 \\ 0 & 0 & 0 & -50 & 174.4 & -124.2 \\ 0 & 0 & 0 & 6.12 & -38.93 & 48.62 \end{bmatrix} \tag{23}$$

The transformed state space representation is

$$\dot{\mathbf{q}}(t) = \overline{A}\,\mathbf{q}(t) + \Delta\overline{A}\,\mathbf{q}(t) + \overline{B}\,\mathbf{u}(t) \tag{24}$$

where

$$\overline{A} = H\,A\,H^{-1}, \tag{25}$$

$$\Delta\overline{A} = H\,\Delta A\,H^{-1}, \tag{26}$$

$$\overline{B} = H\,B, \tag{27}$$

The new state variable representation becomes

$$\begin{bmatrix} \dot{\mathbf{q}}_1(\mathbf{t}) \\ \dot{\mathbf{q}}_2(\mathbf{t}) \end{bmatrix} = \begin{bmatrix} \overline{A}_{11} & \overline{A}_{12} \\ \overline{A}_{21} & \overline{A}_{22} \end{bmatrix} \begin{bmatrix} \mathbf{q}_1(\mathbf{t}) \\ \mathbf{q}_2(\mathbf{t}) \end{bmatrix} + \begin{bmatrix} \Delta\overline{A}_{11} & \Delta\overline{A}_{12} \\ \Delta\overline{A}_{21} & \Delta\overline{A}_{22} \end{bmatrix} \begin{bmatrix} \mathbf{q}_1(\mathbf{t}) \\ \mathbf{q}_2(\mathbf{t}) \end{bmatrix} + \begin{bmatrix} 0 \\ \overline{B} \end{bmatrix} \mathbf{u}(t) \tag{28}$$

In this representation, the m control signals $\mathbf{u}(t)$ act as an input to the lower subsystem only (controlling the m new state variables $\mathbf{q}_2(\boldsymbol{t})$ directly), then, during the sliding phase, $\mathbf{q}_2(\boldsymbol{t})$ acts as an input to the upper subsystem (controlling the remaining $(n-m)$ new state variables $\mathbf{q}_1(\boldsymbol{t})$ by feedback gain):

$$\mathbf{q}_2(t) = -K_\infty \mathbf{q}_1(t) \tag{29}$$

The $m \times (n-m)$ feedback gain matrix K_∞, which is related to the sliding manifold, is to be designed using H_∞ control to provide robustness to parametric uncertainty. Equation (29) and the upper subsystem:

$$\dot{\mathbf{q}}_1(t) = (\overline{A}_{11} + \Delta\,\overline{A}_{11})\mathbf{q}_1(t) + (\overline{A}_{12} + \Delta\,\overline{A}_{12})\mathbf{q}_2(t) \tag{30}$$

have to be represented in the Linear Fractional Transformation (LFT) [42] shown in Figure 5, in which the uncertainty is pulled out in the Δ block, I denotes the identity matrix, B_∞, C_∞, D_∞ are matrices obtained by singular value decomposition, and $\mathbf{z}_\infty$ and $\mathbf{d}_\infty$ represent the disturbed output and input, respectively.

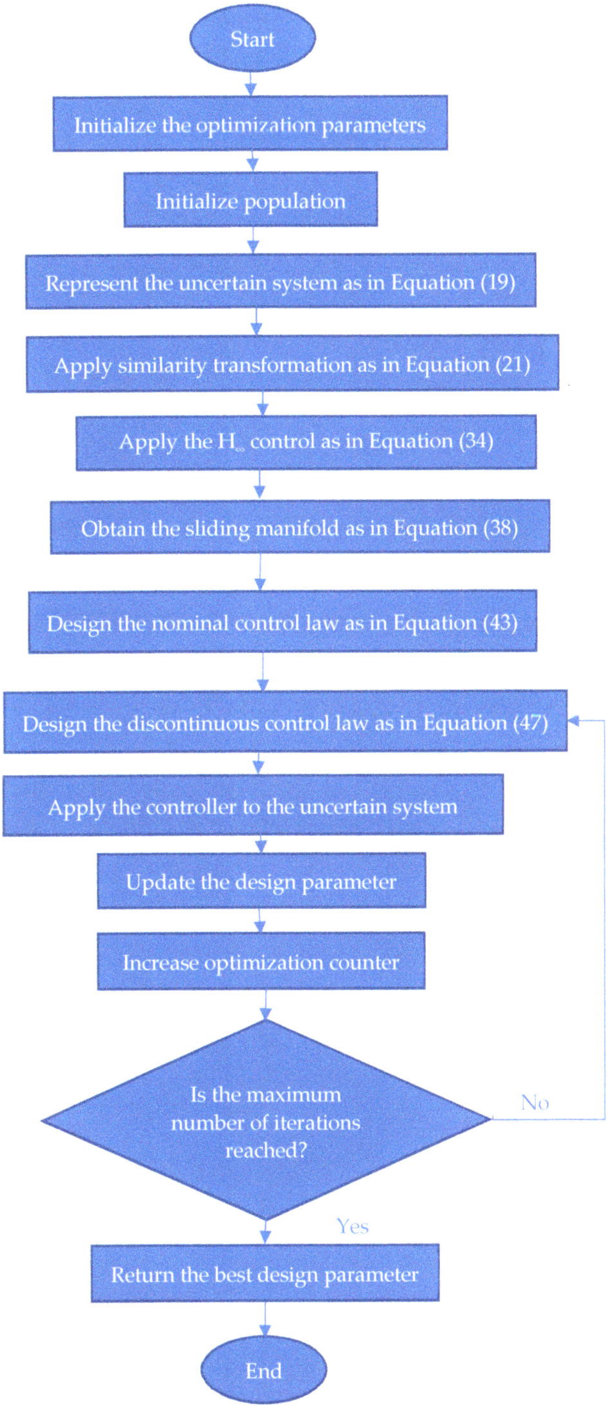

Figure 4. Controller design and optimization.

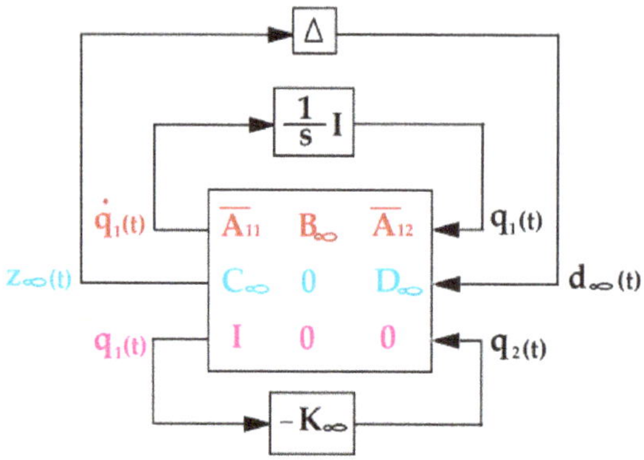

Figure 5. LFT configuration of the system Equations (34) and (36).

Upper LFT is applied to the system, so that the H_∞ problem becomes one of finding the stabilizing feedback gain K_∞ that minimizes $\gamma > 0$ in the quadratic objective function:

$$J_\infty = \frac{1}{2}\int_0^\infty \left[\mathbf{z}_\infty^T(t)\ \mathbf{z}_\infty(t) - \gamma^2\ \mathbf{d}_\infty^T(t)\ \ \mathbf{d}_\infty(t)\right] dt. \tag{31}$$

From Figure 5,

$$J_\infty = \frac{1}{2}\int_0^\infty \left[\mathbf{q}_1^T(t)C_\infty^T\ C_\infty\ \mathbf{q}_1(t) + 2\mathbf{q}_1^T(t)C_\infty^T\ D_\infty\ \mathbf{q}_2(t) + \mathbf{q}_2^T(t)D_\infty^T\ D_\infty\ \mathbf{q}_2(t) - \gamma^2\mathbf{d}_\infty^T(t)\mathbf{d}_\infty(t)\right] dt. \tag{32}$$

The Ricatti-like matrix equation for this linear quadratic problem is

$$P_\infty\left(\overline{A}_{11} - \overline{A}_{12}O_{22}^{-1}O_{12}^T\right) + \left(\overline{A}_{11}^T - O_{12}O_{22}^{-1}\overline{A}_{12}^T\right)P_\infty - P_\infty\left(\overline{A}_{12}O_{22}^{-1}\overline{A}_{12}^T - \gamma^{-2}B_\infty\ B_\infty^T\right)P_\infty + \left(O_{11} - O_{12}O_{22}^{-1}O_{12}^T\right) = 0. \tag{33}$$

where $O_{11} = C_\infty^T C_\infty$, $O_{12} = C_\infty^T D_\infty$, $O_{22} = D_\infty^T D_\infty$. The solution of the above equation (P_∞) is used to find the feedback gain K_∞:

$$K_\infty = O_{22}^{-1}\left(\overline{A}_{12}^T P_\infty + O_{12}^T\right), \tag{34}$$

Thus,

$$K_\infty = \begin{bmatrix} -2447 & -12790 & 14632 & -6474 \\ 392 & 2015 & -2301 & 1024 \end{bmatrix}, \tag{35}$$

which can be used to define the sliding manifold $\mathbf{s}(t)$:

$$\mathbf{s}(t) = K_\infty \mathbf{q}_1(t) + \mathbf{q}_2(t) \tag{36}$$

When the state variables reach and stay on the sliding manifold ($\mathbf{s}(t) = 0$), then the system dynamics is governed by Equation (29). Equation (36) can be written as

$$\mathbf{s}(t) = \begin{bmatrix} K_\infty & I \end{bmatrix}\mathbf{q}(t) \tag{37}$$

From Equation (21), the sliding manifold is expressed in terms of the original states as

$$\mathbf{s}(t) = G\ \mathbf{x}(t) \tag{38}$$

and

$$G = \begin{bmatrix} K_\infty & I \end{bmatrix} H \tag{39}$$

Next, the control law of the SMC is suggested to comprise a nominal term ($\mathbf{u}_{\circ}(t)$) and a discontinuous term ($\mathbf{u_d}(t)$):

$$\mathbf{u}(t) = \mathbf{u}_{\circ}(t) + \mathbf{u_d}(t) \tag{40}$$

The nominal term is responsible for driving the states towards the sliding manifold defined by Equation (38). It is obtained by equating the derivative of the sliding variable to zero, that is

$$\dot{\mathbf{s}}(t) = 0 = G\,\dot{\mathbf{x}}(t) \tag{41}$$

Substituting the nominal terms of Equation (19),

$$\dot{\mathbf{s}}(t) = 0 = GA\,\mathbf{x}(t) + GB\,\mathbf{u}_{\circ}(\mathrm{t}) \tag{42}$$

from which

$$\mathbf{u}_{\circ}(\mathrm{t}) = -(GB)^{-1}GA\,\mathbf{x}(t). \tag{43}$$

provided that the GB matrix is non-singular; thus, it is invertible. To obtain the reaching condition of the SMC, a candidate quadratic Lyapunov function is suggested to be

$$V_L(t) = \frac{1}{2}\,\mathbf{s}^2(t) \tag{44}$$

To ensure the stability of the sliding manifold and the overall system rigorously, the time derivative of $V_L(t)$ has to be negative definite for all $\mathbf{s} \neq 0$, so that $V_L(t)$ keeps decreasing along any trajectory in the state space which, in turn, guarantees the stability of the sliding mode. It implies that

$$\mathbf{s}(t)\dot{\mathbf{s}}(t) < 0 \tag{45}$$

The discontinuous term of the control law ($\mathbf{u_d}(t)$), which ensures the systems converge to the desired equilibrium point by imposing the negative definiteness of $\frac{dV_L(t)}{dt}$ in the presence of parametric uncertainty, disturbance, and noise, can be obtained by substituting Equations (19), (40) and (43) in Equation (45) and replacing the uncertainty matrix by its upper bound Q:

$$\mathbf{s}(t)[G\,Q\mathbf{x}(t) + GB\,\mathbf{u_d}] < 0 \tag{46}$$

$$\mathbf{u_d}(\mathbf{t}) = -\Big(GB)^{-1}[(|G|\,Q|\mathbf{x}(t)| + \eta)sgn(\mathbf{s}(t))\Big] < 0 \tag{47}$$

where the sign function ($sgn()$) is used to bring the states back to the sliding manifold whenever they attempt to leave it due to uncertainty, and η (2×2 diagonal matrix in this system) is a design parameter added to provide protection against disturbance and noise:

$$\eta = \begin{bmatrix} \eta_{11} & 0 \\ 0 & \eta_{22} \end{bmatrix} \tag{48}$$

Instead of using trial and error to find good values of the diagonal elements of η, optimization is suggested, to tune them by searching for their optimal values that lead to both good robustness and performance of the system.

Finally, PSO is applied to enhance the performance of the robust control system in terms of response errors and the overall energy of the needed torques. These characteristics are used to form the cost function

$$J_O = \sum_{i=1}^{3} \int_0^{t_f} t\,|e_i(t)|dt + \sum_{i=0}^{2} \int_0^{t_f} t_{mi}^2(t)dt\,. \tag{49}$$

where J_O represents the optimization cost function, $e_i(t)$ represents the ith link's angle response error, and t_f represents the final time used in simulations.

The ITAE performance index of error is used to penalize the steady-state errors more than transient errors. Since the lower the control energy required, the more efficient the control system, then the ISCS performance index of the control signal is used.

5. Results

This section presents the system and optimization settings and the results of simulating the proposed method for different cases to test its effectiveness. The developed control system is compared with the H_2 control system, whose 3×3 performance weighting matrix is also optimized by PSO. This weighting matrix consists of three second-order transfer functions (each with three coefficients in its numerator and the same in its denominator, and a gain) on its diagonal, and is used for weighting the system's tracking accuracy and robustness to uncertainty. The optimization problem of the two control systems is set as given in Table 2, with the cost function given by Equation (49).

Table 2. Setting of the PSO algorithm.

Parameter	Values for Hybrid H_∞/SMC Control System Optimization	Values for H_2 Control System Optimization
dimension	2	21
lower bound	100	10
upper bound	2000	100
number of search agents	10	10
maximum number of iterations	100	100
$\overline{w}_{PSO}$	0.7	0.7
$\underline{w}_{PSO}$	0.2	0.2
c_{PSO1}	2	2
c_{PSO2}	2	2
r_{PSO1}	random number in the interval (0, 1)	random number in the interval (0, 1)
r_{PSO2}	random number in the interval (0, 1)	random number in the interval (0, 1)

The hybrid H_∞/SMC system (built in MATLAB/Simulink R2023b) used in optimization is given in Figure 6, where, in each iteration, the candidate solutions (η_{11} and η_{22}) are passed from the optimization code (written in MATLAB script) to the controller block (written in MATLAB S-function); the control law is calculated and applied by the actuator to the system block (written in MATLAB S-function), and the states $x_1(t)$, $x_2(t)$, $x_3(t)$ and the actuator signals $t_{m1}(t)$ and $t_{m2}(t)$ are passed back to the optimization code to calculate the cost function of that candidate solution and to update the solutions accordingly. The convergence progress of the optimization algorithm is depicted in Figure 7. The PSO algorithm converges in the 59th iteration to the cost value of 1488.76. The best solution returned by PSO is [η_{11} = 168.4755, η_{22} = 1.0173 $\times 10^3$].

The H_2 control system optimization results in the following performance weighting matrix $W(s)$:

$$W(s) = \begin{bmatrix} 10\frac{51.123s^2+94.763s+10}{10s^2+100s+10} & 0 & 0 \\ 0 & 39.14\frac{17.412s^2+10s+10}{53.782s^2+55.429s+14.368} & 0 \\ 0 & 0 & 10\frac{33.8s^2+100s+100}{100s^2+100s+73.998} \end{bmatrix}$$

The performance index of the controller is 6.102.

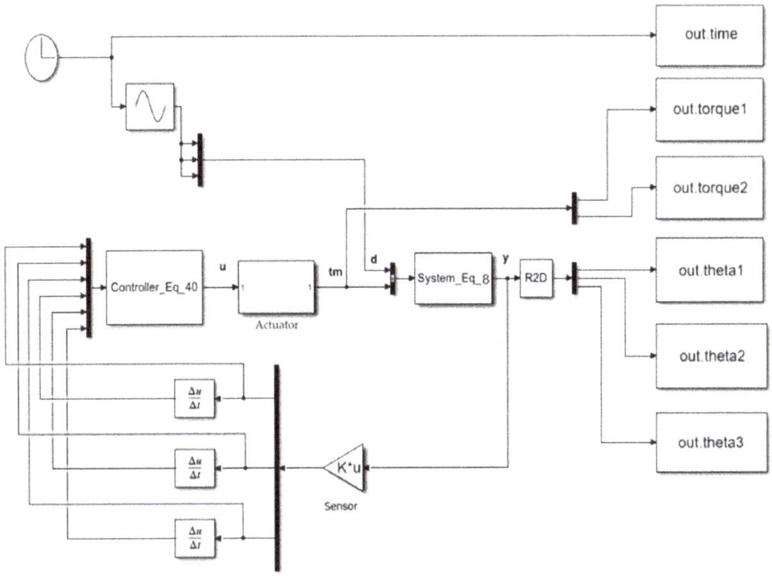

Figure 6. Control system used in optimization.

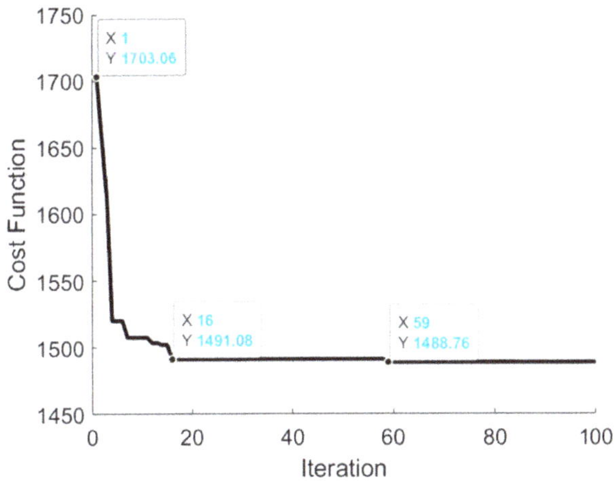

Figure 7. Cost-function convergence of PSO.

5.1. Nominal System Response

First, the developed control system is tested by applying it to the nonlinear model of the triple inverted pendulum given in Equation (3), with the nominal values of its parameters given in Table 1. The initial condition is 0.5 deg, 1 deg, and 2 deg for the first, second, and third angle, respectively. The three angles respond as shown in Figure 8, the phase-plane trajectories of the angles and their derivatives are shown in Figure 9, and the applied torques are shown in Figure 10. Please note that data tips in regular font correspond to hybrid H_∞/SMC system, while data tips in italic font correspond to the H_2 control system in all responses.

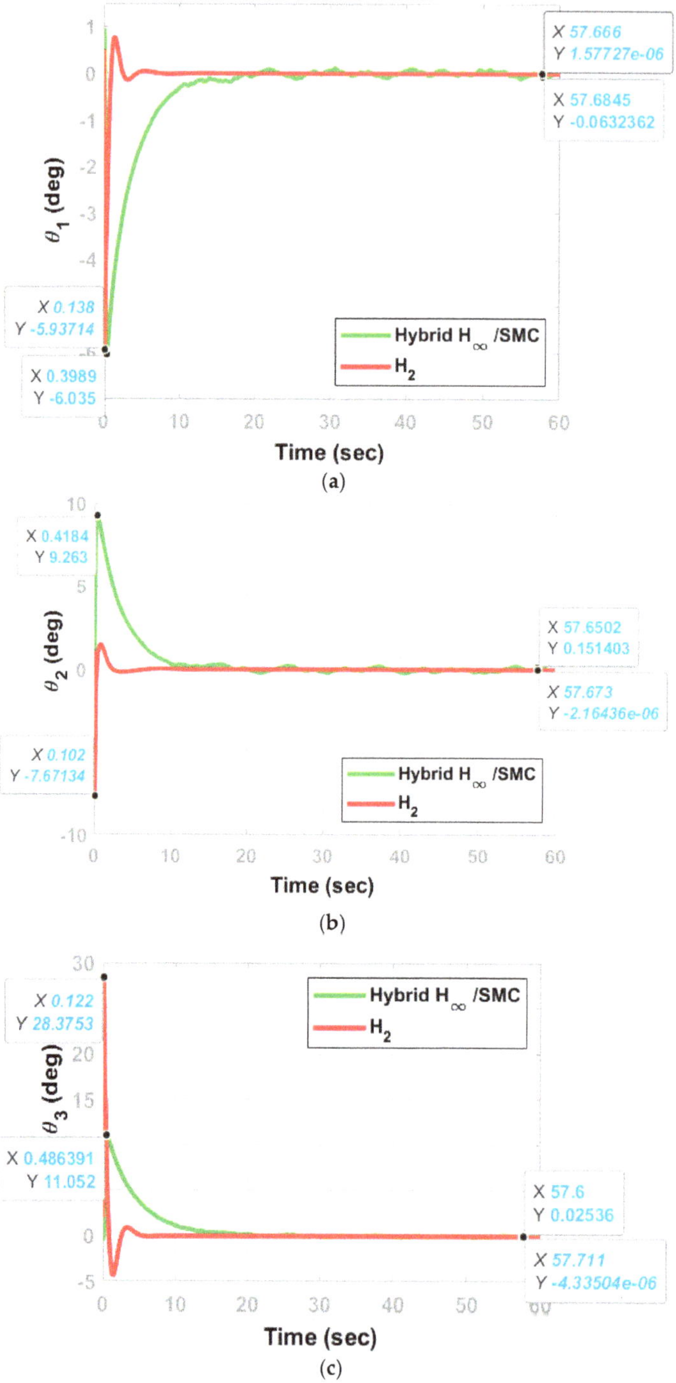

Figure 8. Nominal system response: (**a**) first link's angle; (**b**) second link's angle; (**c**) third link's angle.

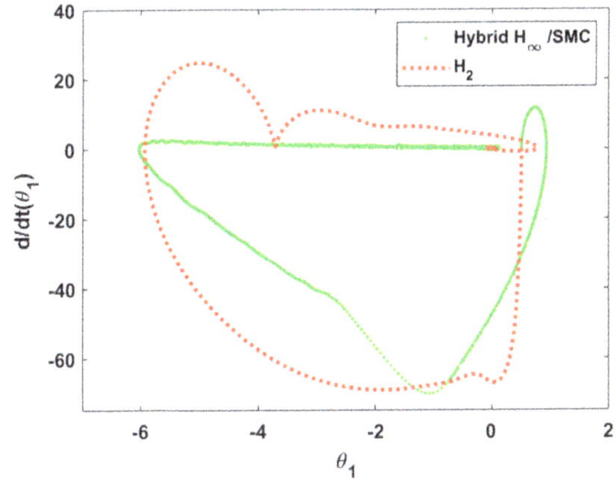

(a)

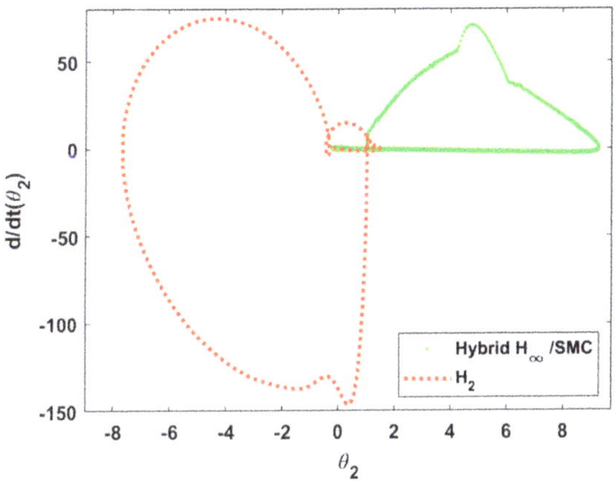

(b)

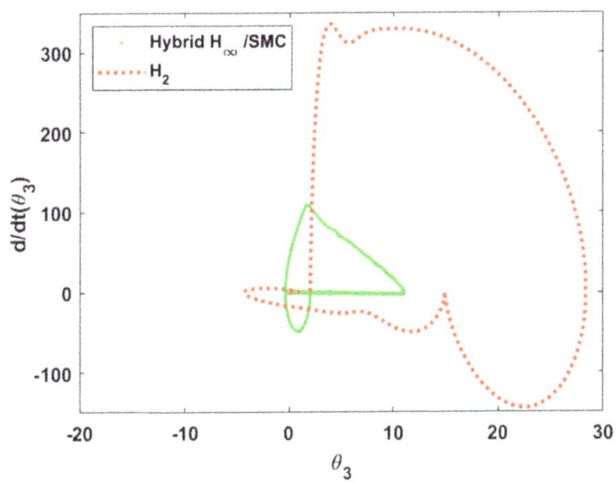

(c)

Figure 9. Nominal system phase-plane trajectory: (**a**) $(\theta_1, \dot{\theta}_1)$; (**b**) $(\theta_2, \dot{\theta}_2)$; (**c**) $(\theta_3, \dot{\theta}_3)$.

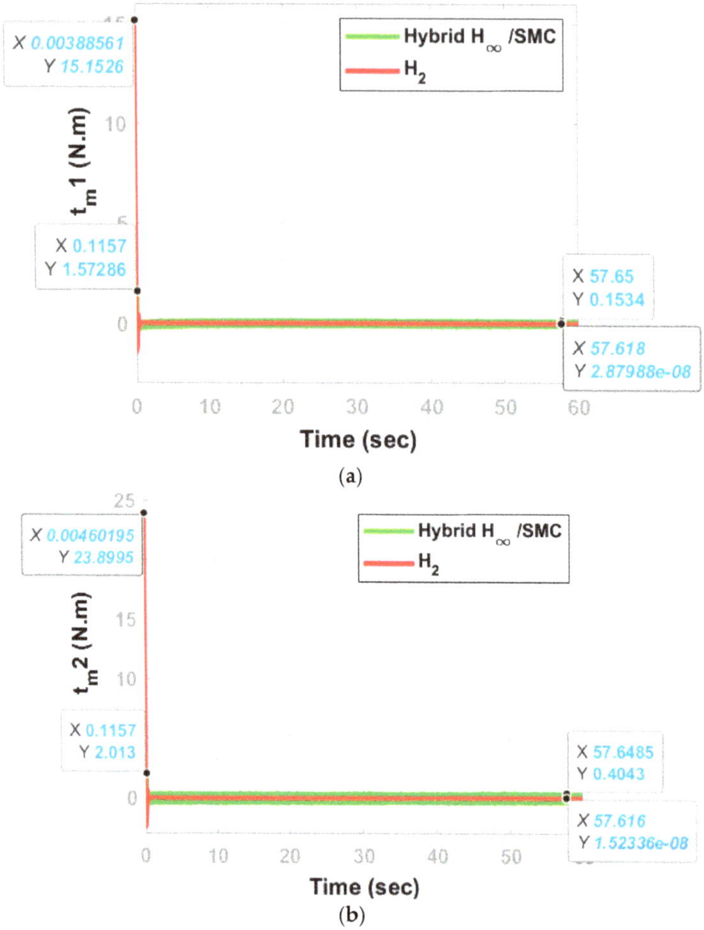

Figure 10. Nominal system torque: (**a**) first motor; (**b**) second motor.

The θ_1 and θ_2 absolute overshoot difference between the two control systems is 0.1 deg and 1.59 deg, respectively, while θ_3 overshoots 17.323 deg higher in the H_2 control system. The difference between the errors of the two systems is small: 0.062, 0.151, and 0.025 deg for θ_1, θ_2, and θ_3, respectively.

The above phase-plane trajectories show the direct sliding nature of the states towards their desired equilibrium point, driven by the hybrid H_∞/SMC control law. The H_2 control system lacks this preference.

The start-up torques t_{m1} and t_{m2} in the H_2 control system are 9.633 and 11.872 N.m times higher, respectively, than in the hybrid H_∞/SMC system. On the other hand, the differences in steady torques are 0.153 and 0.404 N.m.

5.2. Robustness to Parameter Variation

Second, the developed control system is tested by applying it to the nonlinear model of the triple inverted pendulum given in Equation (3), with perturbed uncertain parameters, I_1, I_2, and I_3 = nominal value + 10%, C_1, C_2, C_3, C_{m1}, and C_{m2} = nominal value + 15%. The three angles respond as shown in Figure 11, and the phase-plane trajectories of the angles and their derivatives are shown in Figure 12.

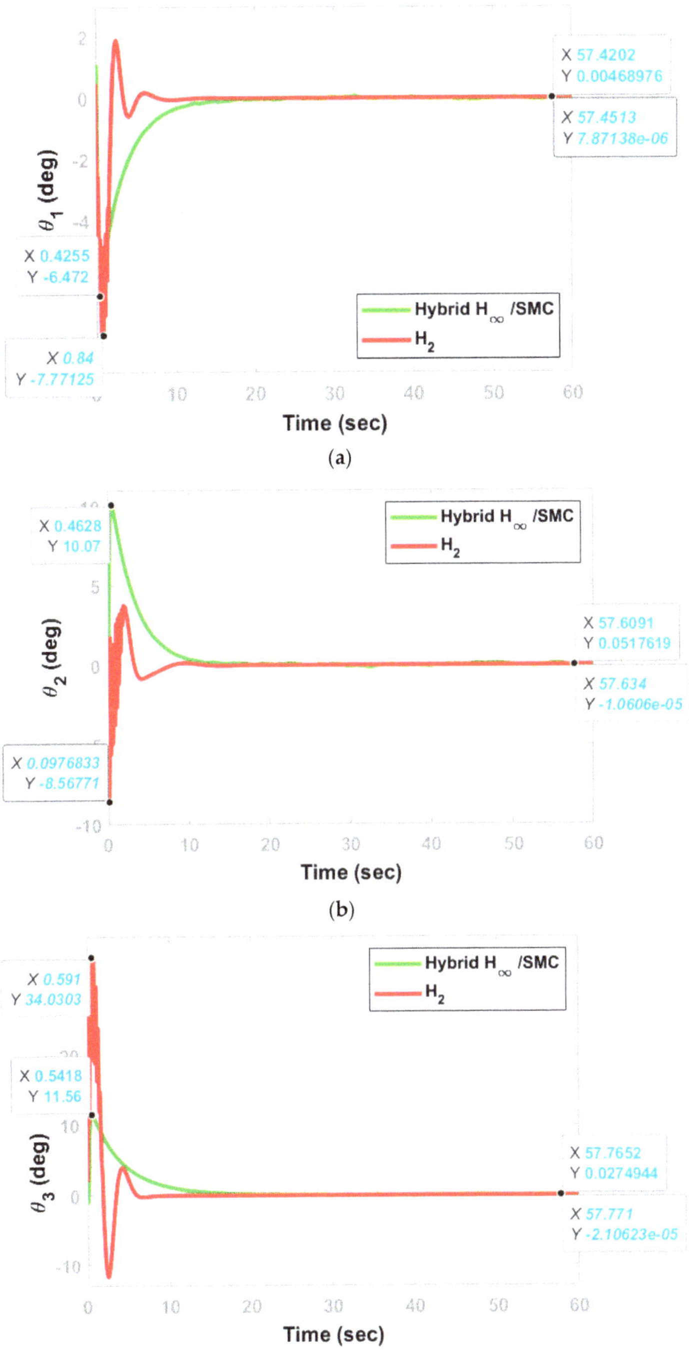

Figure 11. Perturbed by parametric uncertainty-system response: (**a**) first link's angle; (**b**) second link's angle; (**c**) third link's angle.

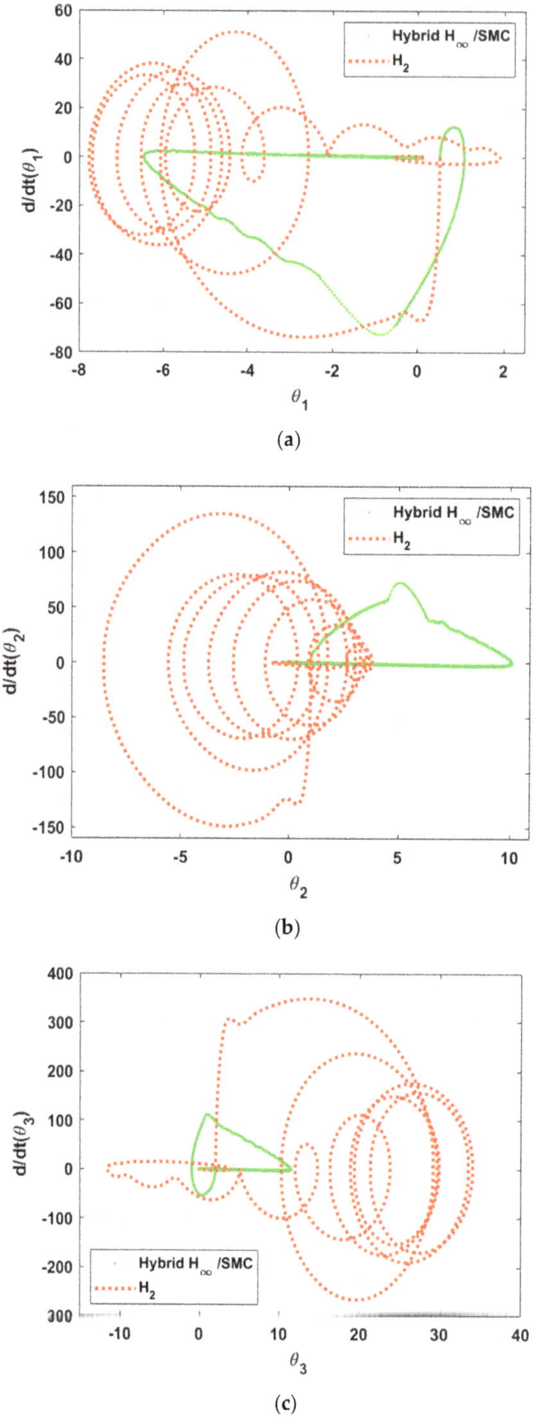

Figure 12. Perturbed by parametric uncertainty-system phase-plane trajectory: (**a**) $(\theta_1, \dot{\theta}_1)$; (**b**) $(\theta_2, \dot{\theta}_2)$; (**c**) $(\theta_3, \dot{\theta}_3)$.

The first and third angles overshoot by 1.299 and 22.47 deg, respectively, in the H_2 control system, more than in the hybrid H_∞/SMC system. Only the second angle overshoots higher in the hybrid H_∞/SMC system, by 1.503 deg. The difference in errors between the two systems are found to be 0.004, 0.05, and 0.027 deg.

The three phase-plane trajectories clearly depict the fluctuations at the start of the response in the H_2 control system.

The applied torques are shown in Figure 13. The start-up torques t_{m1} and t_{m2} in the H_2 control system are again higher than in the hybrid H_∞/SMC system, but by 8.227 and 10.2 N.m, respectively, this time. On the other hand, the differences in steady torques are 0.13 and 0.38 N.m. In this case, the H_2 controller fluctuates significantly at the beginning.

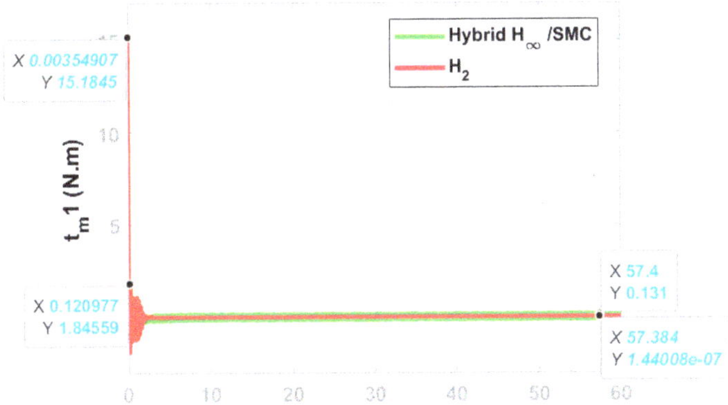

(a)

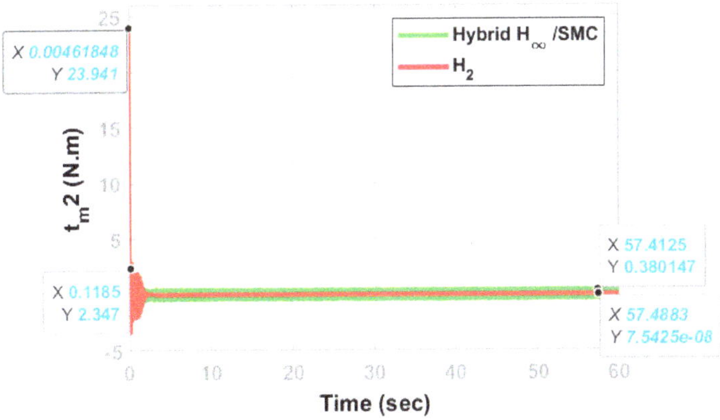

(b)

Figure 13. Perturbed by parametric uncertainty-system torque: (**a**) first motor; (**b**) second motor.

5.3. Robustness to External Disturbance

Third, the developed control system is tested by applying it to the nonlinear model of the triple inverted pendulum given in Equation (3), with external disturbance in the form of a sine wave of 0.1 N.m amplitude applied to each link. The three angles respond as shown in Figure 14; the phase-plane trajectories of the angles and their derivatives are shown in Figure 15.

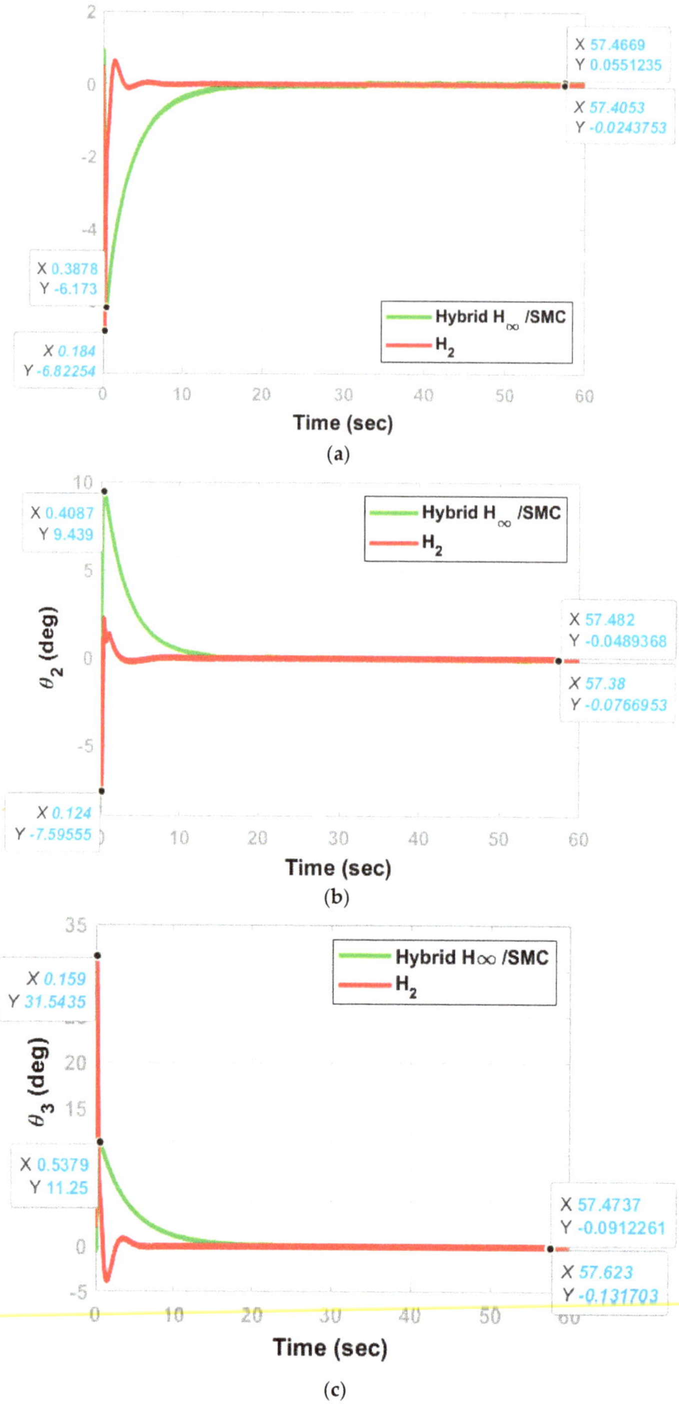

Figure 14. Perturbed by external-disturbance system response: (**a**) first link's angle; (**b**) second link's angle; (**c**) third link's angle.

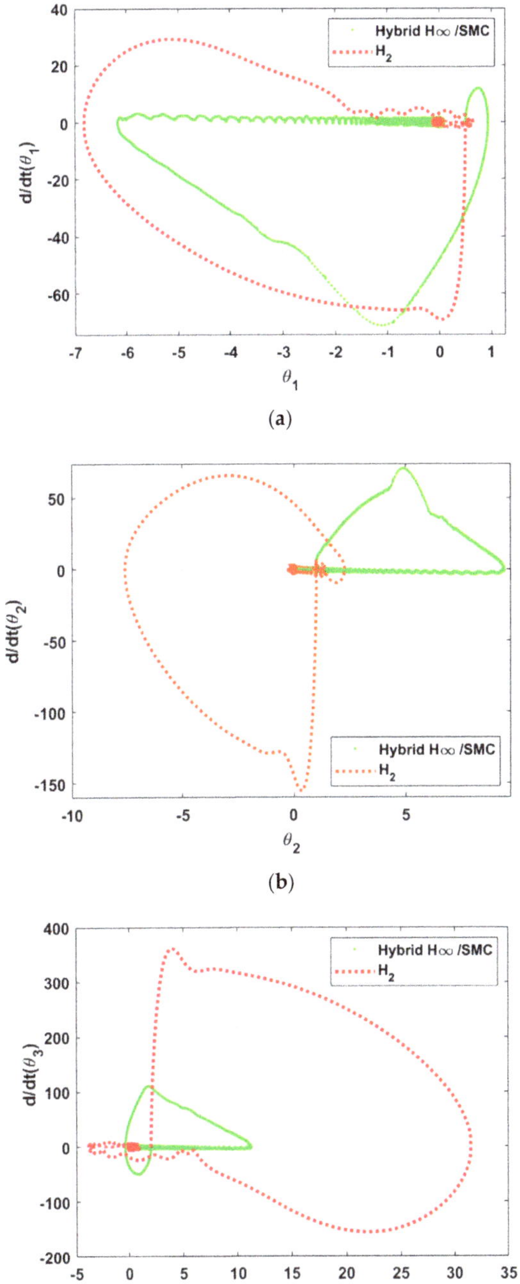

Figure 15. Perturbed by external-disturbance system phase-plane trajectory: (**a**) $(\theta_1, \dot{\theta}_1)$; (**b**) $(\theta_2, \dot{\theta}_2)$; (**c**) $(\theta_3, \dot{\theta}_3)$.

The first, second and third angle responses overshoot slightly higher, moderately lower, and significantly higher, respectively, in the H_2 control system, while the errors are close to each other in all responses.

The phase-plane trajectories show the direct alignment of the systems' states with the desired point in the state space.

Figure 16a shows that t_{m1} at start-up in the H_2 control system is higher by 13.567 N.m, and t_{m2} by 21.839 N.m., while the differences in steady torques of the first and second actuators are only 0.132 and 0.405 N.m, respectively.

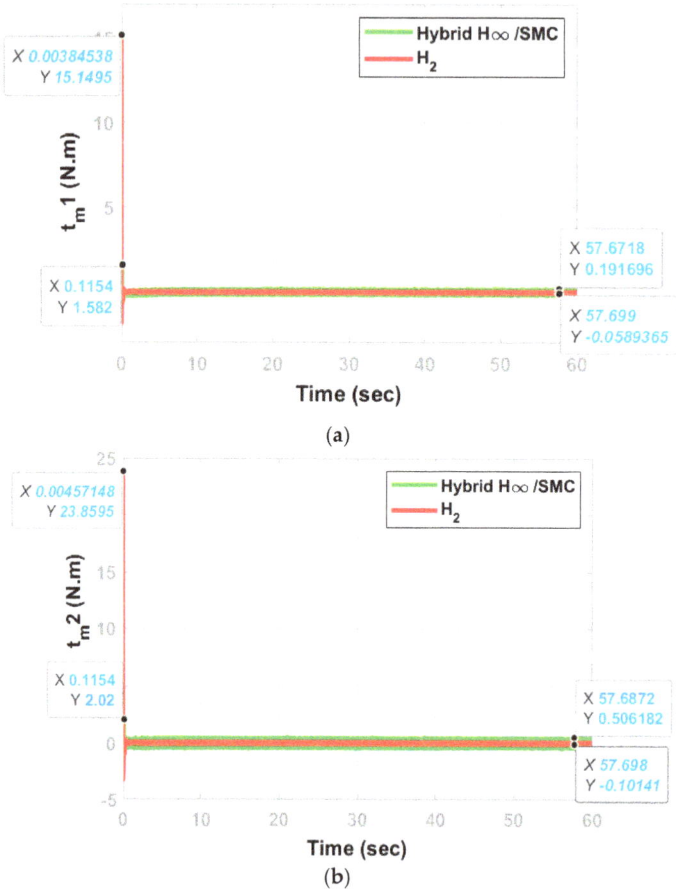

Figure 16. Perturbed by external-disturbance system torque: (**a**) first motor; (**b**) second motor.

5.4. Robustness to Measurement Noise

Finally, the developed control system is tested by applying it to the nonlinear model of the triple inverted pendulum given in Equation (3), with measurement noise in the form of random white noise and with amplitude ±0.4 V applied to each reading. The three angles respond as shown in Figure 17; the phase-plane trajectories of the angles and their derivatives are shown in Figure 18, and the applied torques are shown in Figure 19.

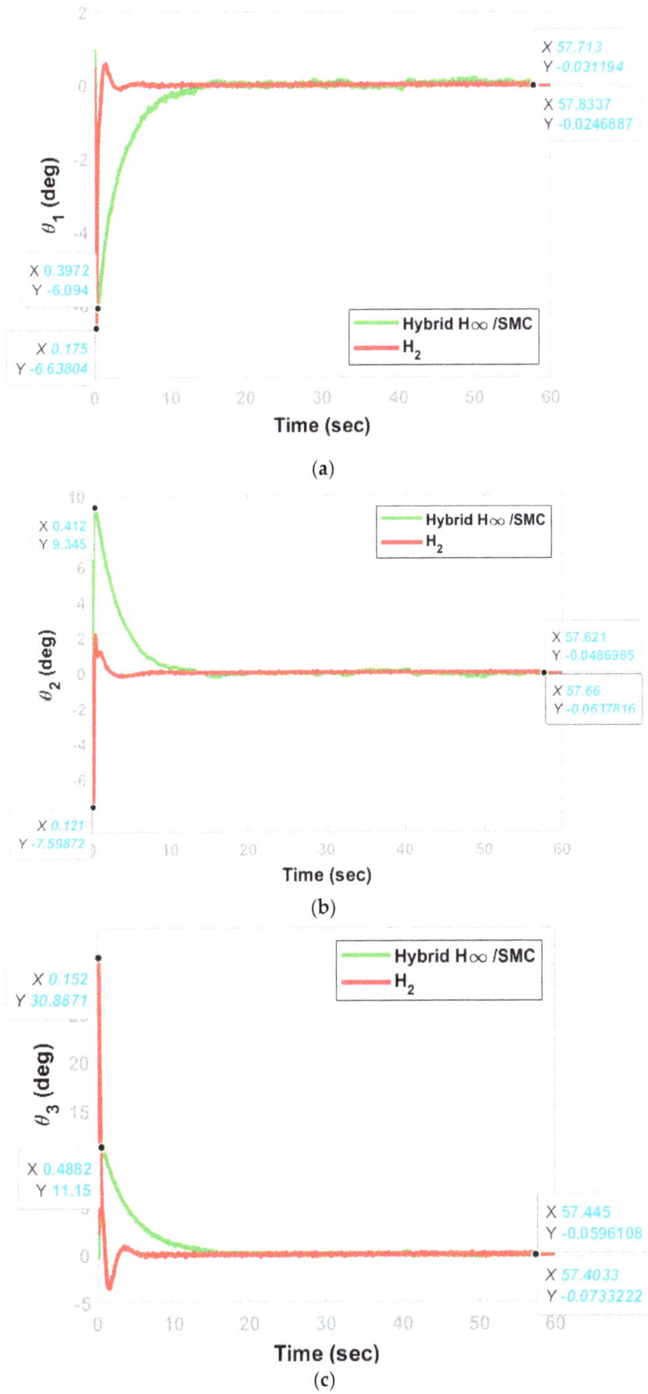

Figure 17. Perturbed by measurement-noise system response: (**a**) first link's angle; (**b**) second link's angle; (**c**) third link's angle.

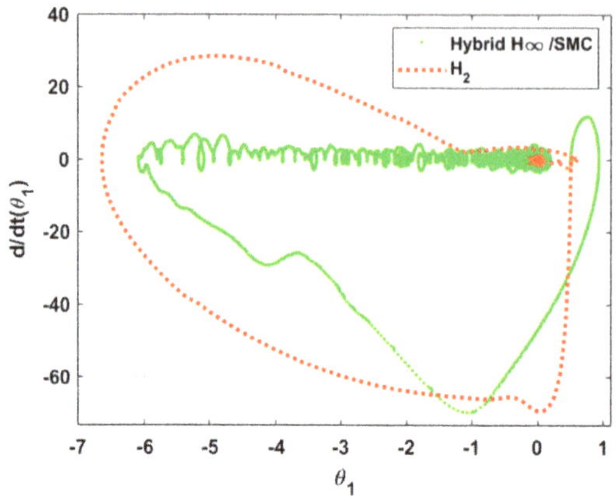

(**a**)

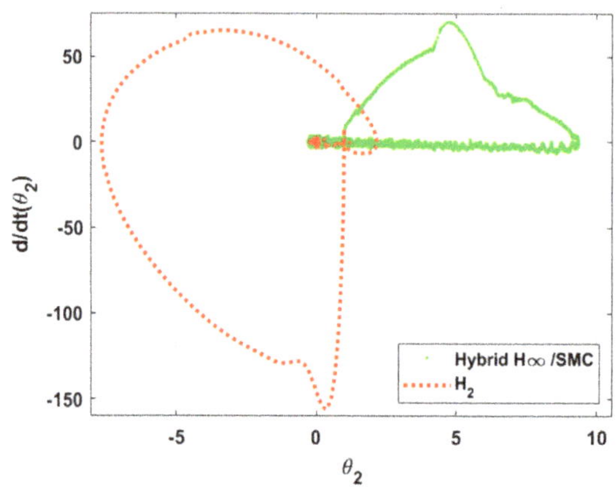

(**b**)

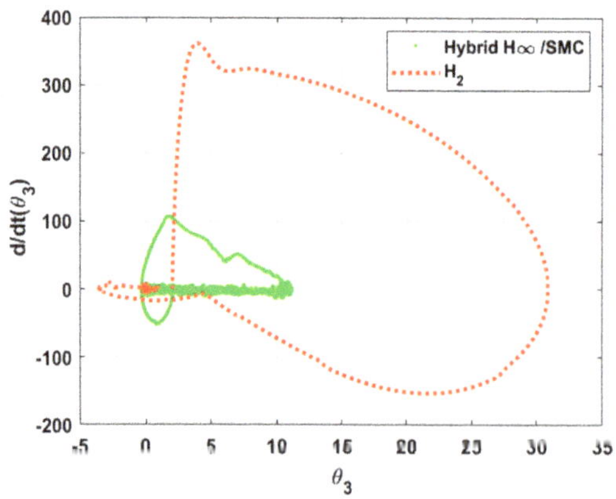

(**c**)

Figure 18. Perturbed by measurement-noise system phase-plane trajectory: (**a**) $(\theta_1, \dot{\theta}_1)$; (**b**) $(\theta_2, \dot{\theta}_2)$; (**c**) $(\theta_3, \dot{\theta}_3)$.

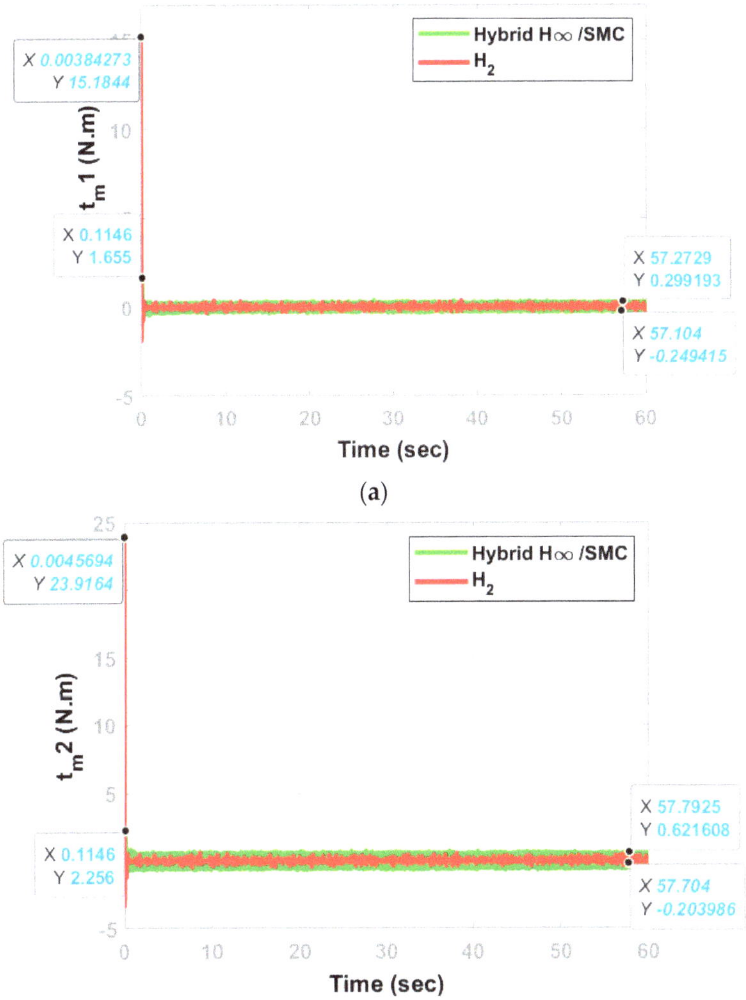

(a)

(b)

Figure 19. Perturbed by measurement-noise system torque: (**a**) first motor; (**b**) second motor.

In this case, the hybrid H_∞/SMC system produces less overshoot in the first and third angles' responses and fewer errors in the second and third angles' responses.

Figure 19 shows that, compared to the H_2 controller, the proposed hybrid H_∞/SMC controller requires much less effort at start-up, and almost the same effort is needed at steady state.

The characteristics of the proposed hybrid H_∞/SMC system performance in all tested cases are summarized in the three charts in Figures 20–22, and will be discussed in the next section.

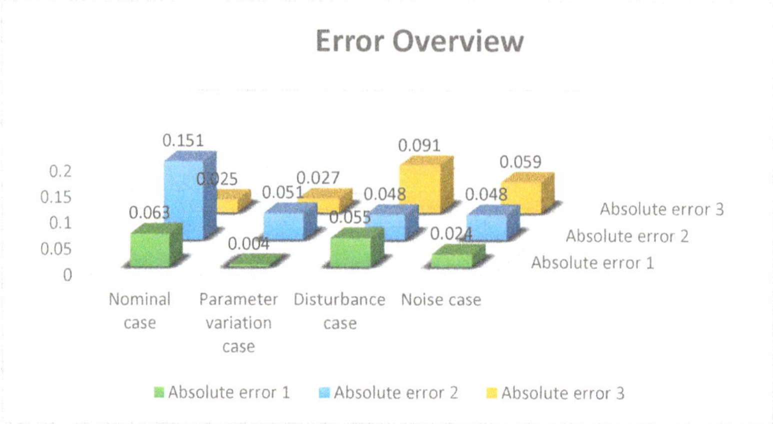

Figure 20. Overview of the three angles' errors in all tested cases.

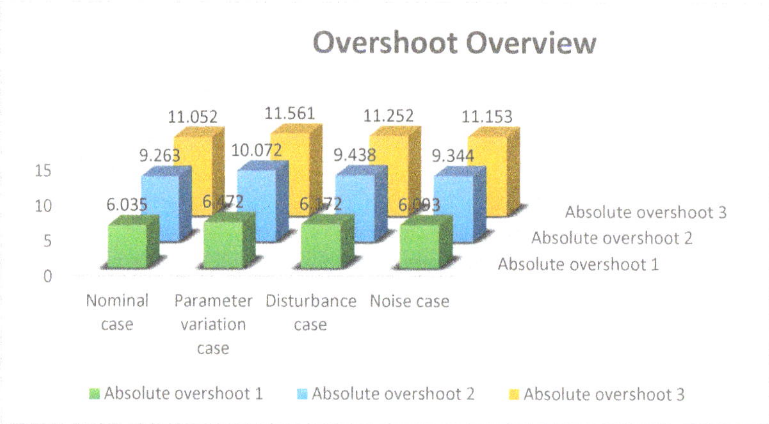

Figure 21. Overview of the three angles' overshoots in all tested cases.

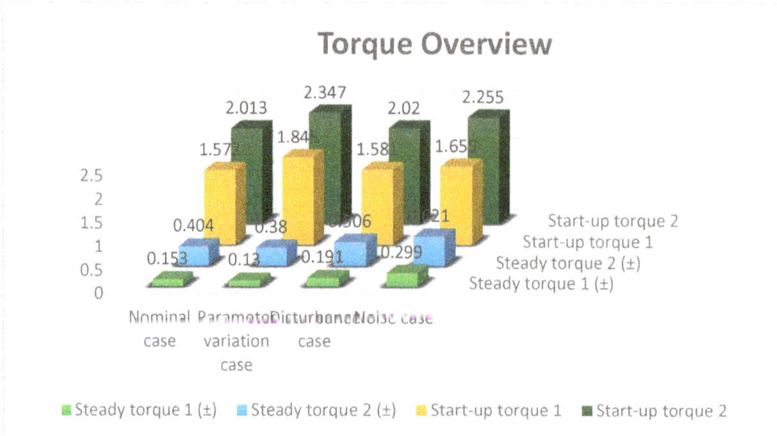

Figure 22. Overview of the two motors' torques in all tested cases.

6. Discussion and Conclusions

This section explores the interpretation and implications of the obtained results. Regarding the optimization aspect, Figure 7 shows that PSO exhibited a very fast convergence rate (majorly within 16 iterations, and completely within 59 iteration). The application of optimization to the controller design helped in improving the control system, through minimizing the response error (as shown in Figures 8, 11, 14 and 17) and the applied torque needed for the system (as shown in Figures 10, 13, 16 and 19).

In relation to the robustness attribute, the developed hybrid H_∞/SMC control system was able to stabilize the nominal, perturbed by parameter variation, perturbed by external disturbance, and perturbed by measurement-noise system. In all cases, the system responded in the same manner (except for very small differences, as shown in Figures 14 and 17).

Concerning performance characteristics, it has been shown in Figure 20 that, in all cases and among all the angles, the error was below 0.152 deg and the average error was 0.053 deg, which is an indication of the system's accuracy. The overshoot in all cases is almost unaffected by the type of uncertainty, as shown in Figure 21. For the first, second, and third angle, the overshoot ranged from 6.035 to 6.472 deg, 9.263 to 10.072 deg, and 11.052 to 11.561 deg, respectively. The least overshoots were in the nominal system case, while the highest were in the perturbed-by-parameter-variation case. Figures 9, 12, 15 and 18 show that the phase plane trajectories of the angular positions and velocities are driven successfully by the developed controller, from the initial condition to a very small neighborhood of the origin, in all cases. Figures 15 and 18 show small fluctuations in the late stage of the trajectory, which come from forcing the states by the controller to reach the origin, in spite of disturbance and noise.

From the system's efficiency point of view, the required steady controller effort in all cases ranged from 0.13 to 0.621 N.m in amplitude, owing to the inclusion of the torque in the objective function of optimization. A higher torque was required only at start-up; the average start-up torque for the first actuator is 1.663 N.m and it is 2.158 N.m for the second actuator. This contributes to a more efficient control system. The small chattering caused by the sign function in Equation (47), which is necessary for driving the states, can be attenuated to obtain a smoother control signal, through using boundary-layer or neural networks.

As compared to the H_2 control system, the proposed hybrid H_∞/SMC system produces fewer overshoots and fewer start-up torques, and smooth states driving in general. The robustness is provided in both control systems with small disparities, as illustrated in the results.

So, instead of having unnecessary conservativeness in the system, which degrades the performance with respect to providing robustness against uncertainties, the application of metaheuristic optimization to the controller design for such an under-actuated nonlinear MIMO system is beneficial in enhancing both performance and efficiency of the control system, besides its robustness.

Another point to be made is that for this specific electro-mechanical control system-design optimization problem, a swarm size of 10 was suitable, as has been presented in the results. A larger swarm size has to be considered in other optimal designs of engineering problems, especially if more parameters are required to be tuned in the application. Also, the adaptation of the cost function can be considered in future, to further enhance the optimization of the design problem, as proposed and described in [43].

Having great impact on control systems' performance and efficiency, metaheuristic optimization algorithms are advised to be applied to enhance other control systems designs, as in the following: modelling predictive control to optimize the control sequence over a specified prediction horizon, hybrid model-based and data-driven control to adapt model parameters and controller gains in real-time, and distributed control systems to optimize the control signals' distribution in large scale systems.

Author Contributions: Conceptualization, Y.A.S. and H.I.A.; methodology, H.I.A.; software, Y.A.S.; validation, Y.A.S., and H.I.A.; formal analysis, H.I.A.; investigation, Y.A.S.; resources, Y.A.S. and H.I.A.; data curation, Y.A.S.; writing—original draft preparation, Y.A.S.; writing—review and editing, H.I.A.; supervision, H.I.A. All authors have read and agreed to the published version of the manuscript.

Funding: This research received no external funding.

Data Availability Statement: Data are contained within the article.

Conflicts of Interest: The authors declare no conflicts of interest.

Nomenclature

Symbol	Meaning
L	Lagrangian
T	Kinetic energy
V	Potential energy
ρ	Position vector in the generalized coordinates
θ	Link angle from the vertical line
I	Link moment of inertia around its center of gravity
m	Link mass
h	Distance from Link bottom to the its center of gravity
l	Link length
$I_{p'}$	Hinge belt–pulley system's moment of inertia
K	Hinge belt–pulley system's ratio of teeth
I_m	Motor moment of inertia
C	Hinge viscous friction coefficient
C_m	Motor viscous friction coefficient
$C_{p'}$	Hinge belt–pulley system's viscous friction coefficient
g	Acceleration of gravity
t_m	Motor control torque
d	Disturbance torque to the link
V_{PSO}	Agent velocity
X_{PSO}	Agent position
w_{PSO}	Inertia weight
$\overline{w}_{PSO}$	Maximum inertia weight
$\underline{w}_{PSO}$	Minimum inertia weight
c_{PSO1}	Cognitive coefficient
c_{PSO2}	Social coefficient
r_{PSO1}	Random number in the interval (0, 1)
r_{PSO2}	Random number in the interval (0, 1)
p_{best}	Best local solution
g_{best}	Best global solution
$\mathbf{x}$	State vector
$\mathbf{u}$	Input vector
t	time
A	State matrix
$\mathbf{y}$	Output vector
ΔA	Uncertainty matrix
Q	Bound matrix of uncertainty
B	Input matrix
$\mathbf{q}$	New state variable
H	Similarity transformation matrix
N	Null space of input matrix
$\overline{A}$	New state matrix
$\Delta\overline{A}$	New uncertainty matrix

Symbol	Meaning
$\overline{B}$	New input matrix
K_∞	Feedback-gain matrix in H_∞ problem
Δ	Diagonal matrix of scalar perturbations
B_∞	Decomposed matrix in LFT
C_∞	Decomposed matrix in LFT
D_∞	Decomposed matrix in LFT
$\mathbf{z}_\infty$	Disturbed output in LFT
$\mathbf{d}_\infty$	Disturbed input in LFT
J_∞	Objective function in H_∞ problem
γ	Performance level of H_∞ problem
O	Abbreviation of two given multiplied matrices
$\mathbf{s}$	Sliding-manifold vector
G	Matrix in sliding-manifold computation
$\mathbf{u}_\circ$	Nominal control term
$\mathbf{u_d}$	Discontinuous control term
V_L	Lyapunov function
η	Design parameter in SMC law
J_O	Optimization cost function
e	Link angle error
t_f	Final time used in simulations
W	Performance weighting matrix

References

1. Gu, D.; Petkov, P.H.; Konstantinov, M.M. *Robust Control Design with MATLAB*, 2nd ed.; Springer: London, UK, 2013; pp. 292–302.
2. Photo by Maximalfocus on Unsplash. Available online: https://unsplash.com/photos/person-in-orange-and-white-robot-costume-eZWGK5sIiBM (accessed on 30 August 2024).
3. Sharma, A.K.; Bhushan, B. Sliding mode control of inverted pendulum with decoupling algorithm. *Int. J. Comput. Appl.* **2018**, *181*, 1–5.
4. Bonifacio, S.R.; Patricio, O.O.; Alexander, P.G. Robust stabilizing control for the electromechanical triple link inverted pendulum system. *IFAC-PapersOnLine* **2018**, *51*, 314–319. [CrossRef]
5. Nguyen, T.T. Sliding mode control-based system for the two-link robot arm. *Int. J. Electr. Comput. Eng.* **2019**, *9*, 2771–2778. [CrossRef]
6. Kharabian, B.; Mirinejad, H. Hybrid sliding mode/H-infinity control approach for uncertain flexible manipulators. *IEEE Access* **2020**, *8*, 170452–170460. [CrossRef]
7. Saif, A.; Fareh, R.; Sinan, S.; Bettayeb, M. Fractional synergetic tracking control for robot manipulator. *J. Control Decis.* **2024**, *11*, 139–152. [CrossRef]
8. Ahmed, S.; Ghous, I.; Mumtaz, F. TDE based model-free control for rigid robotic manipulators under nonlinear friction. *Sci. Iran.* **2024**, *31*, 137–148. [CrossRef]
9. Anjum, Z.; Sun, Z.; Chen, B. Disturbance-observer-based fault-tolerant control of robotic manipulator: A fixed-time adaptive approach. *IET Control Theory Appl.* **2024**, *18*, 1398–1413. [CrossRef]
10. Liu, Z.; He, W.; Li, X.; Liu, C.; Shao, H.; Fu, P. A mismatched composite disturbance observer-based adaptive tracking controller for robotic manipulators. *IET Control Theory Appl.* **2024**, *18*, 1357–1370. [CrossRef]
11. Qiu, J.; Wang, T.; Sun, k.; Rudas, I.J.; Gao, H. Disturbance observer-based adaptive fuzzy control for strict-feedback nonlinear systems with finite-time prescribed performance. *IEEE Trans. Fuzzy Syst.* **2022**, *30*, 1175–1184. [CrossRef]
12. Rigatos, G.; Abbaszadeh, M.; Pomares, J.; Cuccurullo, G. Flatness-based control in successive loops for robotic manipulators and autonomous vehicles. *Int. J. Syst. Sci.* **2024**, *55*, 954–981. [CrossRef]
13. Jabbar, A.; Malik, F.M.; Sheikh, S.A. Nonlinear stabilizing control of a rotary double inverted pendulum: A modified backstepping approach. *Trans. Inst. Meas. Control* **2017**, *39*, 1721–1735. [CrossRef]
14. Siradjuddin, I.; Amalia, Z.; Setiawan, B.; Wicaksono, R.P.; Yudaningtyas, E. Stabilising a Cart Inverted Pendulum System Using Pole Placement Control Method. In Proceedings of the 15th International Conference on Quality in Research (QiR): International Symposium on Electrical and Computer Engineering, Bali, Indonesia, 24–27 July 2017.
15. Pristovani, R.D.; Sanggar, D.R.; Dadet, P. Implementation of push recovery strategy using triple linear inverted pendulum model in "T-FloW" humanoid robot. *J. Phys. Conf. Ser.* **2018**, *1007*, 012068–012081. [CrossRef]
16. Masrom, M.F.; Ghani, N.M.; Jamin, N.F.; Razali, N.A.A. Stabilization Control of aTwo-wheeled Triple Link Inverted Pendulum System with Disturbance Rejection. In Proceedings of the 10th National Technical Seminar on Underwater System Technology, Pahang, Malaysia, 26–27 September 2018.
17. Soltanpour, M.R.; Khooban, M.H. A particle swarm optimization approach for fuzzy sliding mode control for tracking the robot manipulator. *Nonlinear Dyn.* **2013**, *74*, 467–478. [CrossRef]

18. Liu, R.; Li, S. Optimal integral sliding mode control scheme based on pseudospectral method for robotic manipulators. *Int. J. Control* **2014**, *87*, 1131–1140. [CrossRef]
19. Oliveira, J.; Oliveira, P.M.; Boaventura-Cunha, J.; Pinho, T. Chaos-based grey wolf optimizer for higher order sliding mode position control of a robotic manipulator. *Nonlinear Dyn.* **2017**, *90*, 1353–1362. [CrossRef]
20. Soon, C.C.; Ghazali, R.; Jaafar, H.I.; Hussien, S.Y.S. Sliding mode controller design with optimized PID sliding surface using particle swarm algorithm. *Procedia Comput. Sci.* **2017**, *105*, 235–239. [CrossRef]
21. Jibril, M.; Tadese, M.; Tadese, E.A. Comparison of a triple inverted pendulum stabilization using optimal control technique. *Rep. Opin.* **2020**, *12*, 62–70.
22. Hazem, Z.B.; Fatuhi, M.J.; Bingul, Z. A Comparative study of the joint neuro-fuzzy friction models for a triple link rotary inverted pendulum. *IEEE Access* **2020**, *8*, 49066–49078. [CrossRef]
23. Singh, N.; Bhangal, K. Robust control of double inverted pendulum system. *J. Autom. Control Eng.* **2017**, *5*, 14–20. [CrossRef]
24. Shafeek, Y.A.; Ali, H.I. Attaining Robust Stability and Performance for Triple Inverted Pendulum using H-infinity Control. *J. Eur. Des Systèmes Autom.* **2024**, *57*, 443. [CrossRef]
25. Shafeek, Y.A.; Ali, H.I. Balancing of robustness and performance for triple inverted pendulum using μ-synthesis and gazelle optimization. *J. Eur. Des Systèmes Autom.* **2024**, *57*, 443–452. [CrossRef]
26. Ali, H.I. Design of PSO based robust blood glucose control in diabetic patients. *Int. J. Comput. Commun.* **2014**, *14*, 1–9.
27. Qaraawy, S.; Ali, H.I.; Ali, M. Particle Swarm Optimization Based Robust Controller for Congestion Avoidance in Computer Networks. In Proceedings of the 2012 International Conference on Future Communication Networks, Baghdad, Iraq, 2–5 April 2012.
28. Aroniadi, C.; Beligiannis, G.N. Applying particle swarm optimization variations to solve the transportation problem effectively. *Algorithms* **2023**, *16*, 372. [CrossRef]
29. Rasheed, L.T. Optimal tuning of linear quadratic regulator controller using ant colony optimization algorithm for position control of a permanent magnet DC motor. *J. Comput. Commun. Control Syst. Eng.* **2020**, *20*, 29–41.
30. Yasser, Y.A.; Sadiq, A.T.; Alhamdani, W. Generating honeyword based on a proposed bees algorithm. *J. Comput. Commun. Control Syst. Eng.* **2022**, *22*, 177–191.
31. Muncino, D.M.; Damian-Ramirez, E.A.; Cruz-Fernandez, M.; Montoya-Santiyanes, L.A.; Rodriguez-Resendiz, J. Metaheuristic and heuristic algorithms-based identification parameters of a direct current motor. *Algorithms* **2024**, *17*, 209. [CrossRef]
32. Saleem, S.; Hussain, F.; Baloch, N.K. IWO-IGA—A hybrid whale optimization algorithm featuring improved genetic characteristics for mapping real-time applications onto 2D network on chip. *Algorithms* **2024**, *17*, 115. [CrossRef]
33. Kadhim, N.N.; Abood, L.H.; Mohammed, Y.A. Design an optimal fractional order PID controller for speed control of electric vehicle. *J. Eur. Des Systèmes Autom.* **2023**, *56*, 735–741. [CrossRef]
34. Abood, L.H.; Kadhim, N.N.; Mohammed, Y.A. Dual stage cascade controller for temperature control in greenhouse. *Bull. Electr. Eng. Inform.* **2023**, *12*, 51–58. [CrossRef]
35. Nagadurga, T.; Devarapalli, R.; Knypiński, Ł. Comparison of meta-heuristic optimization algorithms for global maximum power point tracking of partially shaded solar photovoltaic systems. *Algorithms* **2023**, *16*, 376. [CrossRef]
36. Rubio, J.D.J.; Hernandez, M.A.; Rosas, F.J.; Orozco, E.; Balcazar, R.; Pacheco, J. Genetic high-gain controller to improve the position perturbation attenuation and compact high-gain controller to improve the velocity perturbation attenuation in inverted pendulums. *Neural Netw.* **2024**, *170*, 32–45. [CrossRef]
37. Rubio, J.D.J. Bat algorithm based control to decrease the control energy consumption and modified bat algorithm based control to increase the trajectory tracking accuracy in robots. *Neural Netw.* **2023**, *161*, 437–448. [CrossRef] [PubMed]
38. Sorcia-Vázquez, F.D.J.; Rumbo-Morales, J.Y.; Brizuela-Mendoza, J.A.; Ortiz-Torres, G.; Sarmiento-Bustos, E.; Pérez-Vidal, A.F.; Rentería-Vargas, E.M.; De-la-Torre, M.; Osorio-Sánchez, R. Experimental validation of fractional pid controllers applied to a two-tank system. *Mathematics* **2023**, *11*, 2651. [CrossRef]
39. Photo by clockinsky on Unsplash. Available online: https://unsplash.com/ (accessed on 2 August 2024).
40. Photo by Lance Anderson on Unsplash. Available online: https://unsplash.com/photos/school-of-gray-fish-G2SDLsJp3rg (accessed on 2 August 2024).
41. Knypiński, Ł. Constrained optimization of line-start PM motor based on the gray wolf optimizer. *Eksploat. I Niezawodn. Maint. Reliab.* **2021**, *23*, 1–10. [CrossRef]
42. Sinha, A. *Linear Systems Optimal and Robust Control*, 1st ed.; CRC Press: Boca Raton, FL, USA, 2007; pp. 377–408.
43. Knypiński, Ł. Adaptation of the penalty function method to genetic algorithm in electromagnetic devices designing. *COMPEL Int. J. Comput. Math. Electr. Electron. Eng.* **2019**, *38*, 1285. [CrossRef]

algorithms

MDPI

Article

Metaheuristic and Heuristic Algorithms-Based Identification Parameters of a Direct Current Motor

David M. Munciño [1], Emily A. Damian-Ramírez [1], Mayra Cruz-Fernández [1,*], Luis A. Montoya-Santiyanes [1,*] and Juvenal Rodríguez-Reséndiz [2]

[1] Industrial Technologies Division, Universidad Politécnica de Querétaro, El Marqués, Querétaro 76240, Mexico; 120037491@upq.edu.mx (D.M.M.); 119034035@upq.edu.mx (E.A.D.-R.)

[2] Facultad de Ingeniería, Universidad Autónoma de Querétaro, Santiago de Querétaro 76010, Mexico; juvenal@uaq.edu.mx

* Correspondence: mayra.cruz@upq.edu.mx (M.C.-F.); luis.montoya@upq.edu.mx (L.A.M.-S.)

Abstract: Direct current motors are widely used in industry applications, and it has become necessary to carry out studies and experiments for their optimization. In this manuscript, a comparison between heuristic and metaheuristic algorithms is presented, specifically, the Steiglitz–McBride, Jaya, Genetic Algorithm (GA), and Grey Wolf Optimizer (GWO) algorithms. They were used to estimate the parameters of a dynamic model that approximates the actual responses of current and angular velocity of a DC motor. The inverse of the Euclidean distance between the current and velocity errors was defined as the fitness function for the metaheuristic algorithms. For a more comprehensive comparison between algorithms, other indicators such as mean squared error (MSE), standard deviation, computation time, and key points of the current and velocity responses were used. Simulations were performed with MATLAB/Simulink 2010 using the estimated parameters and compared to the experiments. The results showed that Steiglitz–McBride and GWO are better parametric estimators, performing better than Jaya and GA in real signals and nominal parameters. Indicators say that GWO is more accurate for parametric estimation, with an average MSE of 0.43%, but it requires a high computational cost. On the contrary, Steiglitz–McBride performed with an average MSE of 3.32% but required a much lower computational cost. The GWO presented an error of 1% in the dynamic response using the corresponding indicators. If a more accurate parametric estimation is required, it is recommended to use GWO; however, the heuristic algorithm performed better overall. The performance of the algorithms presented in this paper may change if different error functions are used.

Keywords: Steiglitz–McBride; Jaya; grey wolf optimizer; genetic algorithm; heuristic; metaheuristic; DC motor

1. Introduction

A direct current (DC) motor has many applications in the industrial sector [1], such as aeronautics, robotics, automation, manufacturing, automotive, etc. Therefore, it is constantly subjected to multiple experiments or investigations in identification, estimation, and parametric control to create solutions in different areas of research that could have applications in real life. For this, the precise control of motor parameters is required to achieve efficient operation [2], and model optimization should help achieve this goal.

Sometimes, models are simulated mathematically, drifting in predictions, control monitoring, and diagnosis. Motor parameters vary with features such as misalignments that can lead to poor motor performance. The design of DC motor controllers could have drawbacks, as they are sensitive to high tolerances [3]. Usually, the parameters specified by manufacturers are somewhat robust, and it is necessary to generate more precise models that help us make a more efficient motor. With precise models, diagnostic systems can be developed, as shown in the research by ref. [4], where a diagnostic method based on a rotor slip was designed to detect faults in the current and speed sensors during the induction motor operation. Another example is from ref. [5], where a technique based on current estimation for fault detection and speed sensor failure for an induction motor is developed. The precise knowledge of mathematical motor models allows for the bypassing of sensors and efficient speed control, as shown in the research by ref. [6], where high-performance field-oriented motor control requires the accurate knowledge of the flow of information and the motor speed. These investigations denote the importance of motor modeling and parameterization, since they allow for better fault-tolerant control, reduce the number of sensors, and even improve control precision.

For example, an improved loss minimization technique based on fuzzy logic for a brushless DC motor (BLDC) drive system is presented in [7]. In addition, the implementation of a platform for testing an algorithm for the closed-loop speed control of a DC motor was sought in [8]. These have not been the only works that have focused on controlling speed through algorithms. Speed control algorithms have also been designed based on the estimation of dynamic error parameters under uncertainty and load variation [9].

Heuristic algorithms solve optimization problems defined by intuitive approximations, and solutions to these problems are intelligently thought out even if the solution is not the best. In recent years, heuristic procedures have been developed to optimize software design problems and component reuse. Some research shows the usefulness of this type of algorithm and its application in parametric evaluation [10]. Some of the proposed limitations of heuristic algorithms have been overcome with more robust methods based on statistics, such as in ref. [11]. However, it is necessary to continue innovating [12]. Investigations such as [13] propose an algorithm that uses a stochastic method to reach optimal points within simple algorithms.

A comparison of heuristic algorithms for the identification of DC motor systems was carried out in [14] through discrete Proportional–Integral (PI) controllers to analyze the system response. Fast and accurate convergence rate results were obtained with the extended Kalman filter (EKF) algorithm. A heuristic method is a well-known set of steps that is used to quickly identify a high-quality solution to a given problem. It consists of intuitive mathematics with a design of static and manual rules that are supposed to give a good solution, although not necessarily the optimal solution, to the problem. For this reason, it is important to compare them with metaheuristic algorithms and observe which offer the highest optimization.

Metaheuristic algorithms have acquired an important role in recent years. The optimization field has been no exception, because metaheuristic algorithms are iterative and intelligently combine the principles of evolution, natural selection, and inheritance, inspired by physical phenomena and different behaviors of animals and even humans. This makes it possible to correctly explore the search space by offering a population of feasible solutions. Due to their simplicity, they are general purpose algorithms and similar phenomena inspire them. In recent years, these algorithms have successfully solved practical problems in different fields. They are easily implementable, as they do not require any particular changes to their structure when applying them across various themes unless optimization is sought. However, there may be limitations, since most of these optimization methods start from random solutions, resulting in approximate, not exact, solutions.

Recent contributions, such as [15], showed that the Artificial Fish Swarm Algorithm (AFSA) has advantages, including high convergence speed, flexibility, fault tolerance, and high accuracy. However, it has been shown that they contain high temporal complexity and a lack of balance in global and local searches. On the other hand, the work by ref. [16] shows that Dolphin Partner Optimization (DPO) offers a quick solution with optimal stability in different function targets. Another area of interest in algorithms is convergence. A study on the hybrid algorithm Particle Swarm–Ant Colony Optimization (PS–ACO) has been presented [17], where ACO was used as a parallel calculation mechanism and showed excellent robustness. Unfortunately, they faced the limitations of stagnation and premature convergence. The Particle Swarm Optimization (PSO) algorithm as an adaptation of reactive power optimization [18] proved to be decisive. Refs. [19,20] use metaheuristic algorithms to optimize a fuzzy controller. Still, it needs predefined parameters for a user-determined issue. These problems can be solved through adaptive adjustments. In [21], the author offers solutions of low algorithmic complexity with optimal convergence results in Proportional–Integral–Derivative (PID) controllers. An example is the Cuckoo algorithm [22] which is a relatively simple method, having minor variation in parameters and containing fast convergence. However, it also has premature convergence defects and low calculation accuracy, which, in the worst case, results in inaccurate solutions.

Genetic algorithms (GA) are another method widely used to optimize a DC motor because of their low complexity of understanding and ease of adaptation. Such is the case in [23], where it is mentioned that GAs are appropriate for the estimation of platform parameters with nonlinear characteristics. Something similar occurs in [24], where it is described that GAs offer superior results in the estimation of parameters. Despite recent articles, GAs also have limitations, since they starts from random solutions. The algorithm requires improvements to optimize it, as shown in ref. [25]. Another example of this is shown in [26], where it is mentioned that although they are good at optimization, the traditional GA still has some shortcomings because most of the time, they require modifications. Therefore, a New Adaptive Genetic Algorithm (NAGA) is proposed to overcome the disadvantages of the traditional one.

Metaheuristic algorithms are sometimes used in combination with other methods to achieve better performance, such as in the case of ref. [27], where they are used for the speed control of a DC motor. Better dynamic and static performance is demonstrated thanks to the implementation of a PID controller with a Backtracking Search Algorithm (BSA) in comparison with a PSO. PID controllers are commonly implemented to drive DC motors and are tuned using different algorithms. This is the case in ref. [28], where a PID controller was tuned through the Jaya optimization algorithm to control the speed of the motor, obtaining better responses in the transitory stage. Speed control has been the subject of many experiments, as in [29], where the optimization of fuzzy rules using GA proved to be an effective method for rate accuracy. In other cases, the original metaheuristic algorithms are modified to optimize their performance. Modifications to the original cuckoo algorithm have been made [30], improving the parametric estimation.

This manuscript presents a parametric estimation of the dynamic model of a DC motor. The study was carried out using the following algorithms: Steiglitz–McBride, Jaya, Genetic Algorithm (GA), and Gray Wolf Optimizer (GWO). These algorithms were used to compare the optimization of the actual responses of current and angular velocity between heuristic and metaheuristic algorithms. Simulations were carried out in MATLAB/Simulink 2010. The evaluation criteria consisted of approximation error, computational cost, convergence time, and dynamic signal behavior. Stabilization times, overshoot values, and the average value of steady-state current were also discussed. This paper is organized as follows. The Section 2 presents the foundations of the DC motor dynamic model, the parameters that govern the development of the metaheuristic algorithms, and the conditions for simulations. The subsequent Section 3 recalls a brief background of the selected algorithms for optimization, diagrams, and conditions for parameter estimation. Sections 4 and 5 show the Results and Discussion, respectively. Here, the parameters estimated by the different algorithms are compared using different metrics. Finally, the Section 6 concludes the present work.

2. Mathematical Modelling

DC Motor Modelling

The dynamic system of the DC motor is composed of two differential equations (an electrical and a mechanical part). The electrical equation has variables such as voltage and current and parameters such as a resistor, which is the internal resistance of the motor, and an inductor, which represents the inductance generated by its windings. In the mechanical equation, there are variables such as speed and torque and mechanical parameters such as the coefficient of friction and inertia, as shown in Figure 1.

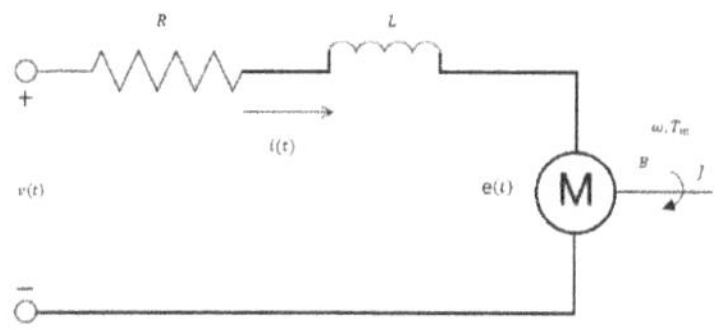

Figure 1. DC motor dynamic circuit diagram.

From the diagram in Figure 1, the following equations are obtained.

$$v(t) = Ri(t) + L\frac{di(t)}{dt} + e(t) \tag{1}$$

$$\tau(t) = J\frac{d\omega(t)}{dt} + B\omega(t) + T_L \tag{2}$$

$$e(t) = K_e\omega(t) \tag{3}$$

$$\tau(t) = K_m i(t) \tag{4}$$

$$K_e \approx K_m = K \tag{5}$$

where R is the resistance expressed in ohms Ω, L the inductance in H, J is the moment of inertia of the rotor, B the coefficient of friction between the rotor and the stator, and T_L the torque load; however, for parameter estimation, a motor free of load is used, which means that $T_L = 0$. $v(t)$ is the voltage induced to the armature, $i(t)$ is the current, $\omega(t)$ is the angular velocity of the rotor, $e(t)$ is the induced electric voltage, and $\tau(t)$ is the torque of the motor. K is the rate of change of the electromotive force with respect to the angular velocity, and the rate of change of the induced voltage is considered to be equivalent to both K_m and K_e. Equation (1) was derived using a simple Kirchhoff's voltage law analysis presented in Figure 1. Equation (2) was obtained by knowing that the sum of moments that make the rotor rotate to the symmetrical axis is the same as that of moments that oppose its movement. K is the rate of change of the electromotive force with respect to the angular velocity, and the rate of change of the induced voltage is considered equivalent to both K_m and K_e. This is because it is a DC motor and the excitation source is considered constant; hence, the previous constants are considered equal. This can be deduced by working with the instantaneous power equation, which relates to the instantaneous power supplied by the excitation source, the instantaneous mechanical power, and the instantaneous power dissipated by the motor windings [31].

The system of differential equations shown in Equations (6) and (7) is obtained from the combination and substitution of Equations (3) and (4) into (1) and (2), respectively.

$$\frac{di(t)}{dt} = \frac{v(t) - Ri(t) - K_e\omega(t)}{L} \tag{6}$$

$$\frac{d\omega(t)}{dt} = \frac{K_m i(t) - B\omega(t)}{J} \tag{7}$$

The equations represent the dynamic model of a DC motor, where Equation (6) corresponds to the mechanical part of the motor, and Equation (7) corresponds to the electrical part. The time response of the dynamic system was simulated using the nominal parameters, as shown in Table 1. Later, simulations were performed with the parameters selected by the algorithms, as described in the following section. It should be noted that the model and nominal parameters may have differences from reality, although they are considered so small as to be negligible.

Table 1. Nominal values of M2 motor.

R	L	K	B	J
0.921042 Ω	0.0077590 H	0.073472	0.000678 $\frac{kg \cdot m^3}{s^2}$	0.000136 $kg \cdot m^2$

The simulation of differential equations was carried out through a block diagram in MATLAB/Simulink 2010. The arrangement consisted of six inputs, which are the five described in Table 1 and the input voltage. The current and angular velocity outputs were delivered as a result. The setup can be seen in Figure 2.

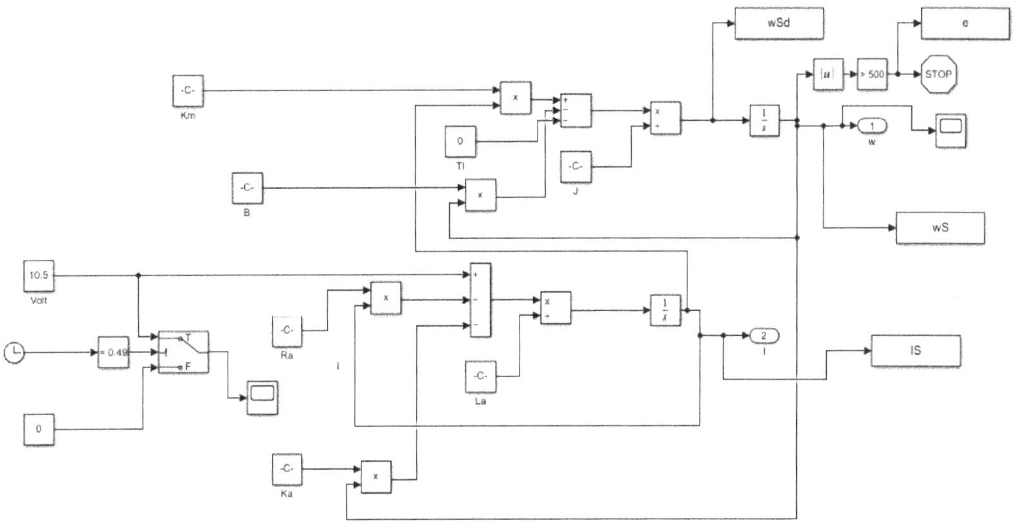

Figure 2. MATLAB/Simulink 2010 setup for the dynamic model.

3. Optimization Algorithms

3.1. Steiglitz–McBride Algorithm as a Parametric Estimator

The Steiglitz–McBride algorithm is a modification of the Minimum Square Recursive (MSR) algorithm. This algorithm was selected because it is a widely used heuristic algorithm equivalent to the GA in metaheuristics. Any heuristic problem can be solved by this algorithm. In the same way, this algorithm has already been subjected to comparisons with metaheuristic algorithms, obtaining favorable results [30]. The most attractive feature of this algorithm is its ability to find the Mean Square Error (MSE) for a linear system by sampling the inputs and outputs in an iterative process, calculating the parameters using a filter defined in the code equivalent to the system. The algorithm takes the results once the filter is applied to the input parameters. It uses them as a starting point in the next iteration, seeking to reduce the error on each iteration. In this way, the best solution is obtained. As mentioned above, the algorithm works in two phases: pre-filtering and the estimation of the parameters. The behavior of the algorithm is better explained in Figure 3, where the two phases of the algorithm, prefiltering and estimation, are shown. In the first phase, the image shows how it uses the input and output of the plant to adjust (in the second phase, it uses the least squares algorithm).

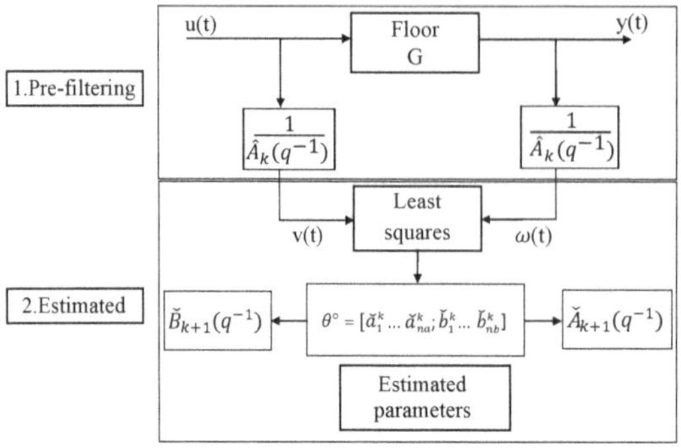

Figure 3. The flowchart of the Steiglitz–McBride algorithm.

To simulate the heuristic Steiglitz–McBride algorithm, the current and angular velocity transfer functions were used to convert the model from discrete to continuous time.

3.1.1. Pre-Filtering

Pre-filtering is implemented on the motor input, i.e., voltage, in addition to two measurable output parameters, where a transfer function defines current and angular velocity.

3.1.2. Estimation

The algorithm uses discrete models. For this reason, they are loaded by samples of signals involved in the estimation. The discrete model must go through the Steiglitz–McBride algorithm to become a continuous model and obtain the estimated motor parameters through calculations. The function changes from a discrete to a continuous model and is simulated in the continuous model, and the error is calculated with the resulting new functions generating a difference between the actual and estimated signals. Each time the coefficients are calculated, the process is repeated, seeking to minimize the error.

3.2. Grey Wolf Optimizer (GWO)

GWO is a metaheuristic algorithm that takes only one inherent parameter to define its search in addition to the general search parameters. It should be noted that this algorithm has already been subjected to research where it has obtained better results than heuristic algorithms. This makes it a suitable algorithm for this research. In other investigations, the authors have managed to obtain good results with the GWO algorithm when adjusting the parameters of nonlinear models for vibration amplitude measurements [32]. Also, there are demonstrations where the estimation of the parameters of a control system can be optimized in real time to improve the performance of robotic manipulators, specifically in terms of tracking precision, robustness to uncertainties, and the smoothness of movement [33].

GWO is based on the social hierarchy that governs decision-making within a pack of gray wolves. The author of [34] mathematically shapes this social hierarchy to consider the best solution as "*alpha*", the second-best solution is called "*beta*", the third-best solution is represented as "*delta*", and the rest of the answers are assumed as "*omegas*". It should be noted that the hierarchy mentioned also influences the main phases of wolf hunting. This is of utmost importance, since the algorithm is responsible for molding or designing a mathematical order to describe each of these hunting phases based on their social hierarchy, since the optimization offered by this algorithm for the solution of different problems is guided by the "*alpha*", "*beta*", and "*delta*" solutions. The main hunting phases of gray wolves are as follows.

3.2.1. Encircling Prey

Mathematically speaking, the cornering of prey tells us that a gray wolf can randomly update its position within the hunting space according to the position of the prey of the best search agents. The author of [34] explains this in the mathematical order, adjusting the vectors of the search agents. This behavior is described mathematically in Equations (8a) and (8b).

$$\vec{X}(t+1) = \vec{X}_p(t) - \vec{A} \cdot \vec{D} \tag{8a}$$

$$\vec{D} = |\vec{C} \cdot \vec{X}_p(t) - \vec{X}(t)| \tag{8b}$$

where $\vec{X}$ is the position of the wolf, t denotes the iteration, $\vec{X}_p$ denotes the position of the prey, and $\vec{D}$ is the distance between wolf and prey. Coefficients $\vec{A}$ and $\vec{C}$ are calculated as

$$\vec{A} = 2\vec{a} \cdot \vec{r}_1 - \vec{a} \tag{9a}$$

$$\vec{C} = 2\vec{r}_2 \tag{9b}$$

r_1 and r_2 are two random values of the vector within the range $[0, 1]$.

3.2.2. Hunting

This phase is mathematically simulated in the algorithm by saving the three best solutions obtained ($alpha, beta, delta$). These solutions will guide the other solutions ($omegas$) to update their position relative to them in the next generation, because the final solution is expected to be at a random place within the circle defined by the three best solutions. We can calculate their updated positions by using alpha, beta, and delta, which are $\vec{X}_\alpha$, $\vec{X}_\beta$, and $\vec{X}_\delta$, as follows:

$$\vec{D}_\alpha = |\vec{C}_1 \cdot \vec{X}_\alpha - \vec{X}|\,,\ \vec{D}_\beta = |\vec{C}_2 \cdot \vec{X}_\beta - \vec{X}|\,,\ \vec{D}_\delta = |\vec{C}_3 \cdot \vec{X}_\delta - \vec{X}| \tag{10a}$$

$$\vec{X}_1 = \vec{X}_\alpha - \vec{A}_1 \cdot (\vec{D}_\alpha)\,,\ \vec{X}_2 = \vec{X}_\beta - \vec{A}_2 \cdot (\vec{D}_\beta)\,,\ \vec{X}_3 = \vec{X}_\delta - \vec{A}_3 \cdot (\vec{D}_\delta) \tag{10b}$$

$$\vec{X}(t+1) = \frac{\vec{X}_1 + \vec{X}_2 + \vec{X}_3}{3} \tag{10c}$$

3.2.3. Attacking Prey

The author of [34] proposes that this behavior can be simulated by decreasing the value $\vec{A}$, a random value in an interval of $[-2a, 2a]$, where a is reduced from 2 to 0 throughout iterations. When the arbitrary values of $\vec{A}$ are at $[1, 1]$, the next position of search agent "a" can be anywhere between its current situation and the position of the dam. With the hunting phases explained so far, the algorithm offers good search agents that converge to attack the dam (optimal solution). However, the algorithm is still prone to stagnation, which brings us to the exploration phase.

3.2.4. Search for Prey (Exploration)

In this stage, the grey wolves diverge from each other to search for prey and converge to attack prey. Some random values are selected to model the divergence to force the search agent to diverge from the prey. The $\vec{C}$ vector contains random values in $[0, 2]$. This is mainly performed to increase the range of exploration and avoid stagnation, perhaps finding a better target. Considering this scheme, the search hyperparameters are selected, and in this case, the prey acts as the vector of engine parameters. The algorithm evaluates each search agent and selects the one with the lowest fitness as the alpha wolf. The algorithm will repeat the process until the selected number of iterations is completed.

The flowchart in Figure 4 is explained as follows. First, the program initializes the population to search for the constants to optimize, within the upper and lower limits. The main program receives the parameters through Simulink in order to calculate the motor current and velocity responses and compare with the actual responses using the fitness function. The algorithm chooses the three best solutions closest to the actual response which become the alpha, beta, and gamma agents, thus emulating the hunting behavior of wolves. The program is initialized again with a new iteration, now creating the agents within the range of solutions of alpha, beta, and gamma agents of the last iteration; in this way, the same cycle is repeated until reaching the maximum iteration, obtaining the best solution.

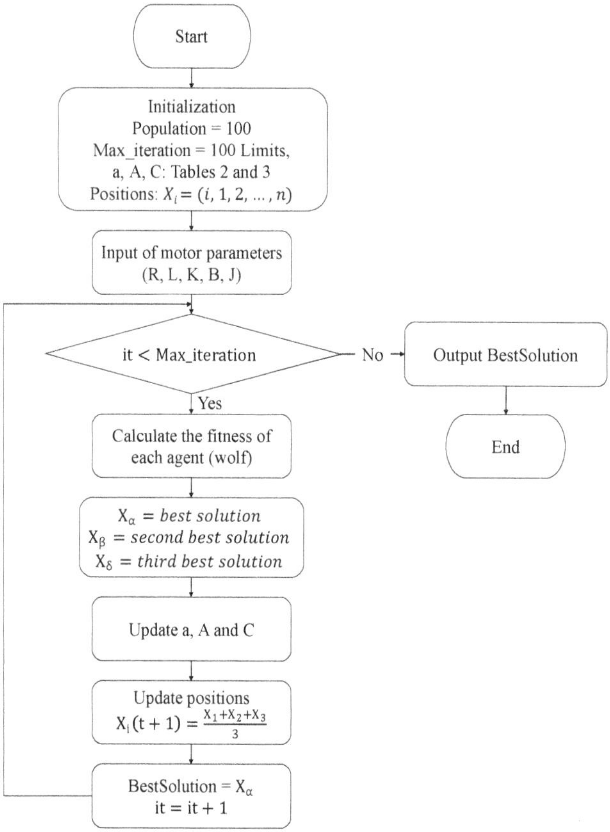

Figure 4. The flowchart of the GWO algorithm.

Table 2. General parameters applied in metaheuristic algorithms.

Parameter	Value	Description
Search agents	100	Number of random solutions proposed of each metaheuristic algorithm.
Upper limits	$R \le 1$, $L \le 0.1$, $K \le 0.1$, $B \le 0.001$, $J \le 0.001$	Upper search limit for each DC motor parameter
Lower limits	$R \ge 0.1$, $L \ge 0.001$, $K \ge 0.005$, $B \ge 0.00001$, $J \ge 0.00001$	Lower search limit for each DC motor parameter
Iterations	100	Repetitions of each metaheuristic algorithm.
Fitness function	$Fitness = \frac{1}{\sqrt{\sum(I-I_s)^2+\sum(\omega-\omega_s)^2}}$	Function used to evaluate the performance of each random solution ($f : i$)

Table 3. Specific parameters applied in metaheuristic algorithms.

Algorithm	Specific Parameters	Description
Genetic Algorithm	Mutation: 20%	An operator that allows the random alteration of a gene to maintain search diversity.
	Selection: Roulette wheel ($s(i) = \frac{f_i}{\sum_{j=0}^{N} f_j}$)	Population with N individuals, for each chromosome i with corresponding fitness value f_i, probability $s(i)$ of selection
	Type: Crossover	A random point is selected to combine the chromosomes.
	Biological pressure: 30%	The percentage of individuals that reproduce
Grey Wolf Optimizer	a, A and C	Intrinsic parameters calculated by original method proposed (see original method in [34]).
Jaya	-	Jaya doesnt contain any specific parameter.

This algorithm has been used in multiple optimization works to estimate the parameters of DC motors. It offers good optimization in robust analyses, as in [35], where it is sought to optimize the fractional-order PID controller in motor speed control.

3.3. Jaya Algorithm

This algorithm is used in optimization and is one of the simplest algorithms to implement. It does not contain any intrinsic or extrinsic search parameters. It only uses general search parameters, which makes it very attractive compared to other metaheuristic algorithms that occupy specific search parameters. The Jaya algorithm is a metaheuristic algorithm proposed by Rao [36] and can be defined as a modification of PSO in which the attraction for the best local functions is removed. Jaya replaces this by moving the worst local solutions away from the next iteration, thus producing optimized solutions for the problem of interest. Compared to other algorithms, Jaya only occupies three input parameters: the size of the population, the number of generations (iterations), and the range of limits. The Jaya optimization process focuses on determining the most optimized solution, leaving aside the lower solutions by disregarding generations that do not offer a better solution than the current generation. When it comes to finding a better solution, the algorithm eliminates the current generation, as it does not keep track of the best solutions. The procedure of the Jaya algorithm is discussed in the following steps [37]. *Step 1*: Initialize population size N and iteration number T. The constrained problem is as follows:

$$\min f(\mathbf{x})$$
$$S.t$$
$$g_j(\mathbf{x}) = c_j \quad \forall j = (1, 2, \ldots, n)$$

where $f(\mathbf{x})$ is the objective function used to calculate fitness value of the solution $x = (x_1, x_2, \ldots, x_D)$, where x_i is a decision variable assigned by a value in the lower and upper limits such that $x_i \in [X_i^{min}, X_i^{max}]$. g_j is the jth equality constraint, and h_k is the k inequality constraint. *Step 2*: Constructing the initial population for Jaya. Note that JM is an augmented matrix of size $N \times D$ corresponding to the Jaya Memory and shown in Equation (11). Conventionally, a solution is randomly constructed: $JM_{i,j} = X_j^{min} + (X_j^{min} - X_j^{max}) \times rnd, \forall i \in (1, 2, \cdots, N) \bigwedge \forall j \in (1, 2, \cdots, N)$. rnd is a uniform function that generates a random value between 0 and 1.

$$\mathbf{JM} = \begin{bmatrix} x_1^1 & x_2^1 & \cdots & x_D^1 \\ x_1^2 & x_2^2 & \cdots & x_D^2 \\ \vdots & \vdots & \cdots & \vdots \\ x_1^N & x_2^N & \cdots & x_D^N \end{bmatrix} \begin{bmatrix} f(\mathbf{x}^1) \\ f(\mathbf{x}^2) \\ \vdots \\ f(\mathbf{x}^N) \end{bmatrix} \tag{11}$$

The objective function $f(\mathbf{x}^i)$ for each solution is also calculated, and the JM solutions are sorted in ascending order. The best solution is $\mathbf{x}^1$ and the worst solution is $\mathbf{x}^N$. *Step 3*:

Jaya evolution process. The decision variables of all solutions in the JM undergo changes using the Jaya operator formulated in Equation (12). Note that $\mathbf{x}_j^{\prime i}$ is the new updated solution, x_j^i is the current solution, x_j^1 is the decision variable j in the best solution, and x_j^N is the decision variable j in the worst solution. r_1 and r_2 are two independent random numbers generated from a uniform distribution $U(0,1)$. *Step 4*: The JM solutions at every iteration will be updated. The objective function value of the new $f(\mathbf{x}^{\prime i})$ is calculated. The current solution $\mathbf{x}^i$ will be replaced by the new solution $\mathbf{x}^{\prime i}$ if $f(\mathbf{x}^{\prime i}) \leq f(\mathbf{x}^i)$. This process will be repeated as many as N times. *Step 5*: Stop rule. The algorithm repeats Step 3 and Step 4 until the maximum number of iterations T is reached.

$$x_j^{\prime i} = x_j^i + r_1 \times (x_j^1 - |x_j^i|) - r_2 \times (x_j^N - |x_j^i|) \quad (12)$$

Figure 5 shows the flowchart for Jaya algorithm.

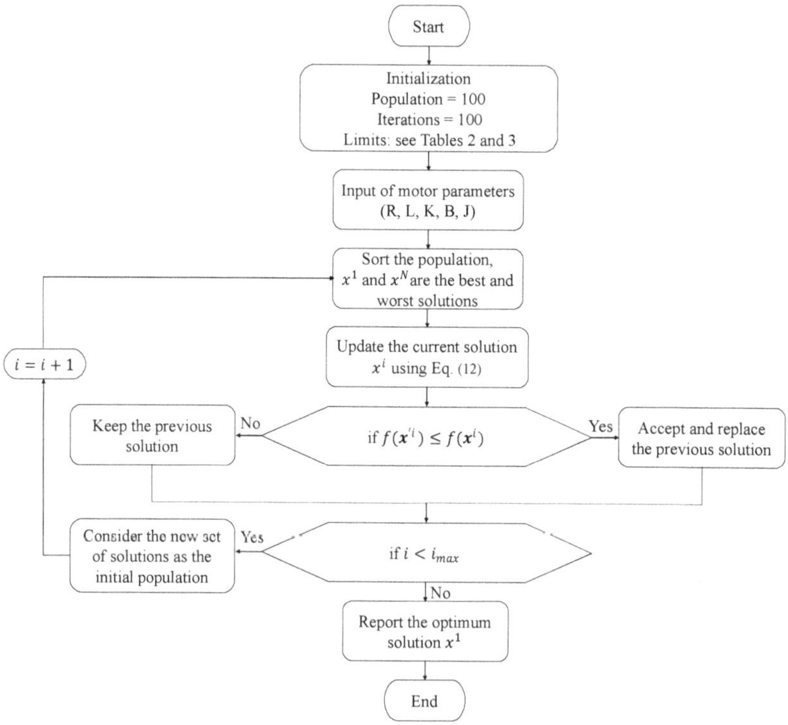

Figure 5. The flowchart of the Jaya algorithm.

3.4. Genetic Algorithm (GA)

GA is the oldest and most widely used metaheuristic algorithm, and it has many antecedents in the field of DC motors. This background extends to the following applications of the Genetic Algorithm: in PID controllers, to tune the parameters of DC motors when they have a time-dependent nature [38]; in the optimization of slotless BLDC motors and the effect on the distribution of magnetic flux and temperature [39,40]; to identify the dynamic state of a DC motor using the least squares error as a metric [41]; in a geared DC motor to improve the actual angular trajectory [23]; and to design a PID controller for a DC shunt motor considering a third-order model [42]. Also, the usefulness of GA for time scheduling and optimization in industrial robotized tasks has been demonstrated, improving efficiency and production time [43].

The main drawback of the algorithm is that it has several specific search parameters. The GA helps us to optimize the search for problem solutions based on natural selection mechanisms and biological and genetic evolution as part of evolutionary computation. The algorithm works through generations, where it first initializes them with random solutions. These are called populations, where each individual or chromosome will be evaluated by the selection function (fitness), which is mathematically molded depending on the problem in question. This process happens in each iteration (generation). Thanks to the selection mechanism, the next generation can be created from the selected chromosomes of the previous generation with the help of the algorithm function called crossing, which modifies the chromosomes through the mutation operator. Thereby, the algorithm finds an optimal solution to the problem in question. Figure 6 shows the operation of the algorithm.

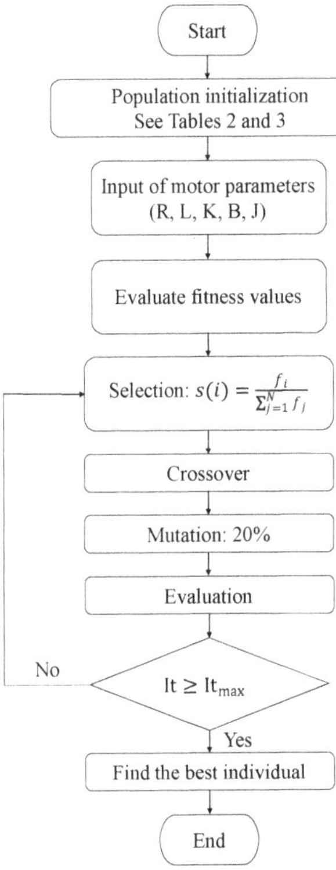

Figure 6. The flowchart of the Genetic Algorithm.

3.5. Implementation of Metaheuristic Algorithms

The parameters used for the development of the algorithms are described here. Much of the optimization is based on a good selection of search parameters that govern the search for acceptable solutions to the problem in question.

The same "fitness function" was applied to each metaheuristic algorithm for a fair comparison. This equation was based on the inverse of the Euclidean distance, as shown in Equation (13), where I_s is the estimated current, I is the actual current, ω_s is the estimated angular velocity, and ω is the actual angular velocity. It should be noted that this function is applied to each iteration of the metaheuristic algorithms.

$$Fitness = \frac{1}{\sqrt{\sum(I - I_s)^2 + \sum(\omega - \omega_s)^2}} \tag{13}$$

The MSE was used as an error evaluation criterion; see Equation (14). This function measures the average of the differences between the real and predicted values and allows us to penalize the largest differences to obtain a more robust model where, if there are large deviations, we can obtain a good approximation of the results. Since the DC motor is a system of differential equations with two variables, both must be included in the error estimate. That is why the fitness function must consider the MSE in current and speed. Since they are two different variables, the Euclidean distance is used to associate them. This statistical method will be used to analyze performance in the parametric estimations of the DC motor. The result will be a percentage of error in each vector of the five parameters [R, L, K, B, J]. It is important to highlight that the error function selected can affect the behavior of the metaheuristic algorithm. In addition to the MSE error, some other error functions have shown similar performance [44,45]; for example, the Integral Squared Error (ISE), the Integral Absolute Error (IAE), the Integral of Time multiplied by the Absolute Error (ITAE), and the Integral of Time multiplied by the Square Error (ITSE). Studying the effects of these error functions is beyond the scope of this manuscript.

$$MSE = \frac{1}{N}\sum_{i=1}^{N}(y_i - \hat{y}_i)^2 \tag{14}$$

This method results in an error calculated by adding the actual values minus the estimated values squared and divided by the number of parameters.

The optimization offered by metaheuristic algorithms depends on the iterations, since the algorithm produces a more optimal solution in each iteration. However, if many iterations are used, the algorithm will require a higher computational cost. For this work, the number of iterations is a fixed parameter aiming for a good performance in both computational cost and parametric estimation, obtaining a good balance between effectiveness and efficiency. The iterations used were 100 for each program.

The range of limits (upper and lower) determines the search space for each motor parameter. The limits adjust each range proportionally to the magnitude of the parameter. Varying these limits increases or reduces the search space. Although an infinite number of values can be selected, it is not essential, as long as the nominal value is within the search range. Thanks to this, the algorithm has a defined search space, where it begins to evaluate random values in search of the most appropriate solution. Mutation and creating a new population are performed within the limits of the GA. In the case of GWO and Jaya, new search agents within the range are checked in each iteration, guaranteeing that the search for all algorithms is performed within the limits. The ranges selected for this work must be positive since, by definition, any parameter takes negative values. On the other hand, the final selection of values is based on the typical values expected for this type of motor. For example, investigations such as refs. [30,46] have successfully searched for values in this range in similar DC motors.

Considering the above, each algorithm uses common general parameters, such as limits and iterations; see Table 2. However, they also use parameters that are specific to each method. The summary with the values of all the parameters used in each metaheuristic algorithm is shown in Table 3.

Simulations were run for 3 s for a constant voltage of 10.5 V. Likewise, a numerical method with a fixed step of 0.001 s was used to adapt to the data acquisition system. In executing the three metaheuristic algorithms, MATLAB was used to model the dynamic system of the DC motor through Simulink, and to obtain the responses of current and angular velocity, Equations (6) and (7) were used, respectively. MATLAB is responsible for executing the heuristic and metaheuristic algorithms; however, to evaluate performance, the Simulink scheme shown in Figure 7 is used, which evaluates the proposed solutions and subsequently returns the current and voltage signals obtained to MATLAB.

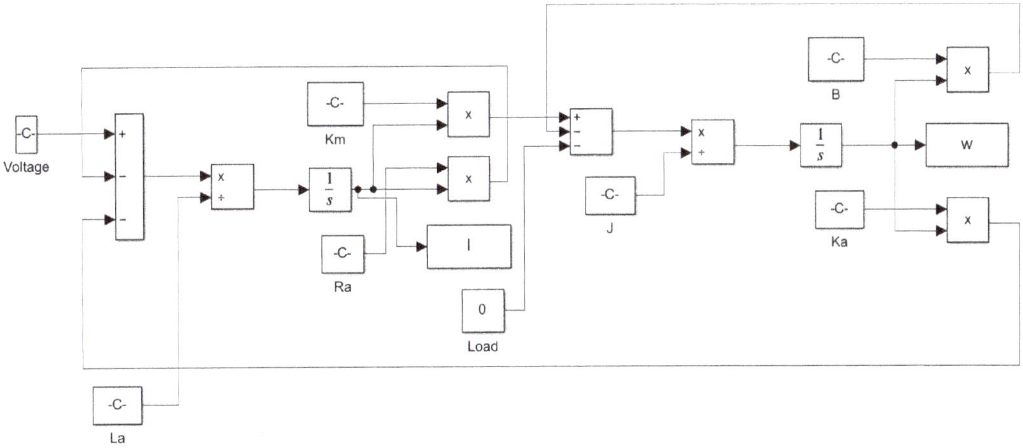

Figure 7. Fitness evolution for Genetic Algorithm.

In the same way, the input parameters of the M2-Robokits motor were the same as in the heuristic algorithm. The same period, voltage, and step mentioned above were used for each simulation.

4. Results

The minimum fitness calculated by each metaheuristic algorithm may vary because it is based on random values. Therefore, cross-validation is required. For this, each metaheuristic algorithm was executed ten times. The best, worst, and average are shown in Table 4. Furthermore, the performance graphs are shown for the GWO algorithm in Figure 8, for the Jaya algorithm in Figure 9, and for the GA in Figure 10.

Table 4. The cross validation for metaheuristic algorithms.

Algorithm	Fitness Value in Fifty Runs		
	Best	Worse	Average
GWO	0.029	0.062	0.042
Jaya	0.35	1.92	0.99
GA	0.044	1.06	0.058

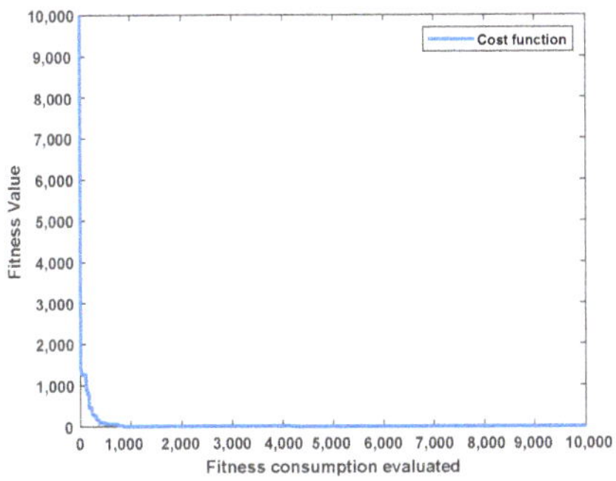

Figure 8. Fitness evolution for Gray Wolf Optimizer.

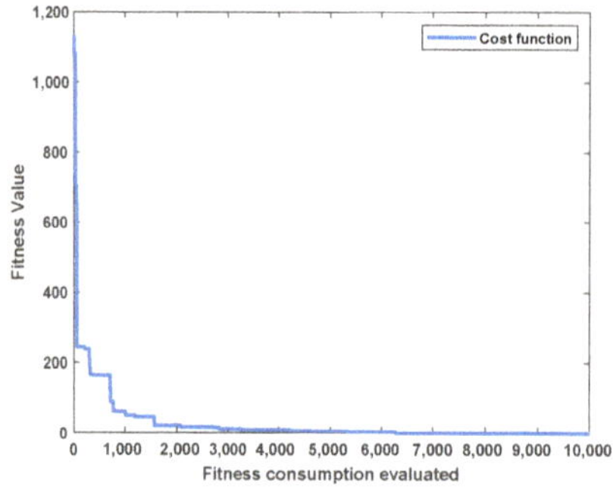

Figure 9. Fitness evolution for Jaya.

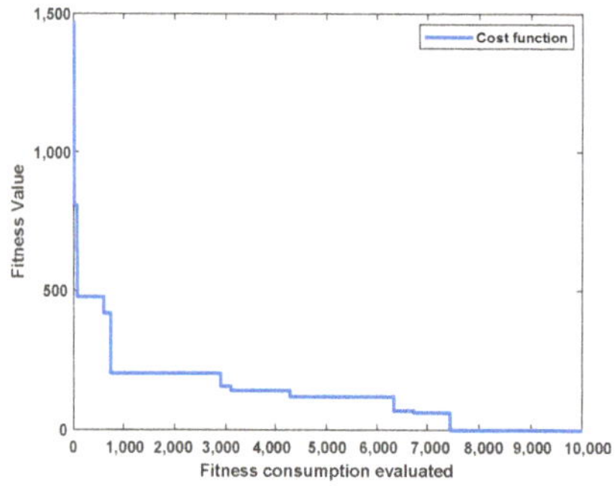

Figure 10. Fitness evolution for Genetic Algorithm.

The parameters calculated for all algorithms are shown in Tables 5 and 6. Two real signals from the M2 Motor were used for this. The current signal was hardware-filtered, and the velocity signal was filtered using a Chebyshev software filter. Both signals are considered real, although there may be variations due to acquisition and filtering.

- Figure 11 shows the comparison between the real current of the motor and the one estimated by the Steiglitz–McBride algorithm. In the same way, Figure 12 shows the comparison of the real velocity of the motor against that estimated by the Steiglitz–McBride algorithm.
- Figure 13 shows the comparison between the real current of the motor and the one estimated by the GWO algorithm. In the same way, Figure 14 shows the comparison of the real velocity of the motor against that estimated by the GWO algorithm.
- Figure 15 shows the comparison between the real current of the motor and the one estimated by the Jaya algorithm. In the same way, Figure 16 shows the comparison of the real velocity of the motor against that estimated by the Jaya algorithm.
- Figure 17 shows the comparison between the real current of the motor and the one estimated by the GA algorithm. In the same way, Figure 18 shows the comparison of the real velocity of the motor with that estimated by the GA algorithm.

Table 5. Comparison of Steiglitz–McBride and GWO algorithms against nominal values.

		Steiglitz–McBride		GWO	
Parameter	**Nominal Value**	**Value**	**MSE**	**Value**	**MSE**
R (Ω)	0.921042	0.914735	00.68%	0.923696	0.42%
L (H)	0.007759	0.008893	14.49%	0.007752	0.96%
K	0.073472	0.073547	00.10%	0.073466	0.08%
$B\left(\frac{\text{kg}\cdot\text{m}^3}{\text{s}^2}\right)$	0.000678	0.000680	00.28%	0.000677	0.16%
$J\ (\text{kg}\cdot\text{m}^2)$	0.000136	0.000135	01.03%	0.000136	0.55%

Table 6. Comparison of Jaya and GA algorithms against nominal values.

		Jaya		GA	
Parameter	**Nominal Value**	**Value**	**MSE**	**Value**	**MSE**
R (Ω)	0.921042	0.996974	08.24%	0.817929	13.44%
L (H)	0.007759	0.008941	15.04%	0.011908	89.00%
K	0.073472	0.000676	01.28%	0.076071	05.75%
$B\left(\frac{\text{kg}\cdot\text{m}^3}{\text{s}^2}\right)$	0.000678	0.291812	01.61%	0.000661	17.07%
$J\ (\text{kg}\cdot\text{m}^2)$	0.000136	0.000128	08.75%	0.000135	17.67%

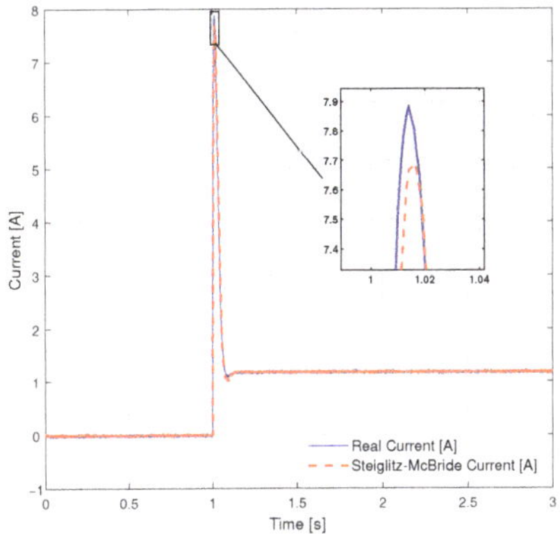

Figure 11. Real current signal vs. Steiglitz–McBride signal.

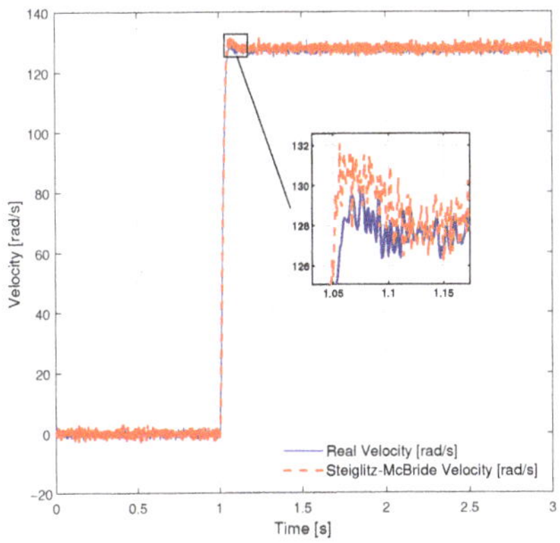

Figure 12. Real velocity signal vs. Steiglitz–McBride signal.

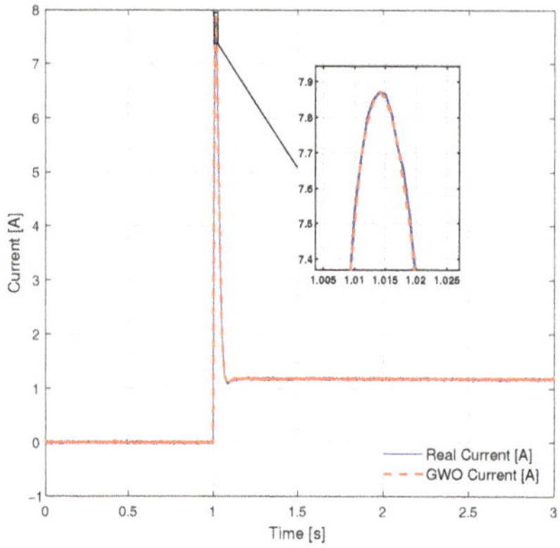

Figure 13. Real current signal vs. GWO signal.

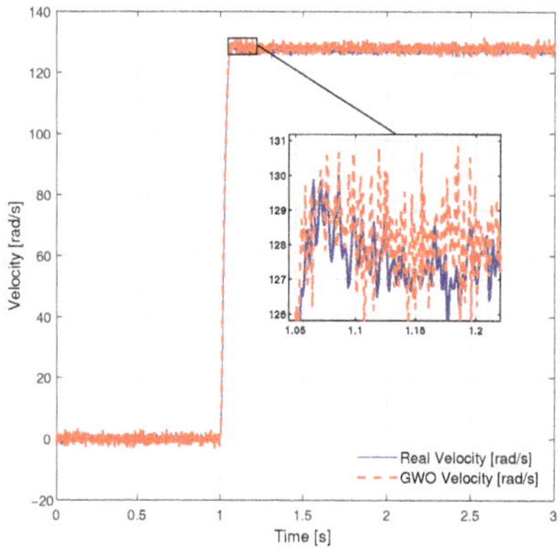

Figure 14. Real velocity signal vs. GWO signal.

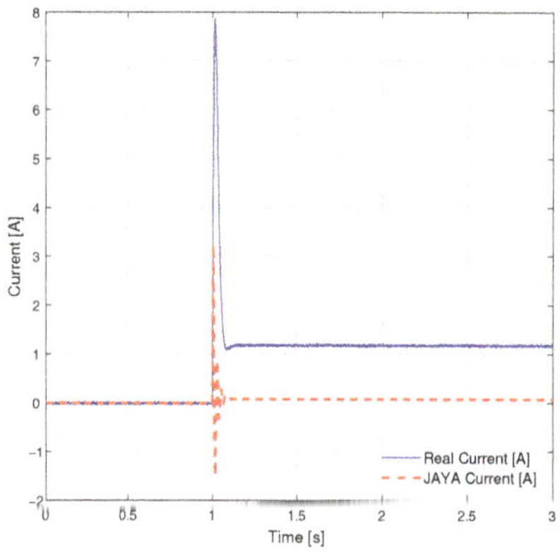

Figure 15. Real current signal vs. Jaya signal.

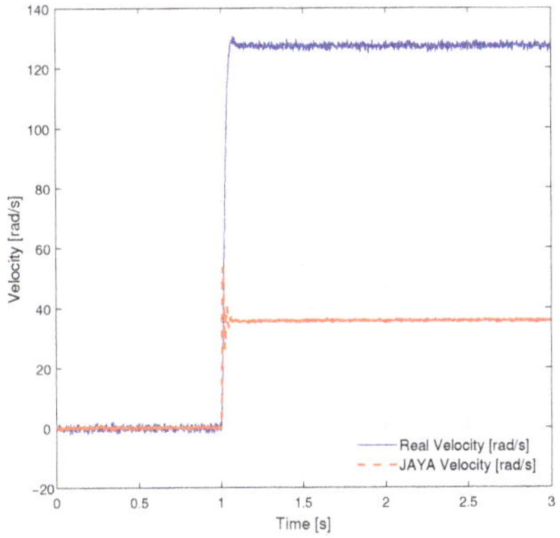

Figure 16. Real velocity signal vs. Jaya signal.

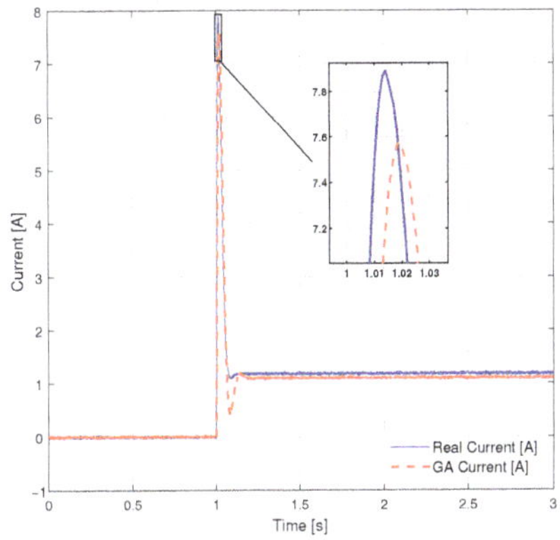

Figure 17. Real current signal vs. GA signal.

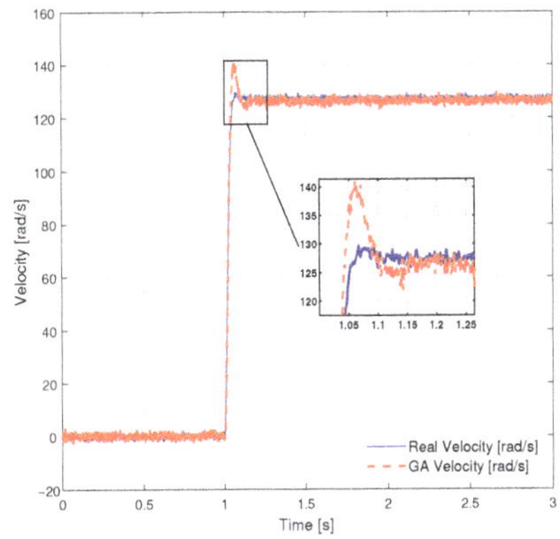

Figure 18. Real velocity signal vs. GA signal.

The computation times for each algorithm to run the 100 iterations are outlined in Table 7. The parameters and errors estimated by the Steiglitz–McBride and GWO algorithms are shown in Table 5 and compared against the nominal values. The parameters and errors estimated with the Jaya and GA algorithms are displayed in Table 6 and compared against the nominal values. As another metric to compare the algorithms, Table 8 includes the standard deviations that resulted from the estimates of each algorithm for each parameter analyzed.

Table 7. Computation time.

Algorithms	Average Time (min)
Steiglitz–McBride	0.102
GWO	49.53
Jaya	51.16
GA	60.23

Table 8. Standard deviation of estimates from each algorithm.

	Algorithms			
Parameters	**Steiglitz–McBride**	**GWO**	**Jaya**	**GA**
R	0.0023	0.0040	0.0086	0.1340
L	6.49×10^{-5}	9.22×10^{-5}	8.69×10^{-4}	0.0114
K	4.53×10^{-5}	7.24×10^{-5}	0.6940	0.0044
B	1.55×10^{-4}	1.15×10^{-6}	1.08×10^{-5}	9.94×10^{-7}
J	3.57×10^{-5}	6.99×10^{-7}	1.20×10^{-5}	5.27×10^{-7}

These results were obtained with real current and velocity signals. Better results are expected from all algorithms with computer-generated signals. However, it is necessary to address the noise inherent in real signals for practical application. The motor model parameterization process is usually performed offline, since the parameters remain constant. However, the parameterization of a motor can allow the use of control schemes that consider the dynamics of the system and not only the error, as in the case of the PID, which can allow for more precise control.

5. Discussion

Table 7 shows the computation performance of each algorithm. In this way, the computation times are compared among themselves to see whether they influence each algorithm's performance, since in works such as ref. [47], it has been reported that the increase in iterations negatively affects the optimization of the solutions. This is contrary to what occurs in our research, where the increased iterations and, consequently, high computational cost improved the parametric estimates. Even though parameterization is an offline process, the response times of the algorithms are too high for the response time expected in a control system. Thus, this type of algorithm is usually avoided in processes that require a high response speed.

As can be seen in Table 7, the Steiglitz–McBride heuristic algorithm has a lower computational cost compared to the metaheuristic algorithms. This is because the algorithm itself works through pure mathematics and is in charge of finding a unique solution. However, among the metaheuristic algorithms, a similar computational cost was observed, with a time variation of 10.7 min.

GWO has been used to estimate the parameters of a DC motor by applying an Integral Squared Error (ISE) object function [3]. It is reported that a smaller number of iterations could be detrimental, as it causes stagnation and local optima problems. Because of this, the authors decided to implement 500 iterations per test and few search agents compared to iterations. The use of 500 iterations in our research significantly increased the computation time. GWO has also been used as a parameter tuner for a PID controller [48], where 50 iterations and 30 search agents were applied. Taking into account the above, the configuration we decided to use in our research consisted of a smaller number of iterations and a larger number of search agents to reduce the computational cost and achieve a low error compared to the literature. This represented an estimation with an average error rate of 1.73%, which was 55% lower than that shown in [3].

The algorithm with the highest average MSE percentage was the GA algorithm, which was 28.59%. Among all the algorithms, the algorithm that showed the lowest average percentage error was the GWO, with 0.43%. However, the average MSE percentage for Steiglitz–McBride, which is not a metaheuristic algorithm, was lower than Jaya and GA; see Tables 5 and 6. Although GWO generated the best parametric estimation in this research, the Steiglitz–McBride algorithm could work as another alternative when the application requires low computational cost and a low standard deviation in the estimation of each parameter; see Table 8. For example, in [49], it has been shown that the Steiglitz–McBride algorithm is useful in the parametric estimation of electrical machines such as DC motors, brush DC, and brushless AC and gear machines. Also, it is frequently used as an optimizer.

The current signals estimated by GWO and Steiglitz–McBride were very similar to the nominal ones. For this reason, a comparison of the real and estimated signals obtained by the algorithms was performed with the following indicators: settling time, overshoot, and steady-state error.

The stabilization times of the current signals were as follows: 1.107 s in the real signal, 1.118 s in Steiglitz–McBride, 1.103 s in GWO, and 1.167 s in GA. However, the current signal estimated by the Jaya algorithm failed to reach the stabilization value; see Figure 15. The current signal obtained by GWO is the closest estimate to the real motor signal; see Figure 13.

The average current values in the steady-state zone of each algorithm were as follows: 1.1821 A for the real signal, 1.1840 A for Steiglitz–McBride, 0.0827 A for Jaya, 1.1805 A for GWO, and 1.0968 A for GA. This means that GWO and Steiglitz–McBride were the algorithms that estimated the closest current response to the real one.

The comparison of the overshoot values in the current signals were as follows: 7.8717 A for the real signal, occurring in 1.014 s; 7.6930 A for the Steiglitz–McBride algorithm, occurring in 1.016 s; 3.2511 A for Jaya, occurring in 1.004 s; 7.8738 A for GWO, occurring in 1.014 s; and, finally, 7.5736 A for GA, occurring in 1.018 s. The GWO algorithm proved to be the most accurate when discussing the maximum peaks, obtaining the same result as the real signal for time. For current, a 0.22% difference was displayed.

In [3], the authors reported that a percentage error comparison cannot be reliable as a measure of the effectiveness of these algorithms. Since the algorithms do not show consistent solutions because they start from random solutions, performing the comparison by means of the standard deviation would be more suitable, which is a more stable metric for random data. Table 8 shows the difference in the standard deviation between the estimated parameters of each algorithm and the nominal values. The algorithms with the most significant deviation were Jaya and GA, with Jaya being the algorithm with the greatest variation in the estimates. On the other hand, Steiglitz–McBride and GWO had less variation in the estimates, where Steiglitz–McBride was the algorithm with the lowest deviation of 37.34% with respect to GWO. This may be due to Steiglitz–McBride having consistent estimated values, because it starts from a mathematical model. In contrast, metaheuristic algorithms start from random solutions which are adjusted to find the most suitable solution during iterations.

6. Conclusions

This work performed a parametric estimation of a DC motor using a heuristic algorithm (Steiglitz–McBride) and three metaheuristic algorithms (Jaya, GWO, and GA). The MSE and standard deviation were used as statistical indicators to evaluate the performance of each algorithm as well as other values of the dynamic response, such as settling time, overshoot, and steady-state error. In this way, this work determined which algorithm provides a parametric estimation closer to the nominal parameters and real DC motor signals. The comparison between metaheuristic and heuristic algorithms is limited to parametric estimation in direct current motors. As the No Free Lunch theorem states, no algorithm is best for all applications. Another limitation is the selection of hyperparameters, because there is no way to guarantee the optimal selection of these parameters in metaheuristic algorithms. There are algorithms for selecting hyperparameters, but they do not guarantee optimal selection. Therefore, the maximum performance of a metaheuristic algorithm may not be found.

According to the aforementioned aspects, the two best-performing algorithms were Steiglitz–McBride and GWO. GWO was the best parametric estimator for this application, having the lowest MSE of all algorithms for the nominal parameters. Likewise, this algorithm had the closest responses to the real ones, with the best overshoot and settling time approximation. However, Steiglitz–McBride also obtained good results as a parametric estimator, even better than Jaya and GA, but with the lowest computational cost of all and a less variable parametric estimation.

Although GWO is a better parametric estimator, it does not mean that it is better than the Steiglitz–McBride algorithm, since implementing the mathematical model in a metaheuristic algorithm can be simpler. However, the search parameters required by metaheuristic algorithms are obtained after many tests, which generate a high computational cost. Although the Steiglitz–McBride is mathematically more complex, if appropriately implemented, it may require fewer tests and fewer corrections within the algorithm, resulting in a lower computational cost.

In this work, the heuristic algorithm produced fewer problems during tests, with lower computational cost and low variation in estimates. However, if a more precise parametric estimation is required, it is recommended to use GWO. Considering these aspects, the heuristic algorithm has more advantages in the parametric estimation of DC motors. The performance of the algorithms presented in this paper may change if different error functions are used, as well as other search parameters. In addition, performance depends on the final application of the algorithm.

Author Contributions: Conceptualization, L.A.M.-S., E.A.D.-R., M.C.-F. and J.R.-R.; methodology, M.C.-F., D.M.M. and E.A.D.-R.; writing—original draft preparation, L.A.M.-S., M.C.-F. and J.R.-R.; writing—review and editing, D.M.M., E.A.D.-R. and L.A.M.-S.; supervision, L.A.M.-S. and M.C.-F. All authors have read and agreed to the published version of the manuscript.

Funding: No funding was received to conduct this study.

Institutional Review Board Statement: Not applicable.

Informed Consent Statement: Not applicable.

Data Availability Statement: The raw data supporting the conclusions of this article will be made available by the authors on request.

Acknowledgments: The authors thank the Polytechnic University of Queretaro (UPQ) for the support.

Conflicts of Interest: The authors declare that they have no conflicts of interest. The datasets generated during and/or analyzed during the current study are available from the corresponding authors on reasonable request.

References

1. Fazdi, M.F.; Hsueh, P.-W. Parameters Identification of a Permanent Magnet DC Motor: A Review. *Electronics* **2023**, *12*, 2559. [CrossRef]
2. Beltran-Carbajal, F.; Tapia-Olvera, R.; Aguilar-Mejia, O.; Favela-Contreras, A.; Lopez-Garcia, I. An online algebraic estimation approach of parameters and variable mechanical torque in shunt DC motors. *Int. Trans. Electr. Energy Syst.* **2018**, *28*, e2474. [CrossRef]

3. Karnavas, Y.L. Application of recent nature-inspired meta-heuristic optimization techniques to small permanent magnet DC motor parameters identification problem. *J. Eng.* **2020**, *2020*, 877–888.
4. Tran, C.D.; Kuchar, M.; Sobek, M.; Sotola, V.; Dinh, B.H. Sensor Fault Diagnosis Method Based on Rotor Slip Applied to Induction Motor Drive. *Sensors* **2022**, *22*, 8636. [CrossRef] [PubMed]
5. Chakraborty, C.; Verma, V. Speed and Current Sensor Fault Detection and Isolation Technique for Induction Motor Drive Using Axes Transformation. *IEEE Trans. Ind. Electron.* **2015**, *62*, 1943–1954. [CrossRef]
6. Merrassi, W.E.; Abounada, A.; Ramzi, M. Advanced speed sensorless control strategy for induction machine based on neuro-MRAS observer. *Mater. Today Proc.* **2021**, *45*, 7615–7621. ISSN 2214-7853. [CrossRef]
7. Nag, T.; Santra, S.B.; Chatterjee, A.; Chatterjee, D.; Ganguli, A.K. Fuzzy logic-based loss minimisation scheme for brushless DC motor drive system. *IET Power Electron.* **2016**, *9*, 1581–1589. [CrossRef]
8. Vikhe, P.S.; Punjabi, N.; Kadu, C.B. DC motor speed control using PID controller in lab view. *Int. J. Innov. Sci. Mod. Eng. (IJISME)* **2015**, 38–41.
9. Song, T.J.; Oh, K.S. Adaptive Velocity Control Algorithm for DC Motors based on Parameter Estimation of Error Dynamics Under Uncertainty and Load Variation. *Trans. Korean Soc. Mech. Eng. A* **2020**, *44*, 83–91. [CrossRef]
10. Majdoubi, R.; Masmoudi, L.; Bakhti, M.; Elharif, A.; Jabri, B. Parameters estimation of BLDC motor based on physical approach and weighted recursive least square algorithm. *Int. J. Electr. Comput. Eng. (IJECE)* **2021**, *11*, 133–145. ISSN 2088-8708. [CrossRef]
11. Knypiński, L.; Reddy, A.V.; Venkateswararao, B.; Devarapalli, R. Optimal design of brushless DC motor for electromobility propulsion applications using Taguchi method. *J. Electr. Eng.* **2023**, *74*, 116–121. [CrossRef]
12. Urra, E.; Cubillos, C.; Cabrera-Paniagua, D.; Mellado, R. hMod: A software framework for assembling highly detailed heuristics algorithms. *Softw.-Pract. Exp.* **2019**, *49*, 971–994. [CrossRef]
13. Zandavi, S.M.; Chung, V.Y.Y.; Anaissi, A. Stochastic dual simplex algorithm: A novel heuristic optimization algorithm. *IEEE Trans. Cybern.* **2021**, *51*, 2725–2734. [CrossRef]
14. Li, Y.; Cherednichenko, A.; Jiang, Z.; Shi, W.; Wu, J. A Novel Generalized Group-Sparse Mixture Adaptive Filtering Algorithm. *Symmetry* **2019**, *11*, 697. [CrossRef]
15. Neshat, M.; Sepidnam, G.; Sargolzaei, M.; Toosi, A.N. Artificial fish swarm algorithm: A survey 590 of the state-of-the-art, hybridization, combinatorial and indicative applications. *Artif. Intell. Rev.* **2014**, *42*, 965–997. [CrossRef]
16. Yang, S.; Jiang, J.; Yan, G. A Dolphin Partner Optimization. In Proceedings of the 2009 WRI Global Congress on Intelligent Systems, Xiamen, China, 19–21 May 2009; Zhou, S.M., Wang, W., Eds.; IEEE Comp Soc; World Res Inst: Xiamen, China, 2009; Volume I, pp. 124–128.
17. Shuang, B.; Chen, J.; Li, Z. Study Hybrid PS-ACO Algorithm. *Appl. Intell.* **2011**, *34*, 64–73. [CrossRef]
18. Wu, E.; Huang, Y.; Li, D. An Adaptive Particle Swarm Optimization Algorithm for Reactive Power Optimization in Power System. In Proceedings of the 2010 8th World Congress on Intelligent Control and Automation (WCICA), Jinan, China, 7–9 July 2010; IEEE: Jinan, China, 2010; pp. 3132–3137.
19. Shi, J.; Mi, Q.; Cao, W.; Zhou, L. Optimizing BLDC motor drive performance using particle swarm algorithm-tuned fuzzy logic controller. *SN Appl. Sci.* **2022**, *4*, 293. [CrossRef]
20. Rodríguez-Abreo, O.; Rodríguez-Reséndiz, J.; García-Cerezo, A.; García-Martínez, J.R. Fuzzy logic controller for UAV with gains optimized via genetic algorithm. *Heliyon* **2014**, *10*, e26363. ISSN 2405-8440. [CrossRef]
21. Serradilla, F.; Canas, N.; Naranjo, J.E. Optimization of the energy consumption of electric motors through metaheuristics and PID controllers. *Electronics* **2020**, *9*, 1842. [CrossRef]
22. Wu, Z.; Du, C. Parameter Identification PMSM Based Improved Cuckoo Algorithm. *Neural Process. Lett.* **2019**, *50*, 2701–2715. [CrossRef]
23. Amiri, M.S.; Ibrahim, M.F.; Ramli, R. Optimal parameter estimation for a DC motor using genetic algorithm. *Int. J. Power Electron. Drive Syst. (IJPEDS)* **2020**, *11*, 1047–1054. [CrossRef]
24. Al-Azzawi, D.S. Evaluation of Genetic Algorithm Optimization in Machine Learning. *J. Inf. Sci. Eng.* **2020**, *36*, 231–241.
25. Cheng, Y.; Lyu, X.; Mao, S. Optimization design of brushless DC motor based on improved Jaya algorithm. *Sci. Rep.* **2024**, *14*, 5427. [CrossRef] [PubMed]
26. Yan, C.; Li, M.-X.; Liu, W. Application of Improved Genetic Algorithm in Function Optimization. *J. Inf. Sci. Eng.* **2019**, *35*, 1299–1309.
27. Ahamed, S.R.; Parumasivam, P.; Lipu, M.S.H.; Hannan, M.A.; Ker, P.J. A comparative evaluation of PID-based optimisation controller algorithms for DC motor. *Int. J. Autom. Control* **2020**, *14*, 634–655. [CrossRef]
28. Achanta, R.K.; Pamula, V.K. DC Motor Speed Control using PID Controller Tuned by Jaya Optimization Algorithm. In Proceedings of the 2017 IEEE International Conference on Power, Control, Signals and Instrumentation Engineering (ICPCSI), Chennai, India, 21–22 September 2017; IEEE: Thandalam, India, 2017; pp. 983–987.
29. Niu, X.; Feng, G.; Jia, S.; Zhang, Y. Control of brushless DC motor based on fuzzy rules optimized by genetic algorithm used in hybrid vehicle. *J. Comput. Methods Sci. Eng.* **2021**, *21*, 951–968. [CrossRef]
30. Rodriguez-Abreo, O.; Hernandez-Paredes, J.M.; Rangel, A.F.; Fuentes-Silva, C.; Velasquez, F.A.C. Parameter Identification of Motors by Cuckoo Search Using Steady-State Relations. *IEEE Access* **2021**, *9*, 72017–72024. [CrossRef]
31. Pillay, P.; Krishnan, R. Modeling, simulation, and analysis of permanent-magnet motor drives. II. The brushless DC motor drive. *IEEE Trans. Ind. Appl.* **1989**, *25*, 274–279. [CrossRef]

32. Rodríguez-Abreo, O.; Rodríguez-Reséndiz, J.; Montoya-Santiyanes, L.A.; Álvarez-Alvarado, J.M. Non-Linear Regression Models with Vibration Amplitude Optimization Algorithms in a Microturbine. *Sensors* **2022**, *22*, 130. [CrossRef]
33. Rezoug, A.; Iqbal, J.; Tadjine, M. Extended grey wolf optimization–based adaptive fast nonsingular terminal sliding mode control of a robotic manipulator. *Proc. Inst. Mech. Eng. Part J. Syst. Control Eng.* **2022**, *236*, 1738–1754. [CrossRef]
34. Mirjalili, S.; Mirjalili, S.M.; Lewis, A. Grey Wolf Optimizer. *Adv. Eng. Softw.* **2014**, *69*, 46–61. [CrossRef]
35. Agarwal, J.; Parmar, G.; Gupta, R.; Sikander, A. Analysis of grey wolf optimizer based fractional order PID controller in speed control of DC motor. *Microsyst.-Technol.-Micro-Nanosyst.-Inf. Storage Process. Syst.* **2018**, *24*, 4997–5006. [CrossRef]
36. Rao, R. Jaya: A simple and new optimization algorithm for solving constrained and unconstrained optimization problems. *Int. J. Ind. Eng. Comput.* **2016**, *7*, 19–34. [CrossRef]
37. Zitar, R.A.; Al-Betar, M.A.; Awadallah, M.A.; Doush, I.A.; Assaleh, K. An Intensive and Comprehensive Overview of JAYA Algorithm, its Versions and Applications. *Arch. Comput. Methods Eng.* **2022**, *29*, 763–792. [CrossRef] [PubMed]
38. Godem Ali, M.I.; Karol, K.; Viliam, F. CAD of Cascade Controllers for DC Drives Using Genetic Algorithm Methods. *Procedia Eng.* **2014**, *96*, 182–189. [CrossRef]
39. Kamal, C.; Thyagarajan, T.; Kalpana, D.; Pragadheeshwaran, T. Multiobjective design optimization and analysis of magnetic flux distribution for slotless permanent magnet brushless DC motor using evolutionary algorithms. *J. Magn. Magn. Mater.* **2019**, *476*, 524–537. [CrossRef]
40. Rahideh, A.; Korakianitis, T.; Ruiz, P.; Keeble, T.; Rothman, M.T. Optimal brushless DC motor design using genetic algorithms. *J. Magn. Magn. Mater.* **2010**, *322*, 3680–3687. [CrossRef]
41. Lankarany, M.; Rezazade, A. Parameter Estimation Optimization Based on Genetic Algorithm Applied to DC Motor. In Proceedings of the 2007 International Conference on Electrical Engineering, Lahore, Pakistan, 11–12 April 2007; pp. 1–6. [CrossRef]
42. Thomas, N.; Poongodi, D.P. Position Control of DC Motor Using Genetic Algorithm Based PID Controller. In Proceedings of the World Congress on Engineering, London, UK, 1–3 July 2009; Volume 2.
43. Baizid, K.; Yousnadj, A.; Meddahi, A.; Chellali, R.; Iqbal, J. Time scheduling and optimization of industrial robotized tasks based on genetic algorithms. *Robot.-Comput.-Integr. Manuf.* **2015**, *34*, 0736–5845. [CrossRef]
44. Soni, Y.K.; Bhatt, R. BF-PSO optimized PID controller design using ISE, IAE, IATE and MSE error criteria. *Int. J. Adv. Res. Comput. Eng. Technol. (IJARCET)* **2013**, *2*, 2333–2336.
45. Mousakazemi, H.; Mohammad, S. Comparison of the error-integral performance indexes in a GA-tuned PID controlling system of a PWR-type nuclear reactor point-kinetics model. *Prog. Nucl. Energy* **2021**, *132*, 103604. [CrossRef]
46. Rodríguez-Abreo, O.; Rodríguez-Reséndiz, J.; Fuentes-Silva, C.; Hernández-Alvarado, R.; Falcón, M.D.C.P.T. Self-Tuning Neural Network PID with Dynamic Response Control. *IEEE Access* **2021**, *9*, 65206–65215. [CrossRef]
47. Afifi, M.; Rezk, H.; Ibrahim, M.; El-Nemr, M. Multi-Objective Optimization of Switched Reluctance Machine Design Using Jaya Algorithm (MO-Jaya). *Mathematics* **2021**, *9*, 1107. [CrossRef]
48. Dutta, P.; Nayak, S.K. Grey Wolf Optimizer Based PID Controller for Speed Control of BLDC Motor. *J. Electr. Eng. Technol.* **2021**, *16*, 955–961. [CrossRef]
49. Jimenez-Gonzalez, J.; Gonzalez-Montanez, F.; Jimenez-Mondragon, V.M.; Liceaga-Castro, U.; Escarela-Perez, R.; Olivares-Galvan, J.C. Parameter Identification of BLDC Motor Using Electromechanical Tests and Recursive Least-Squares Algorithm: Experimental Validation. *Actuators* **2021**, *10*, 143. [CrossRef]

algorithms

MDPI

Article

Comparison of Meta-Heuristic Optimization Algorithms for Global Maximum Power Point Tracking of Partially Shaded Solar Photovoltaic Systems

Timmidi Nagadurga [1], Ramesh Devarapalli [2,*] and Łukasz Knypiński [3,*]

1 Department of Electrical and Electronics Engineering, Lakireddy Bali Reddy College of Engineering, Mylavaram 521230, India; durga.269@gmail.com

2 Department of Electrical/Electronics and Instrumentation Engineering, Institute of Chemical Technology, Indianoil Odisha Campus, Bhubaneswar 751013, India

3 Institute of Electrical Engineering and Electronics, Poznan University of Technology, 60-965 Poznan, Poland

* Correspondence: r.devarapalli@iocb.ictmumbai.edu.in (R.D.); lukasz.knypinski@put.poznan.pl (Ł.K.)

Abstract: Partial shading conditions lead to power mismatches among photovoltaic (PV) panels, resulting in the generation of multiple peak power points on the P-V curve. At this point, conventional MPPT algorithms fail to operate effectively. This research work mainly focuses on the exploration of performance optimization and harnessing more power during the partial shading environment of solar PV systems with a single-objective non-linear optimization problem subjected to different operations formulated and solved using recent metaheuristic algorithms such as Cat Swarm Optimization (CSO), Grey Wolf Optimization (GWO) and the proposed Chimp Optimization algorithm (ChOA). This research work is implemented on a test system with the help of MATLAB/SIMULINK, and the obtained results are discussed. From the overall results, the metaheuristic methods used by the trackers based on their analysis showed convergence towards the global Maximum Power Point (MPP). Additionally, the proposed ChOA technique shows improved performance over other existing algorithms.

Keywords: maximum power point tracking; PV systems; cat swarm optimization; grey wolf optimization; chimp optimization algorithm

1. Introduction

Solar energy has become increasingly important as an energy resource due to its inexhaustible nature, cleanliness, scalability in power generation and low maintenance requirements. As a country with high sunshine, India offers an exceptional opportunity for harnessing solar energy, which can significantly address our pollution problems. In recent times, photovoltaic systems have gained immense popularity due to their numerous advantages. These systems are known for their sustainability, low maintenance requirements, absence of complex parts and extended lifespan, among other benefits [1]. The PV systems are subjected to shading effects along with dynamically changing weather conditions due to the shadow of large buildings, passing clouds, birds' shadows, etc. In the case of Partial Shading Condition (PSC), the PV modules get different irradiations and temperatures, leading to decreasing the power output obtained from the PV array and causing hot spots and loss of tracking efficiency. During shading, power variation with voltage characteristic plot possesses several peaks, one referred to as Global Maximum Power Point (GMPP), and others are well known as local maximum power points, as delineated in Figure 1.

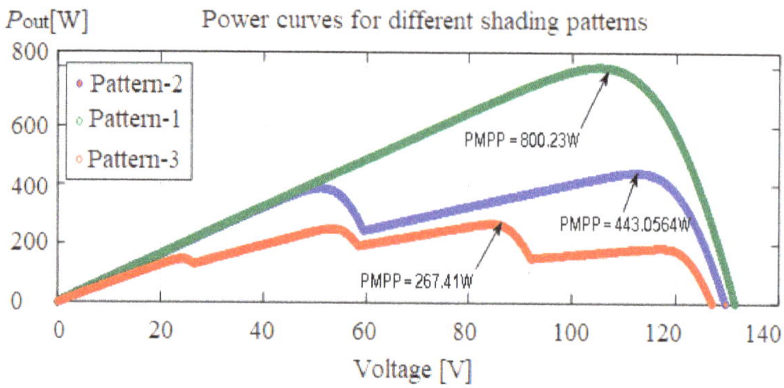

Figure 1. P-V plot of PV module under changing shading conditions.

Most conventional MPPT approaches like Perturb and Observe (P&O), Incremental Conductance (INC) and Constant Voltage (CV) techniques can work efficiently when solar radiation and temperature are uniform but significantly fail when the weather conditions are varying. In order to address the mentioned challenge, researchers have utilized intelligent global search optimization algorithms which effectively tackle the global maximum power tracking (MPPT) problems. In the available literature, diverse MPPT techniques and alternative solutions have been employed to accurately identify the true global maximum power point amidst various local MPPs.

These popular intelligent algorithms include (a) Genetic Algorithm (GA), (b) Particle Swarm Optimization (PSO), (c) Ant Colony Optimization (ACO) and (d) Teaching learning-based optimization. Kulaksiz et al. [2] implemented the GA with an ANN-based tracking method for the standalone PV system. Genetic algorithm optimization was used in this algorithm to optimise the required number of neutrons in a multi-layer perception neural network, thereby tracking required power during uneven irradiance situations. Tajuddin et al. [3] suggested a superior Differential Evolution (DE) algorithm for changing solar environmental conditions to obtain global power output. Thangamani et al. [4] introduced an adaptive differential algorithm that can track MPP under swiftly changing climatic conditions. Ramli et al. [5] surveyed methods resembling artificial intelligence and hybrid algorithm for tracking MPP and summarized the finer points of various techniques understanding to increase the power output of the PV array.

Liu et al. [6] applied the PSO for various operating situations and reported the merits of the PSO algorithm, which includes ease of implementation, system independence and enhanced output. Ishaque et al. [7] discussed the PSO algorithm, which eliminates the need for the direct duty ratio method's proportional-integral (PI) control block. Renaudineau et al. [8] used the PSO algorithm to control the gate pulse given to the boost converter to extract GMPP. Phimmasone et al. [9] used the PSO algorithm by taking additional specific coefficients of the algorithm to progress to the next step and get the global point of the PV. The usage of more specific parameters in PSO algorithms results in decreased tracking efficiency and increased uncertainty of the solution. Therefore, the desired operating point cannot be reached. To overcome these drawbacks, improved optimization algorithms were proposed by the researchers. Chao et al. [10] offered a modified PSO algorithm to harness GMPP from a PV system during a cloudy day. The adjustment practice fallouts to figure out the algorithm-specific parameters for the next step of acceleration. In [11], Chen et al. focused on applying an enhanced PSO algorithm for MPPT to enhance tracking efficiency, particularly under shading conditions. Chowdhury et al. [12] proposed the adaptive PSO to defeat the de-merits of a conventional PSO. Tobon et al. [13] implemented an Improved Pattern Search Method (IPSM) for MPPT, and the solution obtained by the pattern search technique succeeded in their convergence properties. Babu et al. [14] worked on the voltage-band-based improved PSO approach to improve the convergence of the

PSO. Gavhane et al. [15] examined the studies on MPPT on cloudy days by applying Enhanced-Leader-Particle Swarm Optimization (EL-PSO), which was more efficient for implementation. Babu et al. [16] presented the benefits of the modified PSO algorithm, highlighting its nearly zero steady-state oscillations and faster dynamic response compared to traditional approaches. Husain et al. [17] commented on the various parameters of the PSO and elaborated on the numerous procedures for MPP tracking. Kalaiarasi et al. [18] proposed an enhanced PSO method for getting maximum benefits from solar-powered energy integrated through the Z-source inverter and observed various benefits of reduction in the steady state oscillations.

Dileep and Singh [19] gave an inclusive report on different computing methods (PSO and ACO) to get GMPP beneath diverse shading effects. Ahmed et al. [20] applied PSO, ACO, Cuckoo Search (CS) and DE for the simulation of the MPPT system during shading patterns, the provision of operation and merits, and the limitations of each optimization algorithm have been highlighted. Li et al. [21] considered a new GMPP tracking approach which depends on changes in power, besides analysing the nature of the activity. Rezk et al. [22] reviewed the various power extraction algorithms and concluded that PSO and CS algorithms have shown more convergence to the global maximum power tracking. Ahmed et al. [23] investigated the effectiveness of the Particle Swarm Optimization-Support Vector Regression (PSO-SVR) method in reducing the percentage of ripple content and minimizing oscillations in the power waveform around the MPP region. Gangwar et al. [24] studied the cat swarm optimization in a specific pattern well known as phyllotaxy in order to extract more solar power. Mohanty and Tripatty [25] presented a novel Teaching Learning Based Optimization (TLBO) algorithm, that was introduced for the optimal placement of distributed generation in a radial distribution network. This technique offers a unique approach to addressing the problem of determining the best locations for distributed generation sources within the network.

Hegazy and Fathy [26] carried out numerous simulations under different shading patterns using (FLC, PSO, and TLBO) and concluded that TLBO extracts GMPP more dynamically than others. Ahmed and Rezk [27] implemented a novel Mine Blast Algorithm (MBA) and applied different patterns of shadow. The obtained results were compared, thereby declaring MBA as superior to TLBO. Crepinsek et al. [28] observed that the TLBO algorithm showed better performance for maximum power point tracking among the optimisation algorithms. Sundareswaran et al. [29] found that Firefly Algorithm (FA) was one of the efficient approaches to harnessing maximum solar energy beneath partial shading conditions. Javed et al. [30] studied enhancing the FA algorithm; the study aimed to improve MPPT efficiency and mitigate the adverse effects of partial shading. Comparative analysis between the modified FA and PSO algorithms provided valuable insights into the effectiveness of the modified FA algorithm for achieving superior MPPT performance under partially shaded conditions. Mohanty et al. [31] proposed a hybrid maximum power point tracking technique using the P&O method with GWO for better MPPT from PV module strings beneath shading conditions. Eltamay et al. [32] applied the GWO algorithm with fuzzy logic controllers to eliminate the oscillation near GMPP. Mohanty et al. [33] examined the design of the MPPT circuit along with the GWO algorithm that mitigates the shortfalls of conventional tracking methods, such as poor performance in tracking, low efficiency and more oscillations near and around MPP.

The key distinctions among this approach include their effectiveness range, convergence speed, design complexity, sensor requirements, control parameters and hardware implementation costs. Consequently, selecting an appropriate algorithm is crucial when designing a photovoltaic (PV) system, as it depends on the intended application. After the selection of the optimization algorithm, the adaptation for the solved optimization task must be done.

This paper compares widely used MPPT techniques to mitigate the negative impact of partial shading, thereby enhancing maximum power output. It explores the advantages and disadvantages of these techniques and provides a general comparison of various solutions.

This article particularly focuses on discussing different approaches for global maximum power point tracking (GMPPT) under partial shading conditions, with emphasis on recent advancements in the scientific literature.

The main contributions of this paper are as follows:

- A statistical investigation of modern heuristic optimization methods for (GMPPT) in photovoltaic (PV) systems under partial shading conditions has been presented;
- Comprehensive analysis of the challenges associated with heuristic optimization-based GMPPT techniques, focusing on their exploitative and explorative search capabilities;
- Introduction of a novel GMPPT method called Chimp Optimization Algorithm, which effectively balances the exploitative and explorative search capabilities;
- Statistical comparisons of different heuristic optimization-based GMPPT techniques in terms of tracking routines, accumulated energy and tracking efficiency.

Applications of Metaheuristics Algorithm

➢ Metaheuristics algorithms offer a powerful approach to tackling complex optimization problems in diverse domains. Their flexibility, robustness and ability to find satisfactory solutions make them invaluable tools for real world problems as presented below.

➢ Metaheuristics are widely used to tackle problems with a large number of possible combinations, such as the Traveling Salesman Problem, Knapsack Problem or Job Scheduling. Examples of metaheuristics for combinatorial optimization include GA, PSO, ACO and Simulated Annealing (SA).

➢ Metaheuristics are used for optimizing the design parameters of complex systems. For example, they can optimize the shape of an aircraft wing, the layout of an electric circuit, electromagnetic device or the parameters of a chemical process.

➢ Metaheuristics are utilized in optimizing transportation routes, vehicle routing problems and logistics planning. They help find efficient paths for deliveries, minimizing travel time and costs.

➢ Metaheuristics can optimize production schedules, inventory management, and resource allocation in manufacturing processes.

➢ Metaheuristics can be used to create computer programs that can play games effectively by finding near-optimal strategies.

2. Application of CSO-Based MPPT Controller for Solar PV Strings under Partial Shading Conditions

In the CSO algorithm, the natural behaviour of the cats is used to solve the optimization problem [34]. In this algorithm, the behaviour of cats is deciphered in two modes: seeking mode and tracing mode and used to move the virtual cats in the search space. A predetermined ratio, known as the Mixture Ratio (MR), determines how many cats participate in each iteration in tracing and seeking mode.

2.1. Seeking Mode

The virtual cat walks slowly while it follows the searching mode. In this seeking mode, the following four basic criteria are defined as follows:

➢ Seeking memory pool (SMP);

➢ Seeking range dimensions (SRD);

➢ Counts of dimensions to change (CDC);

➢ Self-position consideration (SPC)

Each cat's size in the search pool is determined by the SMP value, which translates the points needed by the cat. SRD announces the boundary condition requirement for updating the dimensions as well as the Mixture Ratio (MR) of the preferred dimensions. The CDC is a key element of the searching mode that reports how many dimensions need to be modified. The SPC value plays a crucial role in determining whether the current candidate of the virtual cat can be deemed as a suitable choice within the search memory. The flow chart interpreting the CSO algorithm is shown in Figure 2. The sequential procedure involved in the Seeking approach is presented below.

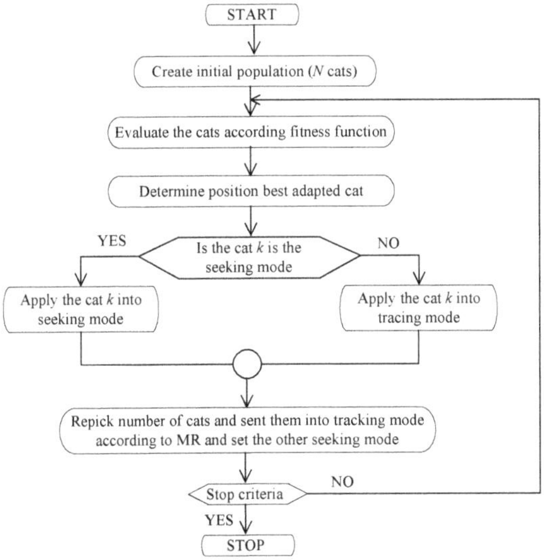

Figure 2. The flow chart of the CSO algorithm.

Step 1: Let us consider "*j*" copy the current position of Cat *k*
where j = SMP if SPC = TRUE
Otherwise,

$$J = \text{SMP} - 1$$

Step 2:

$$\text{XJD}_{new} = (1 + r_1 \cdot \text{SRD}) \cdot \text{XJD}_{old} \tag{1}$$

where r_1 is the random number from range (0, 1)

Step 3: Determines fresh value of fitness for each cat.

Step 4: Find out the selection capability from Equation (2) whenever the fitness function value changes. If not, set each solution's selection probability to 1.

Step 5:

$$P_i = \frac{|FS_i - FS_b|}{FSmin_{max}} \tag{2}$$

where P_i is the probability of selection, FS is the fitness function value.

If the objective function is to minimize, then $FS_b = FS_{max}$. Otherwise,

$$FS_b = FS_{min}$$

where FS_{max} is the largest fitness offered by the candidate, and FS_{max} is the smallest fitness offered by the candidate.

Step 6: Arrange the candidate's cat by P_i and anyone can select to retain the co-ordinate cat k.

2.2. Tracing Mode

The virtual cats can imitate the movement of other cats to track their prey during the tracing mode. The mechanism of the tracing process can be described in Equations (3) and (4).

Step 1: Equation (3) represents the velocity of each cat updated in every dimension.

$$V_{k,d} = V_{k,d} + r_1 \cdot C_1(X_{best,d} - X_{k,d}) \quad (3)$$

where $d = 1, 2, \ldots, M$, r_1 is the random number from range (0, 1), C_1 is the constant number, and X_{best} is the best adapted cat in population.

Step 2: The velocity of each cat is examined to determine if it falls within the permissible limits. In cases where the velocity exceeds the limit, it is adjusted to the maximum allowable velocity value.

Step 3: Determine the position of every cat by adding the current velocity of each cat.

$$X'_{k,d\ new} = X_{k,d\ old} + V_{k,d}/\Delta t \quad (4)$$

where $X_{k,d\ old}$ is the old coordinate of cat k, $X_{k,d\ new}$ is the updated coordinate of cat, and $\Delta t = 1$ is the time step.

3. Application of GWO-Based MPPT Controller for Solar PV String under Shading Conditions

This section presents a mathematical model that describes the social hierarchy, tracking, encircling and attacking behaviour of predators toward their prey. The model aims to capture the dynamics and interactions between predators as they coordinate their actions to maximize their chances of capturing the prey [35].

3.1. Encircling the Prey

The following Equations (5) and (6) represent the behavioural model of wolves while encircling the prey [31].

$$\vec{D} = \left| \vec{C} \cdot \vec{X}_P(t) - X(t) \right| \quad (5)$$

$$\vec{X}(t+1) = \vec{X}_P(t) - \vec{A} \cdot \vec{D} \quad (6)$$

where t is iteration number; D, A and C are the GWO parameters; X_p is the position of the prey.

To determine A and C, the following equations are used.

$$\vec{A} = 2 \cdot \vec{a} \cdot r_1 - \vec{a} \quad (7)$$

$$\vec{C} = 2 \cdot r_2 \quad (8)$$

where r_1 and r_2 are the random numbers from range (0, 1).

The control parameter $\vec{a}$ changes to the grey wolf optimization process, causing omega wolves to either approach or be free from the alpha beta, and delta-dominating wolves. While the iterations are being performed, the operating parameter $\vec{a}$ is set to decrease linearly from a value of 2 to 0.

$$\vec{a} = 2 \cdot \{1 - t/T\} \quad (9)$$

In the above equation, T indicates maximum iteration numbers.

3.2. Hunting Process (Updating of Wolf Position)

The alpha group of wolves, the leaders, usually directs the hunting process. They are followed by beta and delta group wolves, who occasionally engage in hunting. The surrounding wolves in the pack worry the delta and omega wolves. As a result, the alpha wolf is seen as the best option with solid information of where the prey is. The GWO algorithm indicates a superior balance between the exploitation and exploration phases because of these qualities. Moreover, by choosing while C considers the environment, GWO eliminates the local stuck.

Hence, the different wolf groups' positions can be obtained from Equations (10) to (12) as follows:

$$\vec{D}_{\alpha} = \left| \vec{C}_1 \cdot \vec{X}_{\alpha}(t) - \vec{X(t)} \right|, \vec{X}_1 = \vec{X}_{\alpha}(t) - \vec{A}_1 \cdot \vec{D}_{\alpha} \tag{10}$$

$$\vec{D}_{\beta} = \left| \vec{C}_1 \cdot \vec{X}_{\beta}(t) - \vec{X(t)} \right|, \vec{X}_2 = \vec{X}_{\beta}(t) - \vec{A}_2 \cdot \vec{D}_{\beta} \tag{11}$$

$$\vec{D}_{\delta} = \left| \vec{C}_1 \cdot \vec{X}_{\delta}(t) - \vec{X(t)} \right|, \vec{X}_3 = \vec{X}_{\delta}(t) - \vec{A}_3 \cdot \vec{D}_{\delta} \tag{12}$$

Overall, the final grey wolves positioned in the iteration are obtained from Equation (13).

$$\vec{X}_k(t+1) = \frac{\vec{X}_1 + \vec{X}_2 + \vec{X}_3}{3} \tag{13}$$

3.3. Attacking the Prey (Exploitation of Search Process) $\vec{a}$

Grey wolves attack their prey to end their hunt. This method of attack is described mathematically by lowering the value of a, which then modifies the value of $\vec{a}$. A is reduced from 2 to 0 and initially has a random value in the range $[-a, a]$.

3.4. Searching the Prey

Figure 3 represents the flow chart of the GWO algorithm step-by-step process for the present research problem.

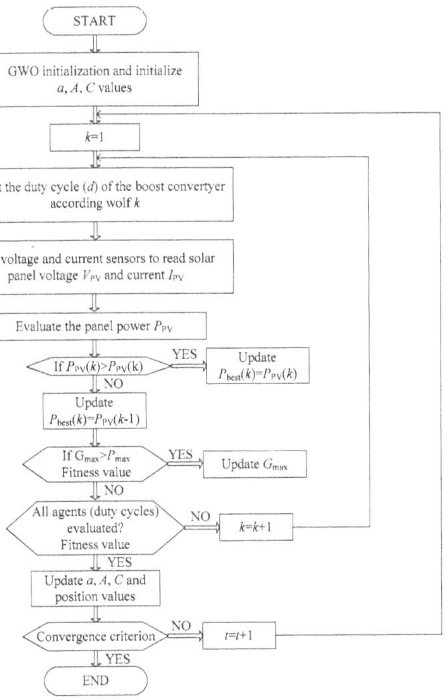

Figure 3. GWO algorithm flow chart.

4. Application of ChOA-Based MPPT Controller for Solar PV Module under Shading Conditions

Exploration and exploitation are the two phases in which the hunting process is completed. Equations (14) and (15) are used for the design process of driving prey and chasing prey as presented below.

$$D = \left| C \cdot X_p - m \cdot X_{Chimp}(t) \right| \quad (14)$$

$$X_{Chimp}(t+1) = X_p(t) - a \cdot D \quad (15)$$

where m is chaotic vector, X_{Chimp} (t) is the chimp position at t-th iteration.

The values of a and C coefficient in each iteration Equations (16)–(18) are used.

$$a = 2 \cdot f \cdot r_1 - f \quad (16)$$

$$C = 2 \cdot r_2 \quad (17)$$

where f is the dynamic vector [36].

In the exploration and exploitation stages of the iteration process, the value of f decreases nonlinearly from 2.5 to 0. Random vectors r_1 and r_2 vary within [0, 1] while regulation vectors are indicated by a and C. The distance between them is denoted as D. The chaotic vector (m) is calculated based on different maps to simulate the sexual motivation of chimps during the search process. In subsection C, a detailed report on the chaotic vector (m) is provided. Unlike conventional swarm intelligent optimization algorithms, where all particles (agents) behave similarly across the entire search space, the mathematical model incorporates various strategies for independent groups of chimps to revise their hunting approach. The behaviour of independent chimp groups in the ChOA is updated using a continuous function. Through the use of constant parameters, the independent process is iteratively refined. All these parameters are explicitly defined throughout the entire process, ensuring that f is decreased in successive iterations of the optimization process. Each self-determining group employs its model to explore both the global and local search spaces. The dynamic coefficient vector with different groups is delineated in Table 1. To enhance the search capability and precisely capture the search nature of the chimp's independent groups, a dynamic coefficient of f is proposed and illustrated in Figure 4. Different curves and slopes are chosen for these dynamic coefficients (f) to facilitate the adjustment of the behaviour of the Chimp Optimization Algorithm's independent groups.

Table 1. Dynamic coefficient vector f.

Type	Barrier	Attacker	Driver	Chaser
$\vec{f}$	$1.95 - 2 \cdot \frac{t^{1/3}}{T^{1/4}}$	$1.95 - 2 \cdot \frac{t^{1/4}}{T^{1/3}}$	$\left(-3 \cdot \frac{t^3}{T^3}\right) + 1.5$	$\left(-2 \cdot \frac{t^3}{T^3}\right) + 1.5$

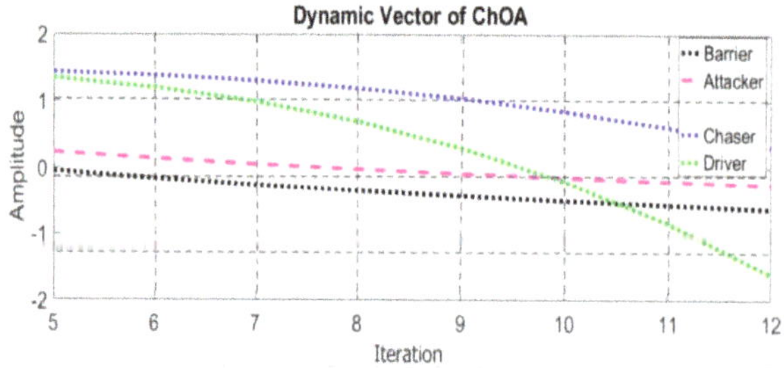

Figure 4. Variation of dynamic coefficients vector with different groups of ChOA.

In the above table, T is the maximum number of iterations, and t is the current iteration.

4.1. Exploration Stage

When chimps are attacking, they are likely to search the prey's local position by the use of diverse stages such as chasing, driving and blocking to enclose the prey. In general, attackers are managed by the chase process. In the initial iteration, since the optimal position of the prey is unknown, the attacker's location is assumed to be the actual location of the prey. Then, the subsequent barrier, driver and chaser locations (denoted by C) are determined relative to the attacker's position. The best position achieved is stored. However, the other chimps need to refine their positions based on the location of the most successful chimp. Mathematical equations are employed to model the process of encircling the prey by the chimps (19), (20) and (21), respectively.

$$\begin{gathered} D_{Attacker} = |C_1 \cdot X_{Barrier} - m_1 \cdot X| \\ D_{Barrier} = |C_2 \cdot X_{Barrier} - m_2 \cdot X| \\ D_{Chacer} = |C_3 \cdot X_{Chaser} - m_3 \cdot X|;\ D_{Driver} = |C_4 \cdot X_{Driver} - m_4 \cdot X| \end{gathered} \tag{18}$$

$$\begin{gathered} X_1 = X_1 X_{Attacker} - a_1 \cdot (D_{Attacker}),\quad X_2 = X_{Barrier} - a_2 \cdot (D_{Barrier}) \\ X_3 = X_{Chaser} - a_3 \cdot (D_{Chaser}), X_4 = X_{Driver} - a_4 \cdot (D_{Driver}) \end{gathered} \tag{19}$$

$$X(t+1) = \frac{X_1 + X_2 + X_3 + X_4}{4} \tag{20}$$

where $X_{Attacher}$ is the best search agent, $X_{Barrier}$ is the second-best search agent, X_{Chaser} is the third-best search agent, and X_{Driver} is the fourth-best search agent.

4.2. Attacking Approach (Exploitation Phase)

The mathematical model represents the attack process, where the value of f decreases linearly from 2.5 to 0. This linear reduction enhances the effectiveness of the attack and the exploitation of the prey's location. To enhance the attacking and exploitation of the prey location, the mathematical model employs a linear reduction of the value of f from 2.5 to 0, representing the process of attack. Further, the scope of the vector is found to decrease like that of function f. The described vector is random within the range of $-2f$ and $2f$. All the time, a random value is selected for the vectors stretch-out within the range of $[-1$ and $1]$, and the value determines the next placement of the chimp, which can be at any location within the available position and conditions of the prey. Although there is mention of projected blocking, chasing and driving mechanisms that highlight the searching limits, there is a concern that the algorithm referred ChOA may get trapped in nearby minima. To address this, an additional operator is deemed necessary to enhance the searching ability during the exploitation period. In this approach, all the chimps embark on hunting the prey, as depicted in Figure 5. The value of "a" coefficient is mathematically modelled to align with the hunting behaviour. Consequently, the disparity among the chimps compels them to wander in pursuit of the prey, ultimately driving them to converge at the location of the prey.

Figure 5. Influence of a on refining the chimp's location mechanism.

4.3. Chaotic Maps (Sexual Motivation)

Introducing chaotic behaviour in the optimization algorithm utilizing chimps helps address two key challenges in solving high-dimensional engineering application problems. Firstly, it helps alleviate the issue of getting trapped in local optima during the final stages of the optimization process. Local optima are suboptimal solutions that can hinder the algorithm from reaching the global optimum. By incorporating chaotic behaviour, the algorithm introduces randomness and exploration, allowing the chimps to venture into unexplored regions of the search space. This exploration enhances the chances of escaping local optima and discovering better globally optimal solutions. The mathematical model for this approach is presented in Equation (22).

$$X_{Chimp}(t+1) = \begin{cases} X_p(t) - a \cdot D, & \text{if } \mu < 0.5 \\ \text{Chaotic\_ value} & \text{if } \mu > 0.5 \end{cases} \tag{21}$$

where μ is random number within the range of [0, 1] [36].

The flow chart of the ChOA is delineated in Figure 6.

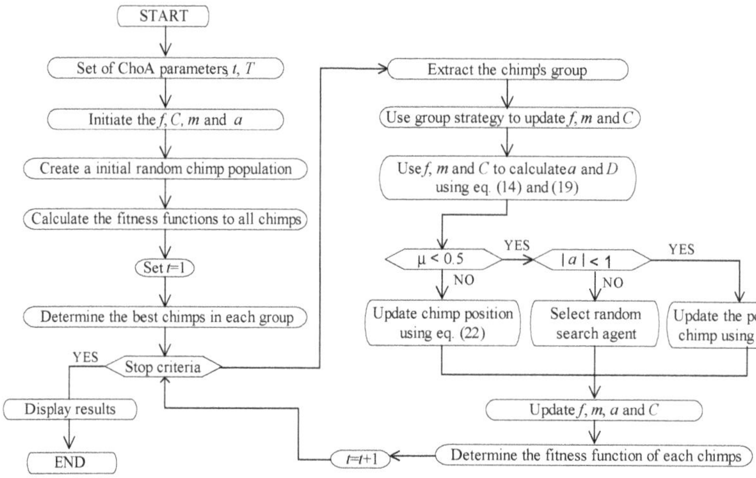

Figure 6. ChOA flow chart.

5. Case Studies

Four cases are considered with four different shading patterns, as shown in Figure 7, and are proposed with different irradiance as mentioned in Table 2 to explore the PV characteristics under shading conditions.

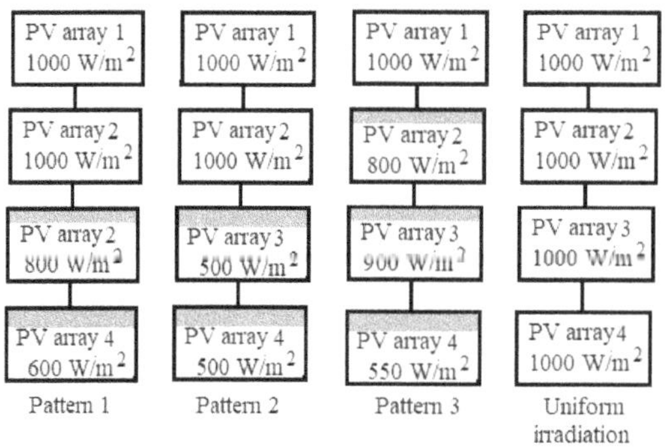

Figure 7. Different irradiation patterns of PV system under shading conditions.

Table 2. PV module with different partial shading patterns.

Case No	Arrangement of PV Modules	No. of PV Modules	Irradiance Level	Temperature
1	Pattern-1	4	1000 W/m^2	25 °C
			1000 W/m^2	
			800 W/m^2	
			600 W/m^2	
2	Pattern-2	4	1000 W/m^2	25 °C
			1000 W/m^2	
			500 W/m^2	
			500 W/m^2	
3	Pattern-3	4	1000 W/m^2	25 °C
			800 W/m^2	
			900 W/m^2	
			550 W/m^2	
4	Pattern-4	4	1000 W/m^2	25 °C
			1000 W/m^2	
			1000 W/m^2	
			1000 W/m^2	

The performance of the PV string is evaluated under partial shading conditions for three cases considering three different shading patterns as shown in Table 2, using various optimization-based MPPT techniques. Figure 8 represents the simulation circuit of four series-connected PV modules by varying the irradiation levels, temperature at 25 °C with the PV module and converter circuit details listed in Table 3.

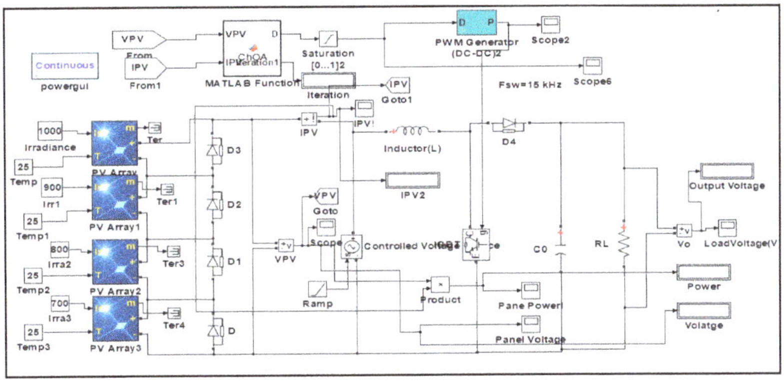

Figure 8. Simulation circuit of four series-connected KC200GT PV module.

All characteristic parameters of the optimization algorithm are carefully selected to provide good convergence to the solved optimization task. The parameters used during simulations for CSO, GWO and ChoA algorithms are listed in Table 4.

Table 3. PV module and converter circuit details.

Parameter	Value
Total cells/modules	54
V_{oc} [V]	33
I_{sc} [A]	8
V_{mpp} [V]	26
I_{mpp} [A]	8
P_{mpp} [W]	200
Specifications of DC-DC Boost Converter	
Input Inductance (L_1)	330 µF
Switching frequency	25 kHz
Output side capacitance (C_{out})	

Table 4. Tuning parameters for metaheuristic algorithms.

Algorithm	Parameter	Value
CSO	Maximum number of iterations (T)	25
	Number of cats	25
	Time step (Δt)	1
	SRD	0.3
	Constant number (C1)	2
	SMP	5
	CDC	1
	MR	0.2
ChoA	Number of chimps	25
	Maximum number of iterations (T)	25
GWO	Number of. agents (wolf)	25
	Maximum number of iterations (T)	25
	Control parameter (a)	2 to 0

In this case, the details of irradiance levels received on the modules in Pattern-1 and their P-V and P-I plots are discussed and presented in Figure 9. The voltage and current at GMPP under this shading condition are 544.12 W, 114.008 V and 4.7771 A, respectively.

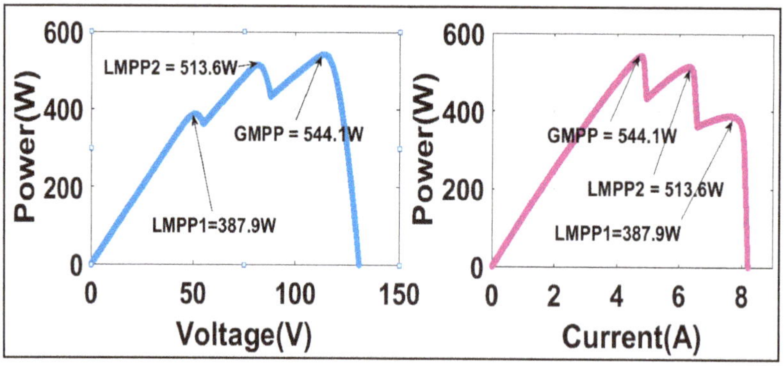

Figure 9. P-V and P-I variations under the first shading pattern on PV string.

According to the simulation findings for Pattern 1 displayed in Figures 10–12, the ChoA shows reduced oscillation during the MPP search process. Specifically, when shading occurs, the PV module power output rapidly converges to the MPP with very minimal fluctuations. However, based on the simulation results depicted in Figure 10 (Simulation results using CSO for Pattern-1), Figure 11 (Simulation results using GWO for Pattern-1) and Figure 12 (Simulation results using ChOA for pattern 1), the average convergence time of the GWO algorithm is relatively high. Even though the suggested chimp algorithm converges sufficiently quickly and catches only after a short period, the ability of ChOA, GWO and CSO algorithms to chase the GMPP under various shading situations is examined. It is observed from the simulation result that the output power obtained from the PV array using CSO is 521.41 W, using GWO is 527.44 W, and 531.198 W when ChoA is used.

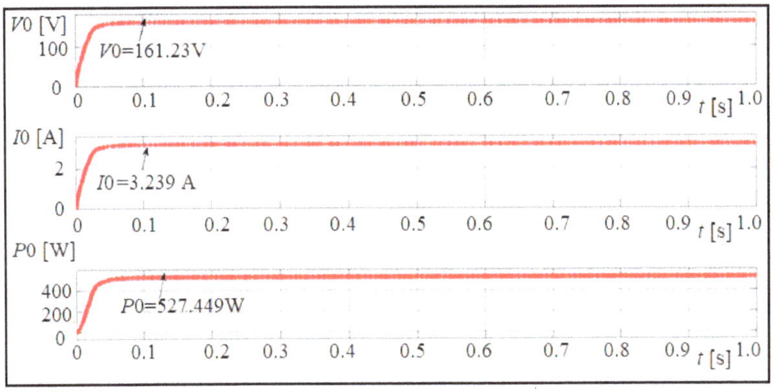

Figure 10. Simulation result using CSO for the Patern-1.

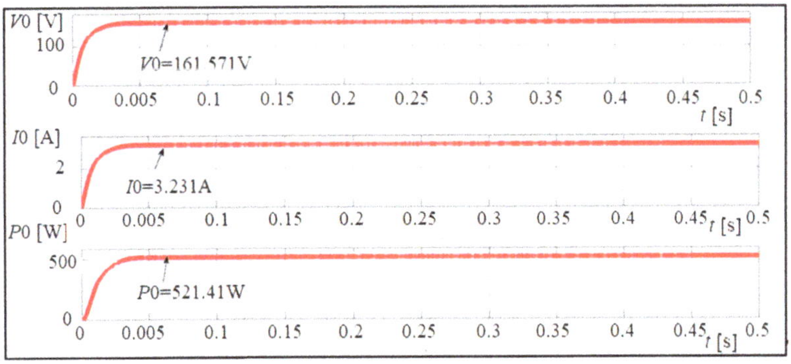

Figure 11. Simulation results using GWO for the Pattern-1.

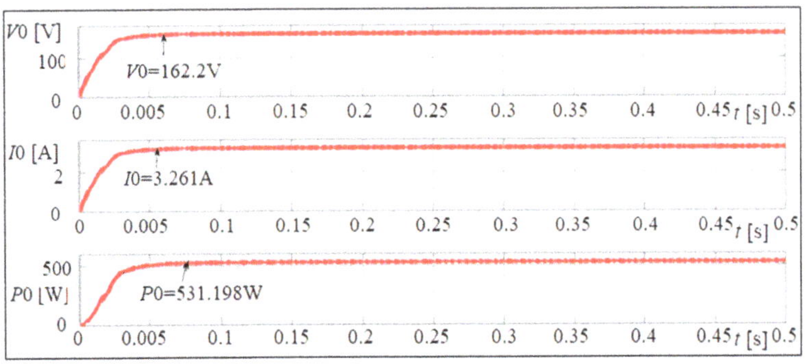

Figure 12. Simulation results using ChoA for the Pattern-1.

Figure 13 presents a performance comparison for the analysed metaheuristic algorithms.

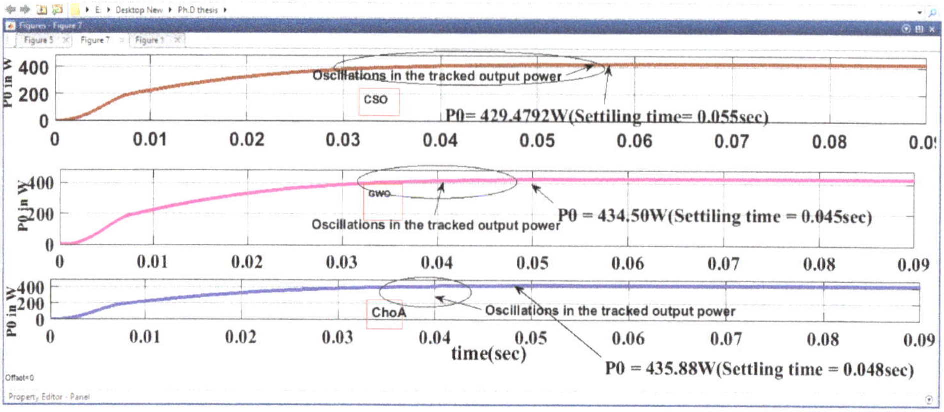

Figure 13. Comparison of settling time for the CSO, GWO and ChoA algorithms.

The above Figure depicts the reduced oscillations and convergence time for the three algorithms, and it clearly shows that the proposed GWO have less oscillation and also less convergence time compared with the other algorithms, while the ChoA is slightly slower than the GWO algorithm.

ChOA, GWO and CSO algorithms are simulated in MATLAB/SIMULINK to validate their expected performance under various partial shade patterns (G_1 to G_6). Table 5 lists the outcomes of the statistical simulation. The variable power output with irradiation with different MPPT techniques under other shading patterns is presented in Figure 14. From Figure 14, G_1 shading pattern will produce the solar photovoltaic system's highest power. Due to its speed and confidence, the simulation results of the various shading situations showed that ChOA performed better than other optimization strategies under partial shading settings. The summarization of statistical simulation results like power, voltage and current of the PV module under other partial shading conditions is presented in Table 5.

Table 5. Summarization of statistical simulation results like power, voltage, and current of PV module.

Other Different Shading Patterns	Parameter	ChOA	GWO	CSO
G_1 = [1000, 900, 800, 700]	Maximum power @GMPP(W)	525.13	525.13	525.13
	Output voltage(V)	115.23	115.47	117.48
	Output current (A)	4.54	4.52	4.412
	Output power (W)	523.14	521.92	518.357
	Conversion efficiency (%)	99.62	99.38	98.71
G_2 = [900, 550, 100, 600]	Maximum power @GMPP(W)	336.61	336.61	336.61
	Output voltage(V)	84.57	83.4V	82.46
	Output current (A)	3.9489	3.86A	3.849
	Output power (W)	333.95	321.92	317.38
	Conversion efficiency (%)	99.21	95.63	94.28

Table 5. *Cont.*

Other Different Shading Patterns	Parameter	ChOA	GWO	CSO
G_3 = [750, 850, 600, 800]	Maximum power @GMPP(W)	340.06	340.06	340.06
	Output voltage(V)	53.67	53.21	82.46
	Output current (A)	6.23	6.21	3.95
	Output power (W)	334.36	329.90	325.71
	Conversion efficiency (%)	98.32	97.01	95.78
G_4 = [600, 800, 400, 200]	Maximum power @GMPP(W)	258.29	258.29	258.29
	Output voltage(V)	56.41	55.41	54.32
	Output current (A)	4.32	4.21	4.123
	Output power (W)	243.69	233.27	223.96
	Conversion efficiency (%)	94.34	90.31	86.70
G_5 = [600, 200, 800, 250]	Maximum power @GMPP(W)	191.22	191.22	191.21
	Output voltage(V)	66.31	65.31	66.21
	Output current (A)	2.873	2.853	2.67
	Output power (W)	188.51	186.13	176.78
	Conversion efficiency (%)	98.58	97.33	92.45
G_6 = [400, 600, 800, 100]	Maximum power @GMPP(W)	232.52	232.52	232.52
	Output voltage(V)	87.54	86.46	85.44
	Output current (A)	2.621	2.61	2.62
	Output power (W)	229.44	225.66	223.86
	Conversion efficiency (%)	98.67	97.04	96.27

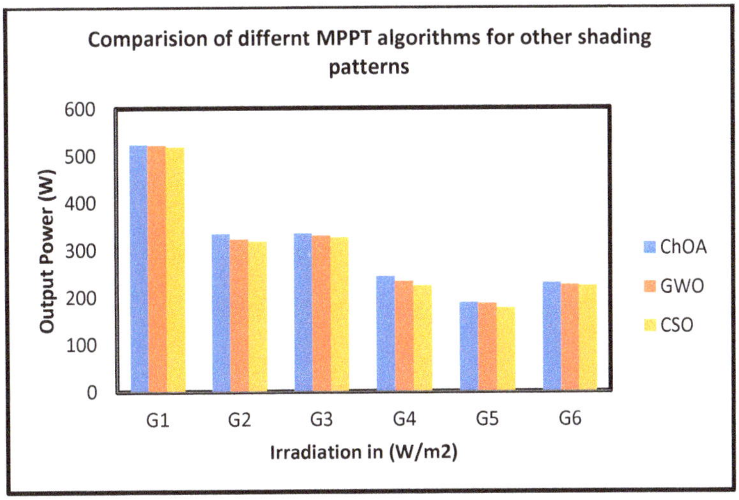

Figure 14. Comparison of output power with different MPPT Algorithms under other shading patterns.

6. Conclusions

This research work presents a comprehensive study that analyses the challenges faced by three heuristic-optimization-based algorithms in the context of GMPPT. The focus is specifically on the search capabilities of these techniques, both in terms of exploitation and exploration. This article introduces a novel method called ChOA GMPPT to address these challenges. This method improves GMPPT performance by utilizing different deterministic starting points for exploitation and exploration. Simulation tests demonstrate that the proposed ChOA-based GMPPT achieves higher tracking accuracy and conversion efficiency compared to other tested methods. Additionally, it reduces computational complexity and is easy to implement. By providing a discussion and analysis of various heuristic optimization methods in the literature, along with the newly suggested ChoA-based GMPPT approach, this current work enables researchers to select the most suitable method based on their preferences and priorities.

In future research, the PSO, CSO, TLBO, GWO and ChOA variants are used to optimize the PV system performance under changing weather conditions. There is a possibility for the usage of recent algorithms as well as hybrid algorithms since a comparative study can be done.

Author Contributions: Conceptualization, T.N. and R.D.; methodology, T.N.; software, T.N. and R.D.; validation, T.N.; investigation resources, T.N. and R.D.; resources, T.N. and Ł.K.; writing—original draft preparation, T.N. and R.D.; writing—review and editing, Ł.K. All authors have read and agreed to the published version of the manuscript.

Funding: This research was supported by the Poznan University of Technology, grant number [0212/SBAD/0594].

Data Availability Statement: Not applicable.

Conflicts of Interest: The authors declare no conflict of interest.

Nomenclature

PV System	Photovoltaic system
MPPT	Maximum Power Point Tracking
GMPP	Global Maximum Power Point
PSC	Partial Shading Condition
CSO	Cat Swarm Optimization
GWO	Grey Wolf Optimization
ChoA	Chimp Optimization algorithm
PSO	Particle Swarm Optimization
TLBO	Teaching Learning Based Optimization
ACO	Ant Colony Optimization
P&O	Perturb and Observe
INC	Incremental Conductance
GA	Genetic Algorithm
DEA	Differential Evaluation Algorithm
FLC	Fuzzy Logic Controller
FA	Firefly Algorithm
SMP	Seeking Memory Pool
MBA	Mine Blast Algorithm

References

1. Available online: https://cea.nic.in/reports/monthly/executive%20summary/2020/exe_summary-02 (accessed on 1 March 2021).
2. Novas, N.; Garcia, R.M.; Camacho, J.M.; Alcayde, A. Advances in Solar Energy towards Efficient and Sustainable Energy. *Sustainability* **2021**, *13*, 6295. [CrossRef]
3. Kulaksız, A.A.; Akkaya, R. A genetic algorithm optimized ANN-based MPPT algorithm for a standalone PV system with induction motor drive. *Solar Energy* **2012**, *86*, 2366–2375. [CrossRef]
4. Tajuddin, M.F.N.; Arif, M.S.; Ayob, S.M.; Salam, Z. Perturbative methods for maximum power point track-ing (MPPT) of photovoltaic (PV) systems: A review. *Int. J. Energy Res.* **2015**, *39*, 1153–1178. [CrossRef]

5. Thangamani, K.; Manickam, M.L.; Chellaiah, C. An experimental study on photovoltaic module with opti-mum power point tracking method. *Int. Trans. Electr. Energy Syst.* **2020**, *30*, e12175. [CrossRef]
6. Ramli, M.A.M.; Twaha, S.; Ishaque, K.; Al-Turki, Y.A. A review on maximum power point tracking for pho-tovoltaic systems with and without shading conditions. *Renew. Sustain. Energy Rev.* **2017**, *67*, 144–159. [CrossRef]
7. Liu, Y.H.; Huang, S.C.; Liang, W.C. A particle swarm optimization –Based maximum power point tracking algorithm for PV systems operating under partially shaded conditions. *IEEE Trans. Energy Convers.* **2012**, *27*, 1027–1035. [CrossRef]
8. Kashiflshaquem, H.; Salam, Z. A direct control based maximum power point Tracking method for photovoltaic system under partial shading conditions using Particle Swarm optimization Algorithm. *Appl. Energy* **2012**, *99*, 414–422. [CrossRef]
9. Renaudineau, H.; Donatantonio, F.; Fontchastagner, J.; Petrone, G. A PSO-Based Global MPPT Technique for Distributed PV Power Generation. *IEEE Trans. Ind. Electron.* **2015**, *62*, 1047–1058. [CrossRef]
10. Phimmasone, V.; Kondo, Y.; Kamejima, T.; Miyatake, M. Evaluation of extracted energy from PV with PSO-based MPPT against various types of solar irradiation changes. In Proceedings of the 2010 International on Electrical Machines and Systems, Incheon, Republic of Korea, 10–13 October 2010; pp. 487–492.
11. Chao, K.H.; Lin, Y.S.; Lai, U.D. Improved particle swarm optimization for maximum power point tracking in photovoltaic module arrays. *Appl. Energy* **2015**, *158*, 609–618. [CrossRef]
12. Roy Chowdhury, S.; Saha, H. Maximum power point tracking of partially shaded solar Photovoltaic arrays. *Sol. Energy Mater. Sol. Cells* **2010**, *94*, 1441–1447. [CrossRef]
13. Tobon, A.; Restrepo, J.; Ceballos, J.P.V.; Ibeas, A. Maximum Power Point Tracking of Photovoltaic Panels by Using Improved Pattern Search Methods. *Energies* **2017**, *10*, 1316. [CrossRef]
14. Babu, T.S.; Sangeetha, K.; Rajasekar, N. Voltage band based improved particle. Swarm optimization technique for maximum power point tracking. in solar photovoltaic System. *J. Renew. Sustain. Energy* **2016**, *8*, 013106. [CrossRef]
15. Gavhane, P.S.; Krishnamurthy, S.; Dixit, R.; Ram, J.P.; Rajasekar, N. EL-PSO based MPPT for solar PV. under partial shaded condition. *Energy Procedia* **2017**, *117*, 1047–1053. [CrossRef]
16. Babu, T.S.; Rajasekar, N.; Sangeetha, K. Modified Particle Swarm optimization technique based Maximum Power Point Tracking for uniform and under partial shading condition. *Appl. Soft Comput.* **2015**, *34*, 13–624. [CrossRef]
17. Husain, M.A.; Tariq, A.; Hameed, S.; Arif, M.S.B.; Jain, A. Comparative assessment of maximum power point tracking procedures for photovoltaic systems. *Green Energy Environ.* **2017**, *2*, 5–17. [CrossRef]
18. Kalaiarasi, N.; Subranshu, D.; Sanjeevikumar, P.; Paramasivam, S. Maximum Power Point Tracking Implementation by DSPACE Controller Integrated through Z-Source Inverter Using PSO Technique for Photovoltaic Applications. *Appl. Sci.* **2018**, *8*, 145. [CrossRef]
19. Dileep, G.; Singh, S.N. Application of soft computing techniques for maximum power point tracking of SPV system. *Sol. Energy* **2017**, *141*, 182–202. [CrossRef]
20. Ahmed, J.; Salam, Z. A critical evaluation on maximum power point tracking methods for partial shading in PV systems. *Renew. Sustain. Energy Rev.* **2015**, *47*, 933–953. [CrossRef]
21. Wen, H.; Chu, G.; Hu, Y.; Jiang, L. A novel power-increment based GMPPT algorithm for PV arrays under partial shading conditions. *Sol. Energy* **2018**, *169*, 353–361. [CrossRef]
22. Rezk, H.; Fathy, A.; Abdelaziz, Y. A Comparision of different global MPPT techniques based on meta-heuristic algorithms for photovoltaic system subjected to partial Shading conditions. *Renew. Sustain. Energy Rev.* **2017**, *74*, 377–386. [CrossRef]
23. Abokhalil, A. Maximum Power Point Tracking for a PV System using Tuned Support Vector Regression by Particle Swarm Optimization. *J. Eng. Res.* **2020**, *8*, 4. [CrossRef]
24. Gangwar, P.; Singh, R.; Tripathi, R.P.; Singh, A.K. Effective solar power harnessing using a few novel solar tree designs and their performance assessment. *Energy Sources Part A Recovery Util. Environ. Eff.* **2019**, *41*, 1828–1837. [CrossRef]
25. Mohanty, B.; Tripathy, S.A. Teaching learning based optimization technique for optimal location and size of DG in distribution network. *J. Electr. Syst. Inf. Technol.* **2016**, *3*, 33–44. [CrossRef]
26. Rezk, H.; Fathy, A. Simulation of global MPPT based on Teaching-Learning based optimization technique for partially shaded PV system. *Electr. Eng.* **2017**, *99*, 847–859. [CrossRef]
27. Fathy, A.; Rezk, H. A novel methodology for simulating maximum power point trackers using Mine blast optimization and teaching learning based optimization algorithm for partially shaded photovoltaic system. *J. Renew. Sustain. Energy* **2016**, *8*, 023503. [CrossRef]
28. Crepinsek, M.; Liu, S.; Mernik, L. A note on teaching learning based optimization algorithm. *Inf. Sci.* **2012**, *212*, 79–93. [CrossRef]
29. Sundareswaran, K.; Peddapati, S.; Palani, S. MPPT of PV Systems Under Partial Shaded Conditions through a Colony of Flashing Fireflies. *IEEE Trans. Energy Convers.* **2014**, *29*, 463–472. [CrossRef]
30. Farzaneh, J.; Keypour, R.; Khanesar, M.A. A New Maximum Power Point Tracking Based on Modified Firefly Algorithm for PV System Under Partial Shading Conditions. *Technol. Econ. Smart Grids Sustain. Energy* **2018**, *3*, 9. [CrossRef]
31. Mohanthy, S.; Subudhi, B.; Ray, P.K. A Grey wolf Assisted Perturb and Observe MPPT Algorithm for a PV system. *IEEE Trans. Energy Conversat.* **2017**, *32*, 340–347. [CrossRef]
32. Eltamaly, A.M.; Farh, H.M.H. Dynamic global maximum power point tracking of the PV systems under variant partial shading using hybrid GWO-FLC. *Sol. Energy* **2019**, *177*, 306–316. [CrossRef]

33. Mohanthy, S.; Subudhi, B.; Ray, P.K. A new MPPT design using Grey Wolf optimization technique for photovoltaic system under partial shading conditions. *IEEE Trans. Sustain. Energy* **2016**, *7*, 181–188. [CrossRef]
34. Songyang, L.; Haipeng, Y.; Miao, W. Cat swarm optimization algorithm based on the information interaction of subgroup and the top-N learning strategy. *J. Intell. Syst.* **2022**, *31*, 489–500. [CrossRef]
35. Knypiński, Ł. Constrained optimization of line-start PM motor based on the gray wolf optimizer. *Eksploat. Niezawodn. Maint. Reliab.* **2021**, *23*, 1–10. [CrossRef]
36. Sadeghi, F.; Larijani, A.; Rostami, O.; Martín, D.; Hajirahimi, P. A Novel Multi-Objective Binary Chimp Optimization Algorithm for Optimal Feature Selection: Application of Deep-Learning-Based Approaches for SAR Image Classification. *Sensors* **2023**, *23*, 1180. [CrossRef] [PubMed]

algorithms

Article

Improved Load Frequency Control in Power Systems Hosting Wind Turbines by an Augmented Fractional Order PID Controller Optimized by the Powerful Owl Search Algorithm

Farhad Amiri [1], Mohsen Eskandari [2,*] and Mohammad Hassan Moradi [1,*]

[1] Electrical Engineering Department, Faculty of Engineering, Bu-Ali Sina University, Hamedan 6516738695, Iran; f.amiri94@basu.ac.ir

[2] The School of Electrical Engineering and Telecommunications, University of New South Wales, Sydney, NSW 2052, Australia

* Correspondence: m.eskandari@unsw.edu.au (M.E.); mhmoradi@basu.ac.ir (M.H.M.)

Abstract: The penetration of intermittent wind turbines in power systems imposes challenges to frequency stability. In this light, a new control method is presented in this paper by proposing a modified fractional order proportional integral derivative (FOPID) controller. This method focuses on the coordinated control of the load-frequency control (LFC) and superconducting magnetic energy storage (SMES) using a cascaded FOPD–FOPID controller. To improve the performance of the FOPD–FOPID controller, the developed owl search algorithm (DOSA) is used to optimize its parameters. The proposed control method is compared with several other methods, including LFC and SMES based on the robust controller, LFC and SMES based on the Moth swarm algorithm (MSA)–PID controller, LFC based on the MSA–PID controller with SMES, and LFC based on the MSA–PID controller without SMES in four scenarios. The results demonstrate the superior performance of the proposed method compared to the other mentioned methods. The proposed method is robust against load disturbances, disturbances caused by wind turbines, and system parameter uncertainties. The method suggested is characterized by its resilience in addressing the challenges posed by load disturbances, disruptions arising from wind turbines, and uncertainties surrounding system parameters.

Keywords: frequency control; FOPD–FOPID controller; wind turbines; developed owl search algorithm

1. Introduction

The use of wind turbines in power systems is growing due to the increasing demand for sustainable and environmentally friendly electrical energy [1–3]. Wind turbines have several advantages as a stable source of electricity in power systems, such as: (1) abundant energy source: for electricity generation, wind turbines require an abundant wind energy source that is usually available day and night; (2) reduced environmental pollution: using wind turbines as a clean energy source reduces environmental pollution; (3) cost reduction in electricity production: the cost of electricity production with wind turbines is lower than that of fossil fuels. Despite these advantages, wind turbines in power systems have some drawbacks, including the complexity of load-frequency control (LFC) [4–6]. Wind turbines rely on wind, which is naturally fluctuating and can complicate LFC in power systems [7]. In power systems, a balance is maintained between generation and consumption [8]. Should any disturbance arise, impeding the smooth functioning of this intricate network, the primary control loop immediately intervenes to restrict any deviations in frequency [9]. Yet it is the secondary control loop, known as the LFC system, which assumes responsibility for restoring frequency to their nominal levels [10]. In an earnest endeavor to enhance frequency stability within the power system, various controllers have been proposed for LFC systems [11–45].

The classic proportional integral derivative (PID) controller is still one of the most popular and widely used controllers in the power industry, which is widely used in power system LFC due to its simplicity, ease of use, fast performance, and stability [11–25]. In LFC systems related to power systems, a number of PID controllers are used in order to improve frequency stability; among these PID controllers [11,12] are: PID controllers whose parameters are optimized using the ICA [13] (see list of abbreviations at the end of the article), PID controllers whose parameters are optimized using the PSO [14], PID controllers whose parameters are optimized using the EHO [15], PID controllers whose parameters are optimized using the ACO [16], fuzzy PID controllers [17], fuzzy PID controllers whose parameters are optimized using the optimization algorithm based on novel HLUS-TLBO [18], fuzzy PID controllers whose parameters are optimized using the DE algorithm [19], fuzzy PID controllers whose parameters are optimized using the HDE-PS algorithm [20], fuzzy PID controllers whose parameters are optimized using the PSO [21], fuzzy PID controllers whose parameters are optimized using ACO [22], fuzzy PID controllers whose parameters are optimized using the HFA-PS algorithm [23], control fuzzy PID controllers whose parameters are optimized using MBA [24], and fuzzy PID controllers whose parameters are optimized using FA [25]. In [26], the fuzzy PID controller whose coefficients are optimized using GA is used to improve the frequency of the power system. The PID controller does not perform effectively against disturbances in the power system and the uncertainty of the parameters related to the power system. Fuzzy PID controllers are somewhat resistant to power system disturbances, but they do not perform well against the uncertainty related to the power system.

The fractional order PID (FOPID) controller has two degrees of freedom compared to the PID controller, and these two degrees of freedom have advantages such as: (1) more accuracy, (2) better stability, and (3) robust performance in systems with disturbance and parameter uncertainty [27]. Due to these advantages, FOPID controllers have been widely used in LFC systems related to the power system in order to improve the frequency stability; among these FOPID controllers [28,29] are: FOPID controllers whose parameters are optimized using the ICA [30], FOPID controllers whose parameters are optimized using the GBMO [31], FOPID controllers whose parameters are optimized using the SCA [32], FOPID controllers whose parameters are optimized using the PSO [33], and FOPID controllers whose parameters are optimized using the JSO algorithm [34]. The performance of FOPID controllers is favorable against disturbances and uncertainty related to system parameters, but their performance is affected by severe disturbances.

Superconducting magnetic energy storage (SMES) systems store electric energy in their magnetic field and release it as needed, making them a significant contributor to improving frequency stability alongside the LFC system in power systems [35], due to their extended lifespan and high storage capacity. In [36], coordinated control of the LFC system and SMES is discussed using an H2/H∞ robust controller. However, designing such a robust controller necessitates an accurate model of the power system, and since some components may be ignored in the model, the controller's optimal performance may be compromised in real-world systems. Another approach to coordinated control is presented in [37], where a neuro-fuzzy controller is employed. Although the neuro-fuzzy controller demonstrates good results in addressing uncertainties related to the power system parameters, it is not robust against disturbances caused by the power system load.

In [38], the coordinated control of the LFC system and SMES using a type-2 fuzzy controller is discussed. However, this controller is also not resistant to severe disturbances in the power system. Similarly, [39] explores the coordinated control of the LFC system and SMES using a PID controller with parameters optimized by the DE algorithm. Unfortunately, this method is also not resistant to disturbances and uncertainties in power system parameters. Reference [40] presents the coordinated control of the LFC system and SMES using a PI controller with parameters optimized using the PSO, taking into account the presence of wind turbines in the power system. However, the performance of this control method is challenging due to the severe disturbances caused by the wind turbine, making the PI controller optimized with the PSO non-resistant to such disturbances. Reference [41] discusses the coordinated control of the LFC system and SMES using a PID controller with parameters optimized by the MSA, considering the presence of wind turbines in the power system. Nevertheless, the performance of this control method is not resistant to load disturbances and disturbances from the wind turbine. In [42], the coordinated control of the LFC system and SMES using a dynamic resistance controller in the presence of wind turbines is discussed. This method exhibits resistance to load disturbances, wind turbine disturbances, and uncertainties related to power system parameters. However, it has certain drawbacks, including: (1) the complexity of the control method, which requires setting numerous parameters and involves heavy calculations, and (2) the necessity for an accurate power system model, as some parameters cannot be accurately modeled, impacting the performance of this control method.

Currently, cascaded controllers are recognized for their superior system control performance compared to single controllers such as PID and FOPID. These cascaded controllers have been employed to enhance frequency stability in power systems [43–46]. In [43], a cascaded FOPI–FOPD controller with optimized parameters using the DSA method is utilized to improve frequency stability in the power system. The FOPI–FOPD controller parameters are set using the ITAE cost function, and the simulation time is set to 10 s. In [44], a PI–TID cascaded controller is introduced, and its parameters are adjusted using the CBO algorithm to enhance frequency stability in the power system. The PI–TID cascaded controller parameters are determined using the *ISE* cost function, and the simulation time is set to 120 s. Furthermore, in [45], a PI–FOPID cascaded controller is investigated for improving frequency stability in the power system. The parameters of the PI–FOPID cascaded controller are adjusted using the GTO technique. The ITAE cost function is employed to set the PI–FOPID cascaded controller parameters, and the simulation time is set to 30 s. In [46], the cascaded FOPDN–FOPIDN controller, whose coefficients are optimized using the CSA, is utilized for automatic generation control of production in the power system. The *ISE* cost function is considered in optimizing the controller coefficients, and the simulation time is set to 100 s.

The main difference between the *ISE* and ITAE cost functions lies in the method of error calculation. Both cost functions are used to evaluate the quality of control, but they differ in how they quantify and integrate the errors over time. In the *ISE* cost function, the squared errors at each time point are integrated over the entire time period. This means that the errors are squared and then summed up. The *ISE* cost function is commonly used for stable control of control systems. On the other hand, the ITAE cost function calculates the absolute errors at each time point, multiplies them by time, and then sums them up. The ITAE cost function captures the process dynamics in the time domain. It emphasizes the errors that occur during the initial response of the system and can be more sensitive to transient behavior. In the context of the paper mentioned, the *ISE* cost function is used to evaluate the frequency stability of the power system in the presence of a wind turbine. The goal is to control the power system effectively. Reference [41] also utilizes the *ISE* cost function to control the frequency of the power system with a wind turbine and to adjust the coefficients of the PID controller. The choice of the *ISE* cost function in this paper allows for comparison with the results obtained using the PID controller in [41].

In this paper, a novel method called the FOPD–FOPID cascaded controller is proposed for the coordinated control of the LFC system and SMES in a power system that includes a wind turbine. The parameters of the FOPD–FOPID cascaded controller are optimized using the developed Owl Search Algorithm (DOSA). The reason for selecting the FOPD–FOPID cascaded controller over other cascaded controllers, such as PI–FOPID, in the power system structure is that the FOPD–FOPID controller, incorporating the FOPD component, provides more accurate and rapid response to frequency changes in the presence of a wind turbine. This characteristic enhances the frequency stability of the power system and ensures robustness against load disturbances, wind turbine disturbances, and uncertainties associated with power system parameters. The DOSA algorithm is employed to optimize the parameters of the cascaded controller within the power system structure. The DOSA algorithm offers several advantages over other meta-heuristic algorithms such as ABC, PSO, MSA, and GTO. These advantages include: (1) robustness in the face of uncertainty related to the objective function, (2) a reduced number of control parameters, and (3) global optimization capability. The paper presents several key innovations, including:

(1) Enhancing the responsiveness of the power system in the presence of a wind turbine using the cascaded FOPD–FOPID controller.
(2) Refining the parameters of the FOPD–FOPID controller through the application of the novel DOSA approach, which has not been previously explored in power system research.
(3) Evaluating and comparing the effectiveness of the proposed algorithm with GTO, MSA, PSO, and ABC algorithms for optimizing the parameters of the FOPD–FOPID controller, employing an objective function based on *ISE*.
(4) Conducting a comprehensive assessment of the performance of the DOSA–FOPD-FOPID controller for improving coordinated control capabilities within both the LFC system and SMES, considering disturbances and uncertain power system variables.

2. The Power System under Scrutiny

In this section, the structure of the power system under scrutiny and the state–space equations of the power system under scrutiny are discussed.

2.1. The Structure of the Power System under Scrutiny

Figure 1a shows the power system configuration incorporating a wind turbine, as described in references [5,42]. Figure 1b shows a single-line diagram of the studied power system [40–42]. This system comprises a hydro power plant, non-reheat power plants, reheat power plants, multiple wind turbines, SMES, and a load [40–42]. The total power generated by the system is 38,000 MW, while the peak load amounts to 29,000 MW [40–42]. Figure 2 illustrates the dynamic model of the power system, taking into account the presence of the wind turbine. The model utilized in this context is a reduced-order (first-order) model, which proves advantageous for analyzing the frequency stability of the power system. The wind turbine model, as shown in Figure 2, is comprehensively described in [40–42]. Figure 2 illustrates the model of the turbine for frequency control [40–42]. In this model, the wind speed is multiplied by a random speed fluctuation, which is derived from the white noise block in MATLAB/SIMULINK. This multiplication allows for the estimation of random fluctuations in the wind output power. Based on Figure 2, the power system incorporates a coordinated control scheme consisting of an FOPD-FOPID cascaded controller. This controller's parameters are optimized using the DOSA method, aiming to enhance the frequency stability of the power system in the presence of a wind turbine. The wind turbine exhibits oscillatory behavior and is influenced by the wind speed. The inclusion of the production rate limit definition has led to an improved accuracy in the dynamic model employed for the power system [40–42]. The production rate limits are set at 0.2 pu MW/min for non-reheat power plants, 0.1 pu MW/min for reheat power plants, and 0.5 pu MW/min for the hydro power plant [40–42].

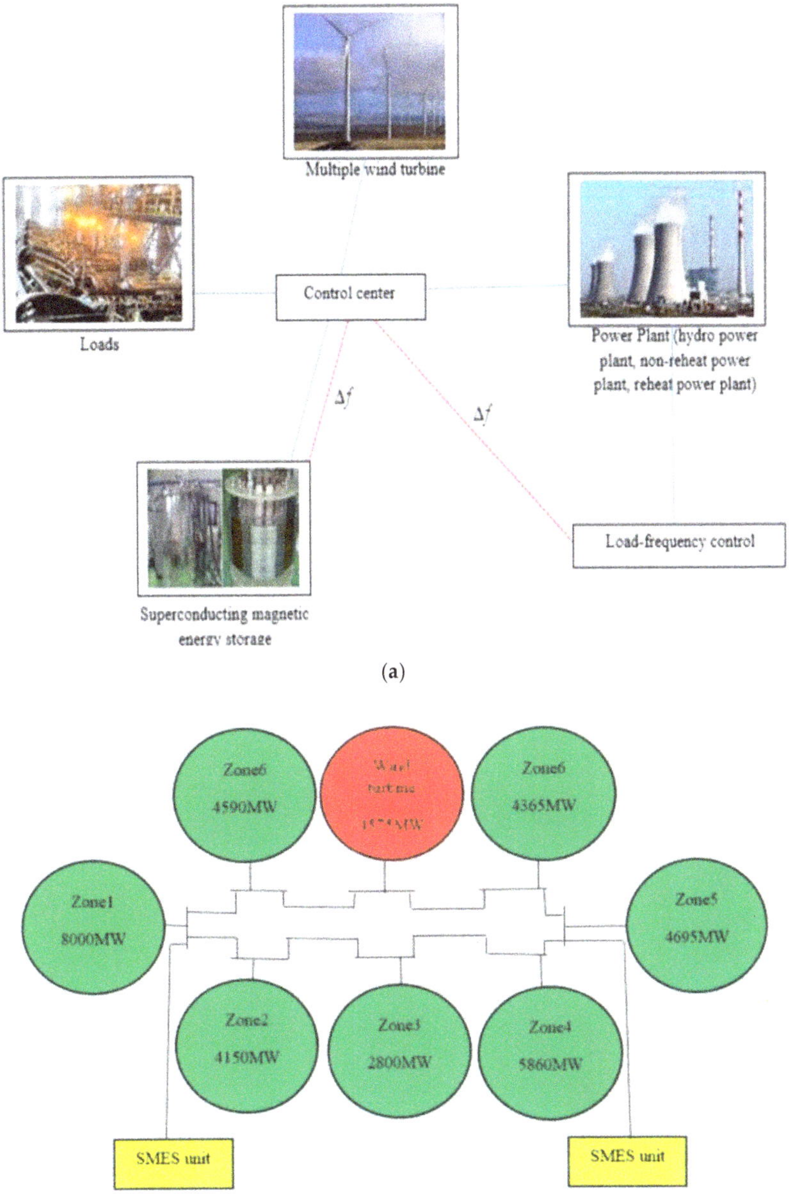

(**b**)

Figure 1. (**a**) Power system configuration incorporating a wind turbine. (**b**) Single-line diagram of the power system [5,42].

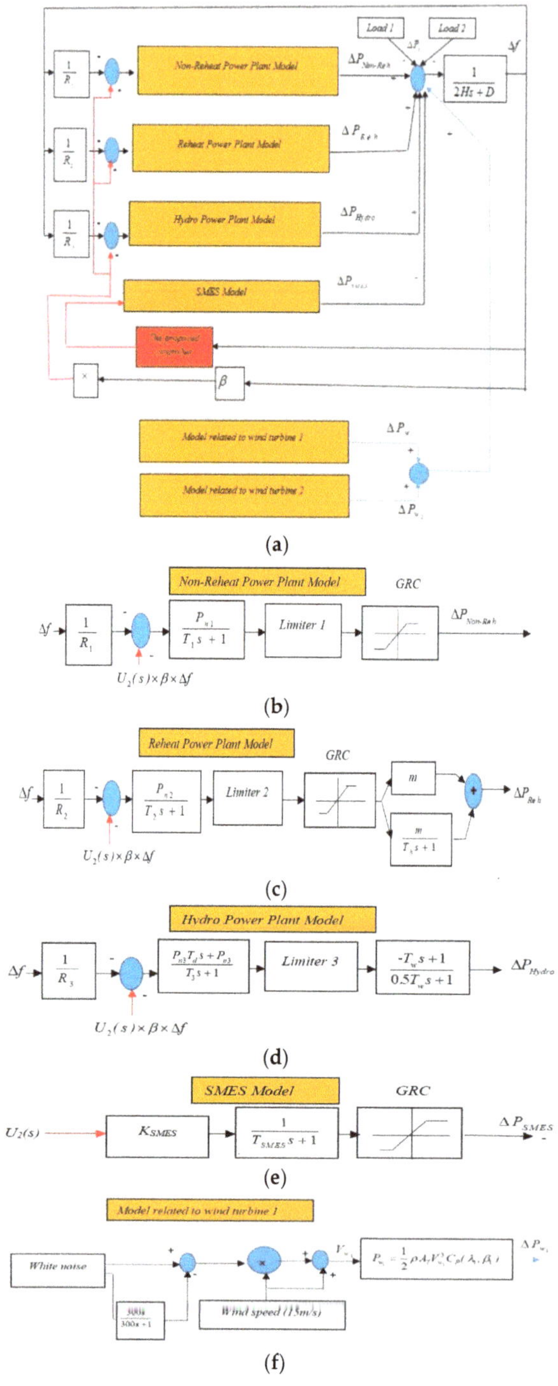

Figure 2. *Cont.*

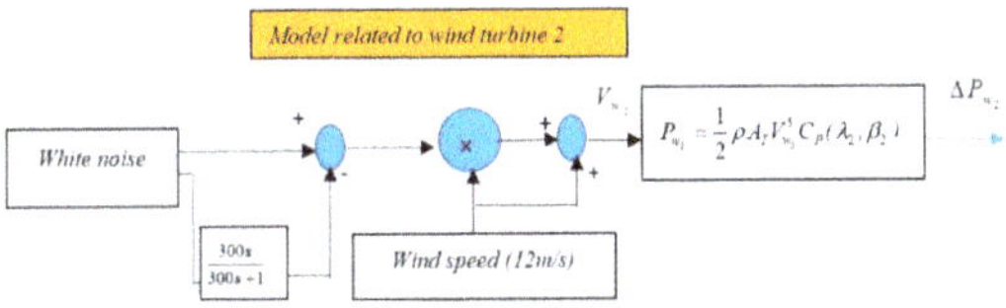

(g)

Figure 2. (**a**) Dynamic model of the power system, taking into account the presence of the wind turbine. (**b**) Non-Reheat power plant model. (**c**) Reheat power plant model. (**d**) Hydro power plant model. (**e**) SMES model. (**f**) Model rated to wind turbine 1. (**g**) Model rated to wind turbine 2.

2.2. The State–Space Equations of the Power System under Scrutiny

The design of the proposed controller for the coordinated control of the LFC system and SMES involves the utilization of state-space equations, as illustrated by Equations (1) and (2). The parameters specific to the analyzed power system are provided in Table 1 [40–42].

$$
\begin{bmatrix} \dot{\Delta f} \\ \Delta \dot{P}_{Non-Reh} \\ \Delta \dot{P}_{Reh} \\ \Delta \dot{P}_{g2} \\ \Delta \dot{P}_{Hydro} \\ \Delta \dot{P}_{g3} \\ \Delta \dot{P}_{WT} \\ \Delta \dot{P}_{SMES} \end{bmatrix} = \frac{2m}{T_h} \begin{bmatrix} \frac{-D}{2H} & \frac{1}{2H} & \frac{1}{2H} & 0 & \frac{1}{2H} & 0 & \frac{1}{2H} & \frac{1}{2H} \\ \frac{-P_{n1}}{T_1 R_1} & \frac{-1}{T_1} & 0 & 0 & 0 & 0 & 0 & 0 \\ \frac{-mP_{n2}}{T_2 R_2} & 0 & \frac{-1}{T_h} & \left(\frac{2m}{T_h} - \frac{m}{T_2}\right) & 0 & 0 & 0 & 0 \\ \frac{-P_{n2}}{T_2 R_2} & 0 & 0 & \frac{-1}{T_2} & 0 & 0 & 0 & 0 \\ \left(\frac{-T_d P_{n3} D}{T_3 R_3 H} + \frac{2P_{n3}}{T_3 R_3}\right) & \left(\frac{T_d P_{n3}}{T_3 R_3 H}\right) & \left(\frac{T_d P_{n3}}{T_3 R_3 H}\right) & 0 & \left(\frac{T_d P_{n3}}{T_3 R_3 H} - \frac{2}{Tw}\right) & \left(\frac{2}{Tw} + \frac{2}{T3}\right) & 0 & 0 \\ \left(\frac{T_d P_{n3} D}{2T_3 R_3 H} - \frac{P_{n3}}{T_3 R_3}\right) & \left(\frac{-T_d P_{n3}}{2T_3 R_3 H}\right) & \left(\frac{-T_d P_{n3}}{2T_3 R_3 H}\right) & 0 & \left(\frac{-T_d P_{n3}}{2T_3 R_3 H}\right) & \frac{-1}{T_3} & 0 & 0 \\ \frac{-D_i}{T_{SMES}} & 0 & 0 & 0 & 0 & 0 & \frac{-1}{T_{WT}} & 0 \\ 0 & 0 & 0 & 0 & 0 & 0 & 0 & \frac{-1}{T_{SMES}} \end{bmatrix} \begin{bmatrix} \Delta f \\ \Delta P_{Non-Reh} \\ \Delta P_{Reh} \\ \Delta P_{g2} \\ \Delta P_{Hydro} \\ \Delta P_{g3} \\ \Delta P_{WT} \\ \Delta P_{SMES} \end{bmatrix} + \begin{bmatrix} 0 \\ \frac{-P_{n1}}{T_1} \\ \frac{-mP_{n2}}{T_2} \\ \frac{-P_{n2}}{T_2} \\ \frac{2P_{n3}}{T_3} \\ \frac{-P_{n3}}{T_3} \\ 0 \\ \frac{-K_{SMES}}{T_{SMES}} \end{bmatrix} [u] + \begin{bmatrix} 0- & \frac{1}{2H} \\ 0 & 0 \\ 0 & 0 \\ 0 & 0 \\ 0 & \frac{-T_d P_{n3}}{T_3 R_3 H} \\ 0 & \frac{T_d P_{n3}}{2T_3 R_3 H} \\ \frac{-1}{T_{WT}} & 0 \\ 0 & 0 \end{bmatrix} \begin{bmatrix} \Delta P_{wi} \\ \Delta P_L \end{bmatrix} \quad (1)
$$

$$
y = [1 \quad 0 \quad 0 \quad 0 \quad 0 \quad 0 \quad 0 \quad 0] \begin{bmatrix} \Delta f \\ \Delta P_{Non-reh} \\ \Delta P_{Reh} \\ \Delta P_{g2} \\ \Delta P_{Hydro} \\ \Delta P_{g3} \\ \Delta P_{WT} \\ \Delta P_{SMES} \end{bmatrix} \quad (2)
$$

Table 1. Parameters specific to the analyzed power system [40–42].

Parameter	Value	Parameter	Value
P_{n2}	0.6107	R_2	2.5
$P_{w,2}$	3000 KW	P_{n1}	0.2529
P_{n3}	0.1364	H	5.7096
T_3	90	T_h	6
T_2	0.4	R_3	1
T_1	0.4	m	0.5
T_d	5	β	1
$P_{w,1}$	750 KW	D	0.028
T_w	1	R_1	2.5
$P_{w,2}$	3000 KW		

3. Design of the Proposed Controller for the Power System

This section provides an overview of the proposed controller's structure, FOPID controller, an analysis of the DOSA, and the design process of the proposed controller utilizing the DOSA.

3.1. Structure of the Proposed Controller

The FOPD–FOPID cascaded controller has been intricately designed to enhance the stability of frequency within power systems that incorporate wind turbines. Its primary objective is to minimize deviations in frequency caused by load disturbances, disruptions from wind turbines, and uncertainties in power system parameters. This elaborate controller consists of two components: the FOPD controller and the FOPID controller. The FOPD controller primarily determines the outcome, while the FOPID controller shapes and guides it by adjusting the reference signal. In this proposed control framework, the FOPD controller is referred to as the main, primary, or external controller, and the FOPID controller is referred to as the internal, secondary, or sub-controller. These components work synergistically, as depicted in Figure 3, to establish cohesive coordination between the LFC system and the SMES. Figure 3 illustrates the suggested setup of the cascaded controller, which facilitates the orchestration of the inner loop dynamics through Equation (3). Equation (3) succinctly represents the transfer function that exclusively encapsulates the inner loop dynamics, as demonstrated in Figure 3:

$$Y_2(s) = M_2(s)U_2(s) \tag{3}$$

In Equation (3), $M_2(s)$ embodies the transfer function of the internal process, whereas $U_2(s)$ signifies the input signal directed towards said process. The principal controller (FOPD), situated in the outer layer, adeptly curtails any oscillations pertaining to frequency and mitigates external interferences. Equation (4) reveals the transfer function characterizing the outer loop's operation:

$$Y(s) = M_1(s)U_1(s) \tag{4}$$

where $M_1(s)$ symbolizes the transfer function of the external process, while $U_1(s)$ stands for the input signal directed towards said process. As Figure 3 illustrates, $N_2(s)$ denotes the FOPID controller found in the inner layer, whereas $N_1(s)$ represents the FOPD controller situated in the outer layer. The transfer functions of both FOPD and FOPID controllers are demonstrated through Equations (5) and (6):

$$N_1(s) = K_{P1} + K_{d1}s^{\mu_1} \tag{5}$$

$$N_2(s) = K_{P2} + K_I s^{-\lambda} + K_{d2}s^{\mu_2} \tag{6}$$

Figure 4 shows the inside structure of the FOPD-FOPID controller.

The settings of the suggested controller are found by making the *ISE* objective function as small as possible by using the DOSA method. Equation (7) shows the main goal of *ISE*. The limits or boundaries of the objective functions are represented by Equation (8):

$$ISE = \int_0^{t_s} (\Delta f)^2 dt \tag{7}$$

$$\begin{aligned}
&K_{P1,\min} \le K_{P1} \le K_{P1,\max}\\
&K_{d1,\min} \le K_{d1} \le K_{d1,\max}\\
&\mu_{1,\min} \le \mu_1 \le \mu_{1,\max}\\
&K_{P2,\min} \le K_{P2} \le K_{P2,\max}\\
&K_{I,\min} \le K_I \le K_{I,\max}\\
&K_{d2,\min} \le K_{d2} \le K_{d2,\max}\\
&\lambda_{\min} \le \lambda \le \lambda_{\max}\\
&\mu_{2,\min} \le \mu_2 \le \mu_{2,\max}
\end{aligned} \tag{8}$$

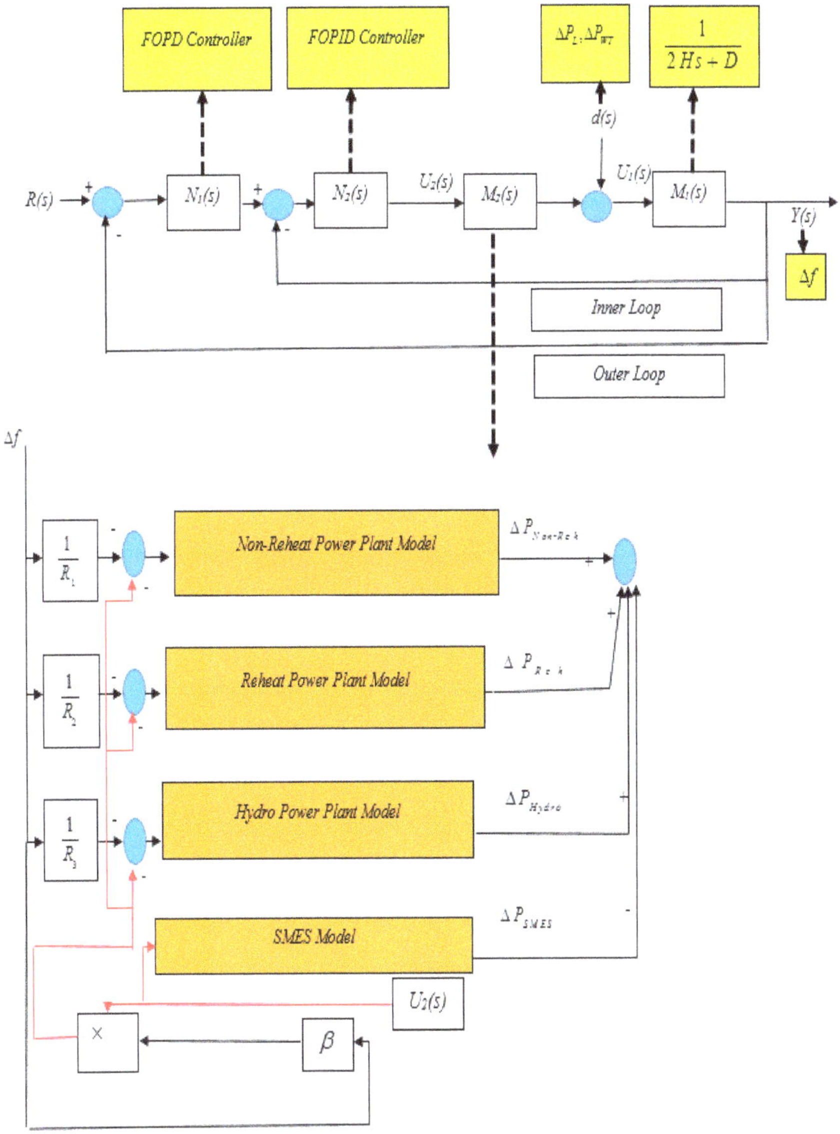

Figure 3. The suggested setup of a cascaded controller manifests itself in order to orchestrate cohesive coordination between the LFC system alongside SMES.

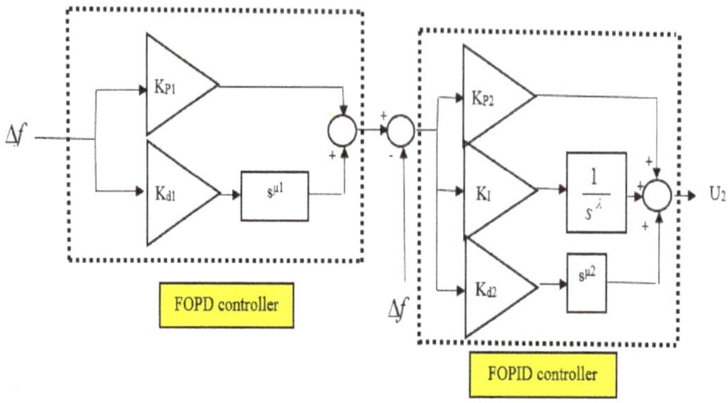

Figure 4. Inside structure of the FOPD-FOPID controller.

3.2. FOPID Controller

Based on the generalized non-integer order fundamental operator, Fractional-order Systems are established (Equation (9)) [27]:

$$aD_t^{\psi} = \begin{cases} \frac{d^{\psi}}{dt^{\psi}} & \psi > 0 \\ 1 & \psi = 0 \\ \int_a^t (dt)^{-\psi} & \psi < 0 \end{cases} \tag{9}$$

The limits of the operation, denoted by a and t, are considered in fractional-order systems. It is typically assumed that $\psi \in R$, although it can also be a complex number. Various definitions of the integral-differential operator have been formulated. The Grünwald–Letnikov definition is commonly employed in the application of FOPID due to its suitability for numerically evaluating fractional order derivatives (Equation (10)) [27]:

$$aD_t^{\psi} f(t) = \lim_{h \to 0} \frac{1}{h^{\psi}} \sum_{j=0}^{k} (-1)^j \begin{pmatrix} \psi \\ j \end{pmatrix} f(t - jh) \tag{10}$$

In Equation (10), a is set to 0, t is equal to kh, where k represents the number of computation steps, and h denotes the step size. Considering zero initial conditions, the Laplace transform of the ψ-order derivative is given by Equation (11) [27]:

$$\int_0^{\infty} e^{-st} {}_0D_t^{\psi} f(t) dt = s^{\psi} F(s) \tag{11}$$

In Equation (11), $\psi \in R^+$, and s is the usual Laplace variable. The PI $^{\lambda}D^{\mu}$ controller, also known as the generalized FOPID controller [27–29], incorporates an integrator of order λ and a differentiator of order μ. The time domain equation for the FOPID controller can be expressed as Equation (12). In Equation (12), the terms $D^{-\lambda}$ $e(t)$ and D^{μ} $e(t)$ denote the fractional-order integral and fractional-order derivative, respectively:

$$u_x(t) = K_p e(t) + K_i D^{-\lambda} e(t) + K_d D^{\mu} e(t) \tag{12}$$

3.3. Developed Owl Search Algorithm (DOSA)

The DOSA is a state-of-the-art meta-heuristic algorithm that offers several advantages over other algorithms such as GTO, MSA, PSO, and ABC. These advantages include: (1) Efficient Search: The DOSA utilizes the owl search behavior, enabling efficient exploration of the search space. The algorithm is designed to strike a balance between exploration and exploitation, facilitating fast convergence towards optimal solutions. (2) Global Optimization: The DOSA is a global optimization algorithm, capable of seeking the global optimum without getting trapped in local optima. This characteristic makes it well-suited for tackling complex optimization problems with multiple solution vertices. (3) Robustness: The DOSA exhibits robustness against noise and uncertainty in the objective function. It adapts effectively to noisy and dynamic environments, making it suitable for real-world applications where the objective function may change over time. (4) Minimal Control Parameters: Compared to other optimization algorithms, the DOSA requires a reduced number of control parameters. This simplifies the tuning process, reduces computational load, and facilitates implementation across various domains. (5) Fast Convergence Speed: The DOSA demonstrates fast convergence in a wide range of optimization problems. It efficiently converges towards near-optimal solutions, significantly reducing the computational time required for finding high-quality solutions. These advantages have been documented in various studies [47–49], highlighting the effectiveness and versatility of the DOSA algorithm.

Similar to the application of other meta-heuristic algorithms in electrical engineering [50–52], its execution is initiated by fortuitously selecting a population. Within this algorithm, said population symbolizes the owls' positioning amidst the forest trees exemplifying the search space. Given that the number of random populations is represented as "n" and the forest is considered as a d-dimensional search space in this algorithm, visually expressing the chance placements of these nocturnal creatures within an n × d matrix can be encapsulated with Equation (13) [47]:

$$x = \begin{bmatrix} x_{1,1} & \cdots & x_{1,d} \\ \cdot & & \cdot \\ \cdot & & \cdot \\ x_{n,1} & \cdots & x_{n,d} \end{bmatrix}_{n\times d} \tag{13}$$

where the component $x_{i,j}$ in the matrix defines the jth variable (dimension) of ith owl. In the owl search algorithm, the term "ith owl" refers to the owl that is being considered or processed in the ith iteration or step of the algorithm. The owl search algorithm is a heuristic search algorithm inspired by the behavior of owls when hunting prey. It uses a combination of local search and global exploration to find an optimal solution. In each iteration, the algorithm evaluates the current owl (ith owl) and updates its position based on certain rules and heuristics. The process continues until a satisfactory solution is found or the search space has been fully explored. To make an introductory state of uniform dissemination, Equation (14) is utilized [47–49]:

$$x_i = x_l + (x_l + x_u)Z(0,1) \tag{14}$$

In Equation (14), X_i could be irregular and uniform numbers between [0, 1], and x_u and x_l are the upper and lower bounds of the ith owl within the jth measurement.

In the realm of mathematical equations, specifically Equation (14), Z(0,1) stands as an intriguing integer that embraces both randomness and uniformity within its numerical essence, constrained between the ethereal boundaries [0,1]. Furthermore, the mysterious confines of x_u and x_l possess a duality and significance in defining not just any owl's position within this enigmatic forest, but rather the profound location it assumes in the jth dimension. As we venture further into comprehending this intricate web of numerics and spatial relations, one is compelled to ponder the cost associated with these owls' chosen abodes in the forest. Illuminatingly explained through Equation (15) [47–50], this particular measure serves as a window into understanding how nature has woven together factors such as distance or resources so crucial to determine what truly befits an owl's dwelling and how they impact its existence:

$$f = \begin{bmatrix} f_1([x_{1,1}, x_{1,2}, \ldots, x_{1,d}]) \\ \cdot \\ \cdot \\ \cdot \\ f_n([x_{n,1}, x_{n,2}, \ldots, x_{n,d}]) \end{bmatrix} \quad (15)$$

The owl's position is contingent upon the magnitude of the sounds it detects through its delicate ears. In this instance, the owl that receives the most intense sound is considered superior because it signifies proximity to the desired goal. The normalized intensity value for each respective owl, denoted by i, will be utilized to revise its position, as derived from Equation (16) [46]:

$$\begin{cases} I_i = \frac{f_i - w}{b - w} \\ b = \max\limits_{m \in 1, \ldots, n} f_m \\ w = \min\limits_{m \in 1, \ldots, n} f_m \end{cases} \quad (16)$$

The measurement of prey distance for each individual owl can be acquired by applying Equation (17) [46–48]:

$$D_i = \sqrt{\sum_i (x_i - L)^2} \quad (17)$$

Equation (17) encompasses the representation of prey positions (L), retrieved through the utilization of the most adept owl. The owl search algorithm postulates the existence of prey amidst the forest as a premise for global optimization. During their pursuit, owls advance meticulously towards their target with gradual aerial movement. The extent to which each ith owl undergoes transformation is delineated in Equation (18) [47–50]:

$$C_i = \frac{I_i}{D_i^2} + R_n \quad (18)$$

In the Equation (18), the variable $4\pi D_i^2$ has been substituted with D_i^2, while R_n represents a stochastic element introduced in order to enhance the model's practicality. As the prey transitions from one location to another, it becomes imperative for the owls to cautiously shift closer towards their target. This algorithm encompasses a mechanism by which the alteration in prey position is determined through probability. Consequently, Equation (19) serves as an update mechanism demonstrating how the new positions of the owls relative to those of their intended prey are depicted:

$$x_i^{t+1} = \begin{cases} x_i^t + \beta C_i \left|\alpha L - x_i^t\right|, & p_{pm} < 0.5 \\ x_i^t - \beta C_i \left|\alpha L - x_i^t\right|, & p_{pm} > 0.5 \end{cases} \quad (19)$$

Equation (19) introduces the concept of p_{pm} as a representation of prey position change probability, while α and β are uniformly distributed random numbers that range from 0 to 0.5 and 0 to 1.9, respectively. This unique characteristic of the owl search algorithm sets it apart from other algorithms, providing superior reliability. While the owl search algorithm is relatively new among optimization algorithms, it occasionally encounters a drawback in becoming trapped within the confines of local optima. Nonetheless, this flaw presents an opportunity for early convergence-inspired solutions. Aware of these limitations, modifications have been implemented to enhance both the performance and effectiveness of the owl search algorithm when confronted with local optima traps. The incorporation of chaos theory has gained traction recently due to its profound influence on modeling effects of nonlinear dynamics; optimization falls within this sphere, susceptible to such influences. By default, in the conventional implementation of the owl search algorithm, only variable β serves as a source for randomness within each iteration. However, including variable β introduces an alternative pathway towards early convergence by expanding its role throughout iterations. In order to avert premature convergence within the system, the implementation of a tumultuous technique known as Singer mapping is employed [53]. This strategy involves treating the obscure variable as a customary equation in conformity with Equation (20):

$$\beta_{i+1} = 1.07(7.9\beta_i - 23.3\beta_i^2 + 28.7\beta_i^3 - 13.3\beta_i^4) \tag{20}$$

Furthermore, an alternative approach to enhance initial convergence in the owl search algorithm entails incorporating Lévy flight. The inclusion of random navigation constitutes an integral facet of this methodology for effectively regulating local search. The mathematical representation of this technique can be found in Equations (21) through (23) [53]:

$$Le(w) \approx w^{-1-\tau} \tag{21}$$

$$w = \frac{A}{|B|^{\frac{1}{\tau}}} \tag{22}$$

$$\sigma^2 = \left\{ \frac{\Gamma(1+\tau)}{\tau\Gamma((1+\tau)/2)} \frac{\sin(\pi\tau/2)}{2^{(1+\tau)/2}} \right\}^{\frac{2}{\tau}} \tag{23}$$

In Equations (21)–(23), τ represents a number that can be between 0 and 2. The letter w stands for a small measurement size. The letter $\Gamma(0)$ represents a mathematical operation called the gamma function. The letters $A/B \approx N(0, \sigma^2)$ indicate that the values are taken from a group where each value is chosen randomly from a bell-shaped curve, with the middle value being zero. The range of values for this group is σ^2. Using Equation (24), we can find the new location of the owls based on the given connections.

$$x_i^{t+1} = \begin{cases} x_i^t + \beta C_i|\alpha L - x_i^t|Le(\delta), & p_{pm} < 0.5 \\ x_i^t - \beta C_i|\alpha L - x_i^t|Le(\delta), & p_{pm} > 0.5 \end{cases} \tag{24}$$

3.4. Design Process of the Proposed Controller Utilizing the DOSA

To improve the performance of the FOPD–FOPID controller for the coordinated control of the LFC system and SMES, we need to follow these steps:

(1) Definition of the objective function: The objective function is a mathematical representation of the goal we want to achieve in this problem. It is determined using Equation (7).
(2) Constraints are rules that help us find the best values for the FOPD–FOPID controller. We define these rules using Equation (8).
(3) Creating the first group of owls: In this step, we create a starting population of owls. Each owl in this group has a different number for each FOPD–FOPID controller setting.
(4) Analyzing the population: The first group of individuals is assessed using a specific measurement called the objective function. We calculate the value of the objective function for every owl.
(5) Choosing the best owls: We select the owls with the highest scores to be part of the next generation.
(6) During this stage, new owls are made for the future generation. This work can be completed by adding or subtracting big owls, or by using random actions.
(7) Assessment of the new group of owls: The new group of owls is judged based on the objective function.
(8) Doing steps 5 to 7 again and again until certain stopping conditions are satisfied, like reaching the desired value of the goal function or finishing a certain number of repetitions.
(9) Choosing the top owl: Once all the rounds are done, the owl with the highest value of the main goal is picked as the best answer. This owl gives the best values for the settings of the FOPD–FOPID controller.

The image depicted in Figure 5 illustrates the utilization of the DOSA to optimize the parameters of the FOPD–FOPID controller. This optimization process is specifically aimed at achieving coordinated control of the LFC system and SMES.

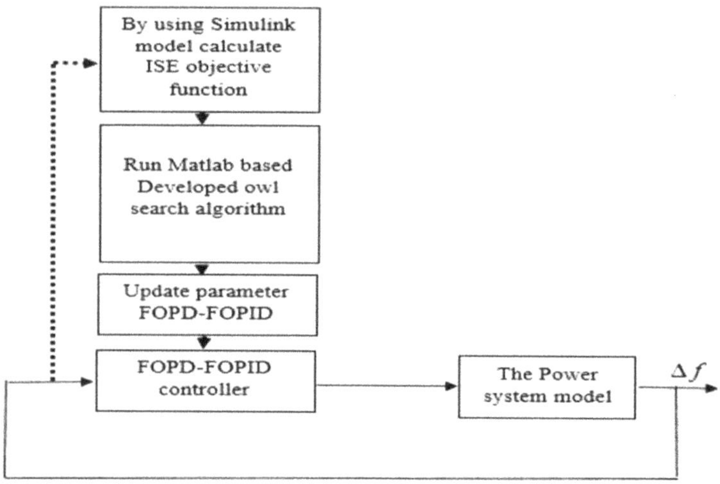

Figure 5. Utilization of the DOSA to optimize the parameters of the FOPD–FOPID controller.

4. Simulation Results and Discussion

In this particular section, the system under study has undergone testing in the presence of a wind turbine across four distinct scenarios. The first scenario (1) is divided into two parts: firstly, an evaluation of the efficiency and adaptability of the DOSA for optimizing the FOPD–FOPID controller parameters takes place alongside a comparison with other algorithms such as MSA, PSO, ABC, and GTO. Secondly, various methods incorporating the proposed DOSA–FOPD–FOPID controller are utilized to compare performance including LFC and SMES based on a robust controller (Controller 1); LFC and SMES based on the MSA–PID controller (Controller 2); LFC based solely on the MSA–PID controller with SMES (Controller 3); and finally, LFC based on the MSA–PID controller without SMES (Controller 4). These comparisons are made while accounting for both load disruptions and wind turbine disruptions. In scenario (2), the proposed method is being compared to several other methods, namely Controller 1, Controller 2, Controller 3, and Controller 4. The aim is to evaluate its performance in handling load disruptions and wind turbine disruptions. Moving on to scenarios (3) and (4), the performance of the proposed method is assessed alongside the mentioned methods. In addition to load disruptions and wind turbine disruptions, these scenarios also consider uncertainties related to power system parameters and disruptions caused by renewable energy sources such as wind turbines.

4.1. Scenario (1)

In this scenario, the power system being analyzed (Figure 2) experiences a disruption in its load with an amplitude of $\Delta P_L = 0.1$ pu at $t = 1$ s, as depicted in Figure 6. The fundamental parameters pertaining to the DOSA and FOPD–FOPID cascaded controller are presented in Table 2. Figure 7 illustrates the optimization process of FOPD–FOPID controller parameters using different algorithms, namely DOSA, MSA, PSO, ABC, and GTO. The objective function utilized is *ISE*. As demonstrated by Figure 7, it can be observed that the DOSA algorithm yields a swifter convergence rate than the others mentioned. In terms of *ISE* values obtained from the respective algorithms (DOSA: 6.8×10^{-6}; GTO: 9×10^{-6}; MSA: 9.1×10^{-6}; ABC: 9.9×10^{-6}; and PSO: 10×10^{-6}), they are displayed in numeric format for easier comprehension. In Table 3, the values of optimized parameters of the FOPD–FOPID controller considering the *ISE* objective function for DOSA, GTO, MSA, ABC, and PSO algorithms are shown. According to Figure 7 and Table 3, the DOSA can be used to optimize the FOPD–FOPID controller parameters. In this scenario, load disruptions and wind turbine disruptions have been applied to the power system according to Figure 8. In Figure 9a–e, the FR of the power system using different control methods to load and wind turbine disruptions is shown. According to Figure 9a, the MFD and ST based on LFC and SMES based on the proposed controller (optimized FOPD–FOPID controller using the DOSA) is equal to 0.0009 Hz and 4.2 s, respectively. The MFD and ST based on controller 1 are equal to 0.0018 Hz and 5 s, respectively; the MFD and ST based on controller 2 are equal to 0.0173 Hz and 19 s, respectively; the MFD and ST based on controller 3 are equal to 0.021 Hz and 38 s, respectively; and the MFD and ST based on controller 4 are equal to 0.0476 Hz and 90 s, respectively (Figure 9b–e). Based on the outcomes of this particular scenario, it is evident that the proposed controller surpasses its counterparts in effectively mitigating power system deviations, and it has also reduced the ST of frequency deviations caused by disruptions on the power system.

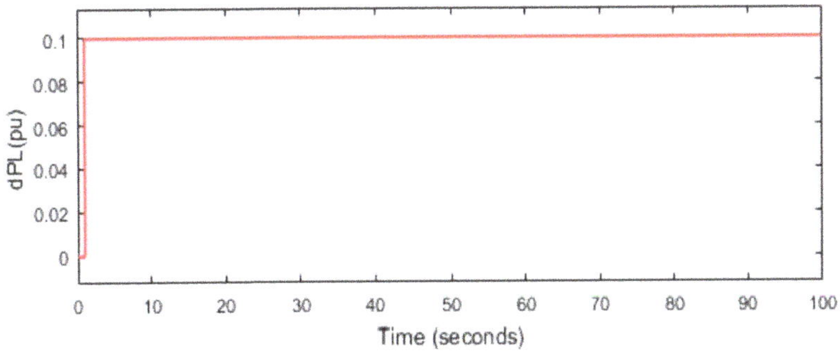

Figure 6. Load disruptions applied to the power system.

Table 2. Fundamental parameters pertaining to the DOSA and FOPD–FOPID cascaded controller.

Parameter	Value	Parameter	Value
Population of owls	100	$K_{P1,min}, K_{d1,min}, K_{P2,min}, K_{I,min}, K_{d2,min}$	0
Forest range for capacity	[0,1000]	$K_{P1,max}, K_{d1,max}, K_{P2,max}, K_{I,max}, K_{d2,max}$	100
α	1	$\lambda_{min}, \mu_{1,2,min}$	0
Iterations (stop criteria)	100	$\lambda_{max}, \mu_{1,2,max}$	1

Table 3. Values of parameters of the FOPD–FOPID controller using different algorithms.

Controller	K_{P1}	μ_1	K_{d1}	K_{P2}	K_I	K_{d2}	λ	μ_2	*ISE*
DOSA–FOPD–FOPID	91.55	0.65	88.91	98.22	91.44	86.35	0.56	0.74	6.8×10^{-6}
GTO–FOPD–FOPID	89.13	0.58	83.66	89.55	83.87	84.18	0.40	0.42	9×10^{-6}
MSA–FOPD–FOPID	85.82	0.62	87.44	92.34	90.56	75.79	0.46	0.40	9.1×10^{-6}
ABC–FOPD–FOPID	68.23	0.49	91.23	86.25	78.45	76.39	0.43	0.38	9.9×10^{-6}
PSO–FOPD–FOPID	70.65	0.47	81.77	75.21	79.92	71.36	0.39	0.48	10×10^{-6}

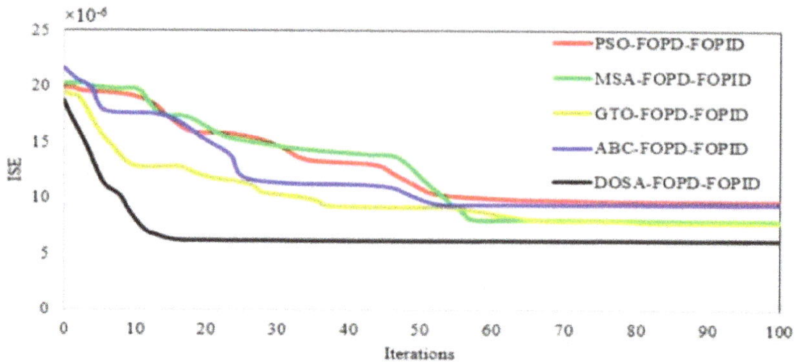

Figure 7. Convergence of DOSA, MSA, PSO, ABC, and GTO algorithms in optimizing FOPD–FOPID controller parameters.

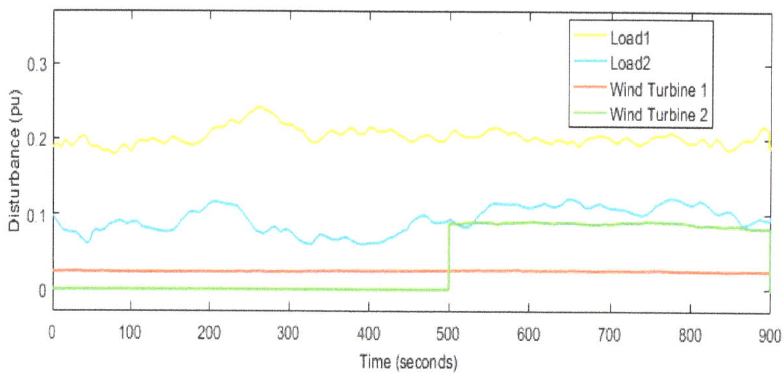

Figure 8. Load disruptions and wind turbine disruptions, scenario 1.

4.2. Scenario (2)

In this scenario, load disruptions and wind turbine disruptions have been applied to the power system according to Figure 10. In Figure 11a–e, the FR of the power system using different control methods to load disruptions and wind turbine disruptions is shown. According to Figure 11a, the MFD and ST based on the proposed controller (optimized FOPD–FOPID controller using the DOSA) are equal to 0.0007 Hz and 3.55 s, respectively. The MFD and ST based on controller 1 are equal to 0.0015 Hz and 4.46 s, respectively; the MFD and ST based on controller 2 are equal to 0.0081 Hz and 21 s, respectively; the MFD and ST based on controller 3 are equal to 0.0129 Hz and 37 s, respectively; and the MFD and ST based on controller 4 are equal to 0.0256 Hz and 45 s, respectively (Figure 11b–e). Based on these results, the proposed controller performs better in reducing the deviations of the power system than the other mentioned controllers, and it also reduces the settling time of the frequency deviations caused by disruptions in the power system.

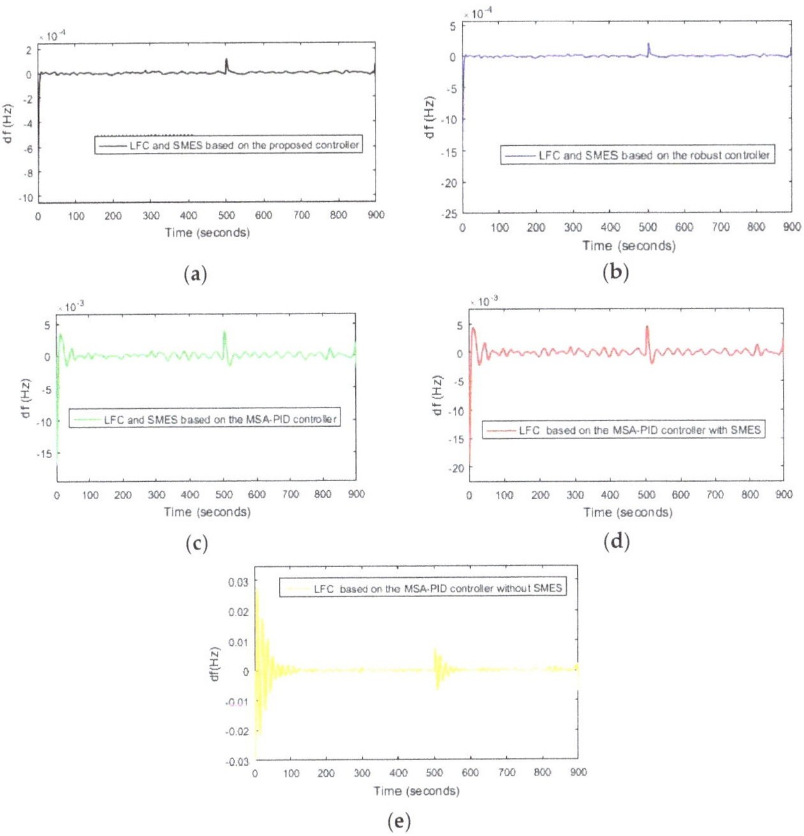

Figure 9. (**a**) The FR of the power system is based on the proposed controller, Scenario (1). (**b**) The FR of the power system using controller 1, Scenario (1). (**c**) The FR of the power system using controller 2, Scenario (1), (**d**) The FR of the power system using controller 3, Scenario (1). (**e**) The FR of the power system using LFC is based on controller 4, Scenario (1).

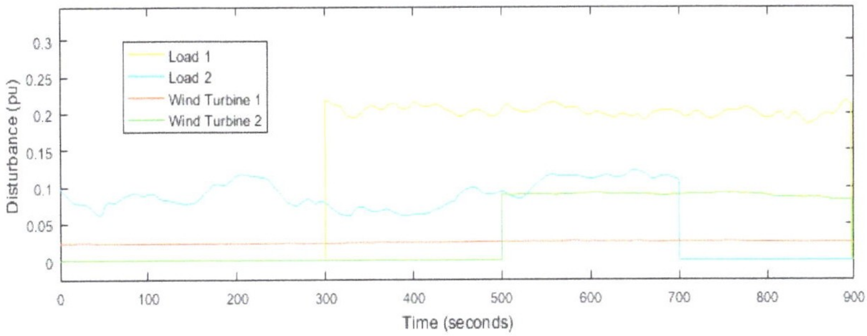

Figure 10. Load disruptions and wind turbine disruptions, scenario 2.

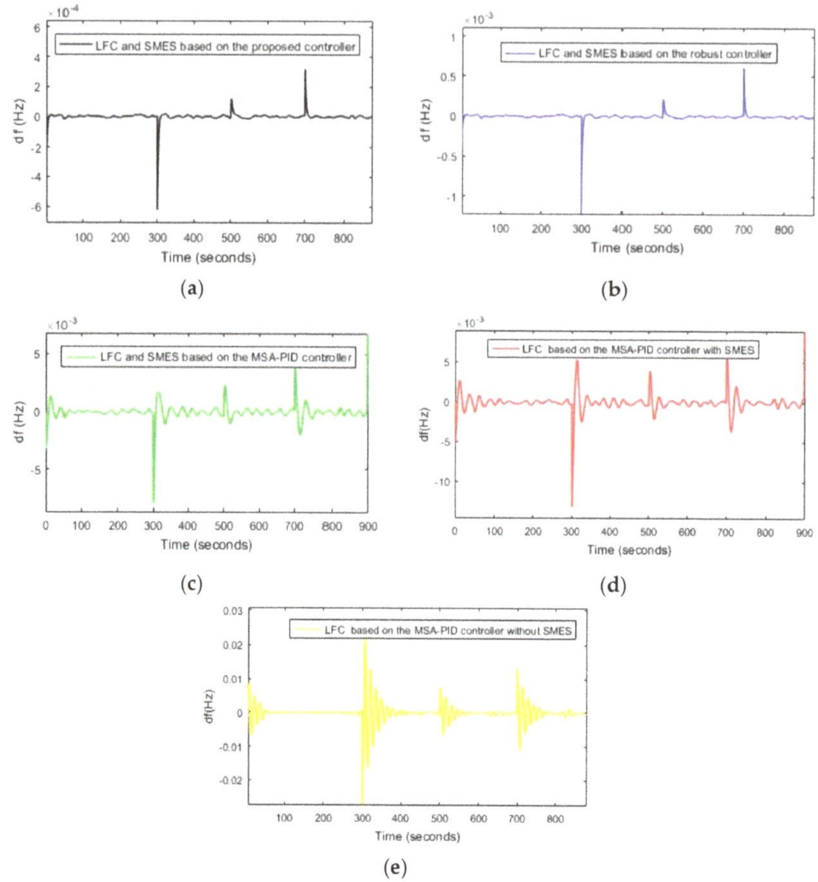

Figure 11. (**a**) The FR of the power system based on the proposed controller, Scenario (2). (**b**) The FR of the power system using controller 1, Scenario (2). (**c**) The FR of the power system using controller 2, Scenario (2), (**d**) The FR of the power system using controller 3, Scenario (2). (**e**) The FR of the power system using LFC based on controller 4, Scenario (2).

4.3. Scenario (3)

In this scenario, load disruptions and wind turbine disruptions have been applied to the power system according to Figure 10. In this scenario, slight uncertainty related to the power system parameters is considered in the system inertia ($H = -25\%$). In Figure 12a–e, the frequency response of the power system to load disruptions, wind turbine disruptions and mild uncertainty related to the system parameters are shown. According to Figure 12a, the MFD and ST based on the proposed controller (optimized FOPD–FOPID controller using the DOSA) are equal to 0.00075 Hz and 3.76 s, respectively. The MFD and ST based on controller 1 are equal to 0.00163 Hz and 4.49 s, respectively; the MFD and ST based on controller 2 are equal to 0.0106 Hz and 24 s, respectively; the MFD and ST based on controller 3 are equal to 0.017 Hz and 42 s, respectively; and the MFD and ST based on controller 4 are equal to 0.0336 Hz and 48 s, respectively (Figure 12b–e). Based on the results obtained in this section, the proposed controller performs better in reducing the deviations of the power system compared to the other mentioned controllers and reduces the ST of the frequency deviations caused by the disruptions in the power system and is resistant to the mild uncertainty related to the system parameters.

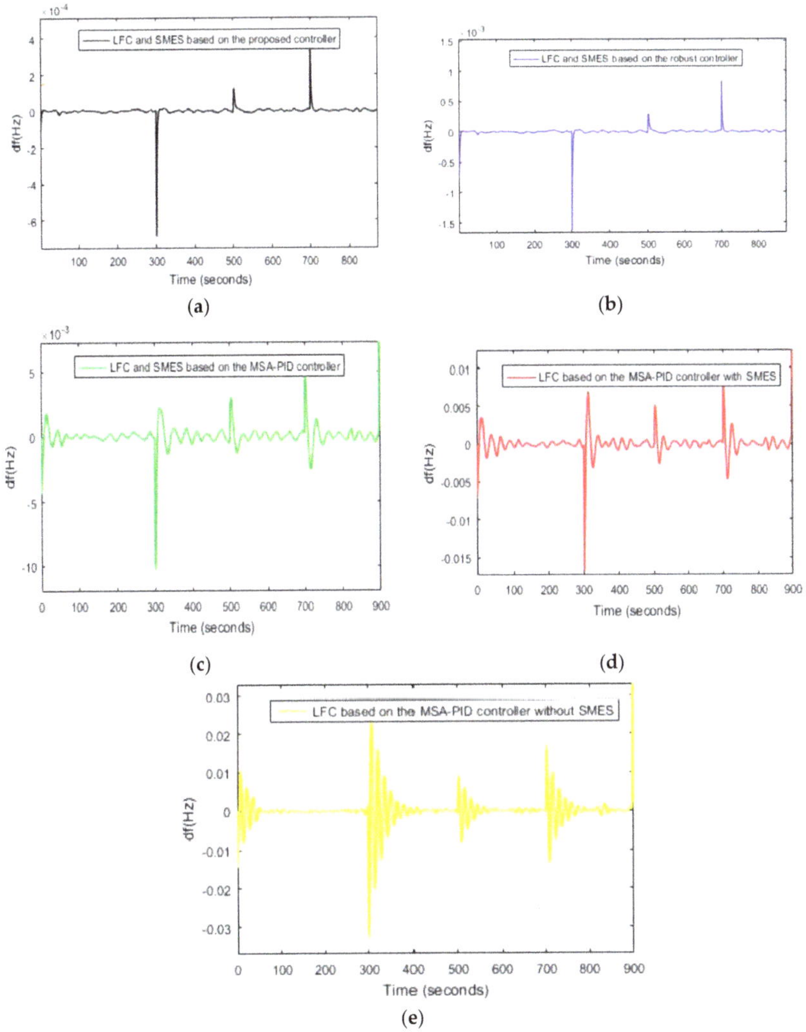

Figure 12. (**a**) The FR of the power system based on the proposed controller, Scenario (3). (**b**) The FR of the power system using controller 1, Scenario (3). (**c**) The FR of the power system using controller 2, Scenario (3), (**d**) The FR of the power system using controller 3, Scenario (3). (**e**) The FR of the power system using LFC based on controller 4, Scenario (3).

4.4. Scenario (4)

In this scenario, load disruptions and wind turbine disruptions have been applied to the power system according to Figure 10. In this scenario, severe uncertainty related to the power system parameters is considered in the system inertia ($H = -50\%$). In Figure 13a–e, the FR of the power system to load disruptions, wind turbine disruptions, and severe uncertainty related to the system parameters are shown. According to Figure 13a, the MFD and ST based on the proposed controller (optimized FOPD–FOPID controller using the DOSA) are equal to 0.00079 Hz and 3.93 s, respectively. The MFD and ST based on controller 1 are equal to 0.00172 Hz and 4.58 s, respectively; the MFD and ST based on controller 2 are equal to 0.0157 Hz and 25 s, respectively; and the MFD and ST based on controller 3 with SMES are equal to 0.0197 Hz and 46 s, respectively (Figure 13b–d).

According to Figure 13e, the FR of the power system is unstable against severe disruptions using LFC based on controller 4, and this control method does not have the ability to maintain frequency stability against severe disruptions related to the parameters of the power system. According to the results of scenario (4), the proposed controller performs better in reducing the deviations of the power system than the other mentioned controllers; it reduces the ST of the frequency deviations caused by disruptions in the power system and is resistant to the severe uncertainty related to the system parameters. In Table 4, performance results of different control methods for 4 scenarios are shown.

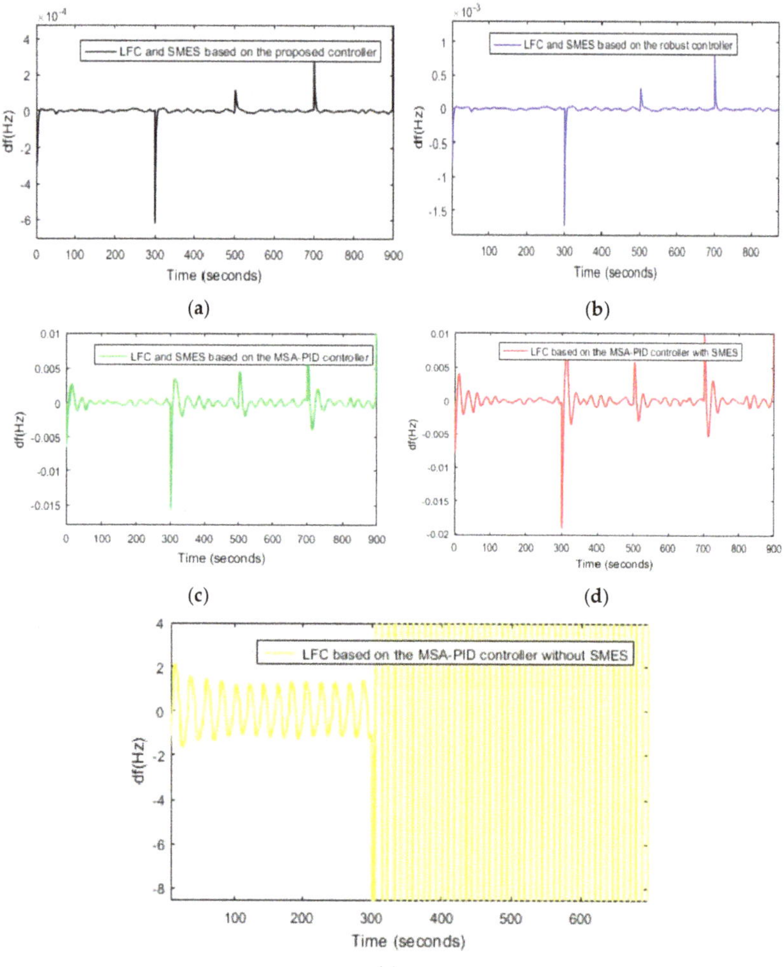

Figure 13. (**a**) The FR of the power system based on the proposed controller, Scenario (4). (**b**) The FR of the power system using controller 1, Scenario (4). (**c**) The FR of the power system using controller 2, Scenario (4), (**d**) The FR of the power system using controller 3, Scenario (4). (**e**) The FR of the power system using LFC based on controller 4, Scenario (4).

Table 4. Performance results of different control methods for four scenarios.

Controller		Scenario (1)	Scenario (2)	Scenario (3)	Scenario (4)
Proposed controller	MO (Hz)	0.0001	0.00035	0.00037	0.00038
	MU (Hz)	0.0009	0.0007	0.00075	0.00079
	ST (s)	4.2	3.55	3.76	3.93
Controller 1	MO (Hz)	0.0004	0.0008	0.0009	0.0010
	MU (Hz)	0.00184	0.00152	0.001631	0.001724
	ST (s)	5.05	4.461	4.492	4.492
Controller 2	MO (Hz)	0.00421	0.00341	0.0053	0.0092
	MU (Hz)	0.01734	0.00816	0.01066	0.01578
	ST (s)	19.03	21.24	24.53	25.22
Controller 3	MO (Hz)	0.00643	0.00582	0.0127	0.0146
	MU (Hz)	0.0214	0.01291	0.01708	0.01975
	ST (s)	38.12	37.39	42.11	46.25
Controller 4	MO (Hz)	0.0476	0.023	0.0367	---
	MU (Hz)	0.04667	0.02565	0.03363	---
	ST (s)	90.1	45.03	48.21	---

5. Conclusions

In this paper, a robust control method has been designed for the coordinated control of LFC and SMES using the FOPD–FOPID controller. The DOSA algorithm, which has many advantages over other optimization algorithms, was used to adjust the parameters of the proposed controller. The proposed method in this paper is compared with other methods presented in the field of power system frequency control in the presence of wind turbines. The proposed method was able to improve the MFD and ST related to frequency deviations in the power system caused by load disturbances and wind turbine disturbances by 50% and 17%, respectively, compared to other methods presented in this field (frequency control). The proposed method has been able to improve the MFD and ST related to frequency deviations in the power system caused by load disturbances, wind turbine disturbances, and slight uncertainty of parameters (H = −%25) by 45% and 18%, respectively (compared to other methods presented in the field of frequency control). The proposed method improved the MFD and ST related to frequency deviations in the power system caused by load disturbances, wind turbine disturbances, and extreme uncertainty of parameters (H = −%50) by 45% and 18%, respectively (compared to other methods presented in the field of frequency control). To continue the work of this paper in the future, several suggestions can be made, including: (1) Combining the FOPD–FOPID controller with a neural network and using it in different parts of the power system, (2) Investigating the possibility of using the FOPD–FOPID controller in power systems Smart, (3) checking the performance of the FOPD–FOPID controller in case of faults in the power system and defects in other system components.

Author Contributions: F.A.: methodology, resources investigation, writing—original draft, conceptualization, methodology, visualization. M.E.: investigation, review and editing, conceptualization, formal analysis. M.H.M.: validation, supervision. All authors have read and agreed to the published version of the manuscript.

Funding: This work was supported by the Australian Research Council, DP190102501.

Data Availability Statement: Data are contained within the article.

Acknowledgments: National Elites Foundation.

Conflicts of Interest: The authors declare no conflict of interest.

Abbreviations

ABC	Artificial bee colony	ΔP_{SMES}	Changes in power production of the SMES system
ACO	Ant colony optimization	ΔP_{WT}	Changes in power production of the wind turbine
CBO	Chaotic butterfly optimization	$\Delta P_{non\text{-}Reh}$	Changes in power output of the gas power plant
CSA	Crow search algorithm	ΔP_{Reh}	Changes in power output of the reheat power plants
DE	Differential evolution	D	System damping coefficient of the area (pu MW/Hz)
DSA	Dragonfly search algorithm	H	Equivalent inertia constant (pu s)
EHO	Elephant herding optimization	T_1	Valve time constant of the non-reheat plant (s)
FA	Firefly algorithm	T_2	Steam valve time constant of the reheat plant (s)
GA	Genetic algorithm	T_3	Water valve time constant of the hydro plant (s)
GBMO	Gases Brownian motion ptimization	T_d	Dashpot time constant of the hydro plant speed governor (s)
HDE-PS	Hybrid differential evolution and pattern search	T_h	Time constant of the reheat thermal plant (s)
HFA-PS	Hybrid firefly algorithm–pattern search algorithm	T_w	Water starting time in the hydro intake (s)
HLUS-TLBO	Hybrid local unimodal sampling and teaching learning based optimization	β	Frequency bias factor (pu MW/Hz)
ICA	Imperialist competitive algorithm	m	Fraction of turbine power (intermediate pressure section)
JSO	Jellyfish search optimizer	R_1	Governor speed regulation of the non-reheat plant (Hz/pu MW)
MBA	Mine blast algorithm	R_2	Governor speed regulation of the reheat plant (Hz/pu MW)
MO	Maximum overshoot	R_3	Governor speed regulation of the hydro plant (Hz/pu MW)
MU	Maximum undershoot	P_{n1}	Nominal rated power output for the non-reheat plant (MW pu)
MFD	Maximum frequency deviation	P_{n2}	Nominal rated power output for the reheat plant (MW pu)
MSA	Moth swarm algorithm	P_{n3}	Nominal rated power output for the hydro plant (MW pu)
PSO	Particle swarm optimization	FOPIDN	FOPID with filter
ST	Settling time	ρ	Air density (kg/m^3)
SCA	Sine–cosine algorithm	A_T	Rotor-swept area (m^2)
TID	Tilt-integral-derivative		
Δf	Changes in power system frequency	$C_r(\lambda_1,\beta_1)$	Power coefficient of the rotor blades (wind turbine 1)
$\Delta P_{non\text{-}Reh}$	Changes in power output of the non-reheat power plants	$C_r(\lambda_2,\beta_2)$	Power coefficient of the rotor blades (wind turbine 2)
ΔP_{g2}	Changes in power output of governor 2	$P_{w,1}$, $P_{w,2}$	Production power of wind turbines 1 and 2
ΔP_{g3}	Changes in power output of governor 3	ISE	Integral of squared error
ΔP_{Hydro}	Changes in power output of the hydro power plant	ITAE	Integral time absolute error
ΔP_L	Changes in load		

References

1. Ukaegbu, U.; Tartibu, L.; Lim, C.W. Multi-Objective Optimization of a Solar-Assisted Combined Cooling, Heating and Power Generation System Using the Greywolf Optimizer. *Algorithms* **2023**, *16*, 463. [CrossRef]
2. Yang, C.; Yi, H.; Chai, C.; Huang, B.; Zhang, Y.; Chen, Z. Predictive current control of boost three-level and T-type inverters cascaded in wind power generation systems. *Algorithms* **2018**, *11*, 92. [CrossRef]
3. Zheng, C.; Eskandari, M.; Li, M.; Sun, Z. GA—Reinforced Deep Neural Network for Net Electric Load Forecasting in Microgrids with Renewable Energy Resources for Scheduling Battery Energy Storage Systems. *Algorithms* **2022**, *15*, 338. [CrossRef]
4. Eskandari, M.; Rajabi, A.; Savkin, A.V.; Moradi, M.H.; Dong, Z.Y. Battery energy storage systems (BESSs) and the economy-dynamics of microgrids: Review, analysis, and classification for standardization of BESSs applications. *J. Energy Storage* **2022**, *55*, 105627. [CrossRef]

5. Amiri, F.; Eskandari, M.; Moradi, M.H. Virtual Inertia Control in Autonomous Microgrids via a Cascaded Controller for Battery Energy Storage Optimized by Firefly Algorithm and a Comparison Study with GA, PSO, ABC, and GWO. *Energies* **2023**, *16*, 6611. [CrossRef]
6. Amiri, F.; Hatami, A. Load frequency control for two-area hybrid microgrids using model predictive control optimized by grey wolf-pattern search algorithm. *Soft Computing* **2023**, *27*, 18227–18243. [CrossRef]
7. Paducel, I.; Safirescu, C.O.; Dulf, E.-H. Fractional Order Controller Design for Wind Turbines. *Appl. Sci.* **2022**, *12*, 8400. [CrossRef]
8. Lu, L.; Saborío-Romano, O.; Cutululis, N.A. Reduced-Order-VSM-Based Frequency Controller for Wind Turbines. *Energies* **2021**, *14*, 528. [CrossRef]
9. Abumeteir, H.A.; Vural, A.M. Design and Optimization of Fractional Order PID Controller to Enhance Energy Storage System Contribution for Damping Low-Frequency Oscillation in Power Systems Integrated with High Penetration of Renewable Sources. *Sustainability* **2022**, *14*, 5095. [CrossRef]
10. Idir, A.; Canale, L.; Bensafia, Y.; Khettab, K. Design and Robust Performance Analysis of Low-Order Approximation of Fractional PID Controller Based on an IABC Algorithm for an Automatic Voltage Regulator System. *Energies* **2022**, *15*, 8973. [CrossRef]
11. Tan, W. Tuning of PID load frequency controller for power systems. *Energy Convers. Manag.* **2009**, *50*, 1465–1472. [CrossRef]
12. Hussain, I.; Das, D.C.; Latif, A.; Sinha, N.; Hussain, S.S.; Ustun, T.S. Active power control of autonomous hybrid power system using two degree of freedom PID controller. *Energy Rep.* **2022**, *8*, 973–981. [CrossRef]
13. Shabani, H.; Vahidi, B.; Ebrahimpour, M. A robust PID controller based on imperialist competitive algorithm for load-frequency control of power systems. *ISA Trans.* **2013**, *52*, 88–95. [CrossRef] [PubMed]
14. Bahgaat, N.K.; El-Sayed, M.I.; Hassan, M.M.; Bendary, F.A. Load frequency control in power system via improving PID controller based on particle swarm optimization and ANFIS techniques. In *Research Methods: Concepts, Methodologies, Tools, and Applications*; IGI Global: Hershey, PA, USA, 2015; pp. 462–481.
15. Sambariya, D.K.; Fagna, R. A novel Elephant Herding Optimization based PID controller design for Load frequency control in power system. In Proceedings of the 2017 International Conference on Computer, Communications and Electronics (Comptelix), Jaipur, India, 1–2 July 2017; pp. 595–600.
16. Bernard, M.; Musilek, P. Ant-based optimal tuning of PID controllers for load frequency control in power systems. In Proceedings of the 2017 IEEE Electrical Power and Energy Conference (EPEC), Saskatoon, SK, Canada, 22–25 October 2017; pp. 1–6.
17. Osinski, C.; Leandro, G.V.; da Costa Oliveira, G.H. Fuzzy PID controller design for LFC in electric power systems. *IEEE Lat. Am. Trans.* **2019**, *17*, 147–154. [CrossRef]
18. Sahu, B.K.; Pati, T.K.; Nayak, J.R.; Panda, S.; Kar, S.K. A novel hybrid LUS-TLBO optimized fuzzy-PID controller for load frequency control of multi-source power system. *Int. J. Electr. Power Energy Syst.* **2016**, *74*, 58–69. [CrossRef]
19. Sahu, R.K.; Sekhar, G.C.; Panda, S. DE optimized fuzzy PID controller with derivative filter for LFC of multi source power system in deregulated environment. *Ain Shams Eng. J.* **2015**, *6*, 511–530. [CrossRef]
20. Sahu, R.K.; Panda, S.; Yegireddy, N.K. A novel hybrid DEPS optimized fuzzy PI/PID controller for load frequency control of multi-area interconnected power systems. *J. Process Control* **2014**, *24*, 1596–1608. [CrossRef]
21. Joshi, M.; Sharma, G.; Bokoro, P.N.; Krishnan, N. A fuzzy-PSO-PID with UPFC-RFB solution for an LFC of an interlinked hydro power system. *Energies* **2022**, *15*, 4847. [CrossRef]
22. Chen, G.; Li, Z.; Zhang, Z.; Li, S. An improved ACO algorithm optimized fuzzy PID controller for load frequency control in multi area interconnected power systems. *IEEE Access* **2019**, *8*, 6429–6447. [CrossRef]
23. Sahu, R.K.; Panda, S.; Pradhan, P.C. Design and analysis of hybrid firefly algorithm-pattern search based fuzzy PID controller for LFC of multi area power systems. *Int. J. Electr. Power Energy Syst.* **2015**, *69*, 200–212. [CrossRef]
24. Fathy, A.; Kassem, A.M.; Abdelaziz, A.Y. Optimal design of fuzzy PID controller for deregulated LFC of multi-area power system via mine blast algorithm. *Neural Comput. Appl.* **2020**, *32*, 4531–4551. [CrossRef]
25. Sekhar, G.C.; Sahu, R.K.; Baliarsingh, A.K.; Panda, S. Load frequency control of power system under deregulated environment using optimal firefly algorithm. *Int. J. Electr. Power Energy Syst.* **2016**, *74*, 195–211. [CrossRef]
26. Tammam, M.A.; Aboelela, M.A.S.; Moustafa, M.A.; Seif, A.E.A. Fuzzy like PID controller tuning by multiobjective genetic algorithm for load frequency control in nonlinear electric power systems. *Int. J. Adv. Eng. Technol.* **2012**, *5*, 572.
27. Sorcia-Vázquez, F.D.J.; Rumbo-Morales, J.Y.; Brizuela-Mendoza, J.A.; Ortiz-Torres, G.; Sarmiento-Bustos, E.; Pérez-Vidal, A.F.; Osorio-Sánchez, R. Experimental Validation of Fractional PID Controllers Applied to a Two-Tank System. *Mathematics* **2023**, *11*, 2651. [CrossRef]
28. Sondhi, S.; Hote, Y.V. Fractional order PID controller for load frequency control. *Energy Convers. Manag.* **2014**, *85*, 343–353. [CrossRef]
29. Kumar, A.; Pan, S. Design of fractional order PID controller for load frequency control system with communication delay. *ISA Trans.* **2022**, *129*, 138–149. [CrossRef]
30. Taher, S.A.; Fini, M.H.; Aliabadi, S.F. Fractional order PID controller design for LFC in electric power systems using imperialist competitive algorithm. *Ain Shams Eng. J.* **2014**, *5*, 121–135. [CrossRef]
31. Zamani, A.; Barakati, S.M.; Yousofi-Darmian, S. Design of a fractional order PID controller using GBMO algorithm for load-frequency control with governor saturation consideration. *ISA Trans.* **2016**, *64*, 56–66. [CrossRef]
32. Babaei, F.; Safari, A. SCA based fractional-order PID controller considering delayed EV aggregators. *J. Oper. Autom. Power Eng.* **2020**, *8*, 75–85.

33. Ahuja, A.; Narayan, S.; Kumar, J. Robust FOPID controller for load frequency control using Particle Swarm Optimization. In Proceedings of the 2014 6th IEEE Power India International Conference (PIICON), Delhi, India, 5–7 December 2014; pp. 1–6.
34. Daraz, A.; Malik, S.A.; Basit, A.; Aslam, S.; Zhang, G. Modified FOPID controller for frequency regulation of a hybrid interconnected system of conventional and renewable energy sources. *Fractal Fract.* **2023**, *7*, 89. [CrossRef]
35. Zaid, S.A.; Kassem, A.M.; Alatwi, A.M.; Albalawi, H.; AbdelMeguid, H.; Elemary, A. Optimal Control of an Autonomous Microgrid Integrated with Super Magnetic Energy Storage Using an Artificial Bee Colony Algorithm. *Sustainability* **2023**, *15*, 8827. [CrossRef]
36. Shayeghi, H.; Jalili, A.; Shayanfar, H.A. A robust mixed $H_2/H\infty$ based LFC of a deregulated power system including SMES. *Energy Convers. Manag.* **2008**, *49*, 2656–2668. [CrossRef]
37. Pappachen, A.; Fathima, A.P. Load frequency control in deregulated power system integrated with SMES-TCPS combination using ANFIS controller. *Int. J. Electr. Power Energy Syst.* **2016**, *82*, 519–534. [CrossRef]
38. Sudha, K.R.; Santhi, R.V. Load frequency control of an interconnected reheat thermal system using type-2 fuzzy system including SMES units. *Int. J. Electr. Power Energy Syst.* **2012**, *43*, 1383–1392. [CrossRef]
39. Biswal, A.; Dwivedi, P.; Bose, S. DE optimized IPIDF controller for management frequency in a networked power system with SMES and HVDC link. *Front. Energy Res.* **2022**, *10*, 1102898. [CrossRef]
40. Magdy, G.; Shabib, G.; Elbaset, A.A.; Mitani, Y. Optimized coordinated control of LFC and SMES to enhance frequency stability of a real multi-source power system considering high renewable energy penetration. *Prot. Control Mod. Power Syst.* **2018**, *3*, 39. [CrossRef]
41. Magdy, G.; Mohamed, E.A.; Shabib, G.; Elbaset, A.A.; Mitani, Y. SMES based a new PID controller for frequency stability of a real hybrid power system considering high wind power penetration. *IET Renew. Power Gener.* **2018**, *12*, 1304–1313. [CrossRef]
42. Amiri, F.; Moradi, M.H. Coordinated Control of LFC and SMES in the Power System Using a New Robust Controller. *Iran. J. Electr. Electron. Eng.* **2021**, *17*, 1–17.
43. Çelik, E. Design of new fractional order PI-fractional order PD cascade controller through dragonfly search algorithm for advanced load frequency control of power systems. *Soft Comput.* **2021**, *25*, 1193–1217. [CrossRef]
44. Bhuyan, M.; Das, D.C.; Barik, A.K.; Sahoo, S.C. Performance assessment of novel solar thermal-based dual hybrid microgrid system using CBOA optimized cascaded PI-TID controller. *IETE J. Res.* **2022**, 1–18. [CrossRef]
45. Ali, M.; Kotb, H.; Aboras, K.M.; Abbasy, N.H. Design of cascaded pi-fractional order PID controller for improving the frequency response of hybrid microgrid system using gorilla troops optimizer. *IEEE Access* **2021**, *9*, 150715–150732. [CrossRef]
46. Babu, N.R.; Chiranjeevi, T.; Devarapalli, R.; Knypiński, Ł.; Garcìa Màrquez, F.P. Real-time validation of an automatic generation control system considering HPA-*ISE* with crow search algorithm optimized cascade FOPDN-FOPIDN controller. *Arch. Control. Sci.* **2023**, *33*, 371–390.
47. Ren, X.; Wu, Y.; Hao, D.; Liu, G.; Zafetti, N. Analysis of the performance of the multi-objective hybrid hydropower-photovoltaic-wind system to reduce variance and maximum power generation by developed owl search algorithm. *Energy* **2021**, *231*, 120910. [CrossRef]
48. Lai, G.; Li, L.; Zeng, Q.; Yousefi, N. Developed owl search algorithm for parameter estimation of PEMFCs. *Int. J. Ambient Energy* **2022**, *43*, 3676–3685. [CrossRef]
49. Cao, Y.; Wang, Q.; Wang, Z.; Jermsittiparsert, K.; Shafiee, M. A new optimized configuration for capacity and operation improvement of CCHP system based on developed owl search algorithm. *Energy Rep.* **2020**, *6*, 315–324. [CrossRef]
50. Moradi, M.H.; Eskandari, M.; Hosseinian, S.M. Operational strategy optimization in an optimal sized smart microgrid. *IEEE Trans. Smart Grid* **2014**, *6*, 1087–1095. [CrossRef]
51. Eskandari, M.; Li, L.; Moradi, M.H. Decentralized optimal servo control system for implementing instantaneous reactive power sharing in microgrids. *IEEE Trans. Sustain. Energy* **2017**, *9*, 525–537. [CrossRef]
52. Eskandari, M.; Li, L.; Moradi, M.H.; Siano, P.; Blaabjerg, F. Optimal voltage regulator for inverter interfaced distributed generation units part I: Control system. *IEEE Trans. Sustain. Energy* **2020**, *11*, 2813–2824. [CrossRef]
53. Yang, D.; Li, G.; Cheng, G. On the efficiency of chaos optimization algorithms for global optimization. *Chaos Solitons Fractals* **2007**, *34*, 1366–1375. [CrossRef]

algorithms MDPI

Article

A Self-Adaptive Meta-Heuristic Algorithm Based on Success Rate and Differential Evolution for Improving the Performance of Ridesharing Systems with a Discount Guarantee

Fu-Shiung Hsieh

Department of Computer Science and Information Engineering, Chaoyang University of Technology, Taichung 413310, Taiwan; fshsieh@cyut.edu.tw

Abstract: One of the most significant financial benefits of a shared mobility mode such as ridesharing is cost savings. For this reason, a lot of studies focus on the maximization of cost savings in shared mobility systems. Cost savings provide an incentive for riders to adopt ridesharing. However, if cost savings are not properly allocated to riders or the financial benefit of cost savings is not sufficient to attract riders to use a ridesharing mode, riders will not accept a ridesharing mode even if the overall cost savings is significant. In a recent study, the concept of discount-guaranteed ridesharing has been proposed to provide an incentive for riders to accept ridesharing services through ensuring a minimal discount for drivers and passengers. In this study, an algorithm is proposed to improve the performance of the discount-guaranteed ridesharing systems. Our approach combines a success rate-based self-adaptation scheme with an evolutionary computation approach. We propose a new self-adaptive metaheuristic algorithm based on success rate and differential evolution for the Discount-Guaranteed Ridesharing Problem (DGRP). We illustrate effectiveness of the proposed algorithm by comparing the results obtained using our proposed algorithm with other competitive algorithms developed for this problem. Preliminary results indicate that the proposed algorithm outperforms other competitive algorithms in terms of performance and convergence rate. The results of this study are consistent with the empirical experience that two people working together are more likely to come to a correct decision than they would if working alone.

Keywords: shared mobility; ridesharing; optimization; self-adaptive; evolutionary computation; metaheuristic

1. Introduction

Shared mobility is a paradigm of transport modes that enables the reduction of vehicles, traffic congestion, consumption of energy and emission of greenhouse gas in cities. Due to the potential benefits of shared mobility, different sharing models have emerged in the past years. These include ridesharing, car sharing and bike sharing. As all of these transport models are helpful for sustainability issues, relevant issues and problems have attracted researchers' and practitioners' attention in academia and industry. In particular, ridesharing has been implemented in university campuses [1], by companies [2] and by transport service providers such as Uber [3], Lyft [4] and BlaBlaCar [5].

In the literature, early studies of ridesharing focused on the problem of meeting the transport requirements of drivers and passengers. A lot of the work from the early ridesharing literature can be found in [6,7]. In these early works, the goal was to optimize total cost savings or total travel distance through matching drivers and passengers based on their itineraries. The ways to achieve this goal can include building a simulation environment to simulate the application scenarios or formulating an optimization model to solve the ridesharing problems. Due to the wide variety of ridesharing problems, different models were proposed and studied in the past years. Optimization methods were applied to formulate the ridesharing problems. The challenges and opportunities for solving ridesharing problems with optimization models can be found in [8,9]. A review on variants of shared mobility, problems and solution approaches is available in [10].

In addition to the issues of optimizing total cost savings or total travel distance [11], there are recent works on other issues in ridesharing systems. For example, to promote ridesharing, the optimization of monetary issues and non-monetary issues in ridesharing systems has been studied. As mentioned in [12–14], dealing with these issues often requires the modeling of more complex constraints that are highly nonlinear. These constraints may lead to a more complex solution space and make it difficult to find a solution for the problem. Therefore, these monetary issues and non-monetary issues in ridesharing systems pose new challenges in the development of effective methods to solve relevant ridesharing problems.

One prominent financial benefit of a shared mobility mode such as ridesharing is cost savings. For this reason, a lot of studies focus on maximization of cost savings in shared mobility systems. Cost savings provide an incentive for riders to adopt ridesharing. However, if cost savings are not properly allocated to riders or the financial benefit of cost savings is not sufficient to attract riders to use ridesharing mode, riders will not accept ridesharing mode even if the overall cost savings is significant [15,16]. In a recent study [13], the concept of discount-guaranteed ridesharing has been proposed to provide an incentive for riders to accept ridesharing services through ensuring a minimal discount for drivers and passengers. In [13], several algorithms have been applied to solve the Discount-Guaranteed Ridesharing Problem (DGRP). With the advances in computing technology, it is possible to develop a more effective algorithm to solve the problem. In this study, we will propose an algorithm to improve the performance of the discount-guaranteed ridesharing systems and the convergence rate to find a solution for the DGRP. Neighborhood search has been recognized as an effective mechanism to improve solutions in an evolutionary computation approach. The concept of self-adaptation has been widely used in meta-heuristic algorithms to identify better search strategies through learning and to apply them in the solution-finding processes to improve convergence rate. In this paper, a success rate-based self-adaptation mechanism and neighborhood search are used jointly to develop an effective algorithm to solve the DGRP.

One of the challenges in solving the DGRP arises from the large number of constraints and discrete decision variables. To tackle the constraints effectively, we adopt a method that discriminates feasible regions from infeasible regions in the solution space [17] by designing a proper fitness function. To deal with the discrete decision variables, we use a transformation approach that transforms the real values of decision variables into discrete values. The contributions of this paper include the development of a new self-adaptive neighborhood search algorithm for solving the DGRP and the assessment of its effectiveness by comparing with existing methods.

The rest of this paper is organized as follows. In Section 2, we will provide a literature review of ridesharing problems and relevant solution methods. We will present the problem formulation and the model of the DGRP in Section 3. In Section 4, the details about the development of a solution algorithm based on self-adaptation and neighborhood search will be presented. In Section 5, the results obtained by applying the proposed algorithm and other competitive algorithms will be presented. We will discuss the results of experiments and conclude this study in Section 6.

2. Literature Review

In this section, we briefly review existing studies relevant to this paper. As we concentrate on the development of an effective algorithm for the DGRP based on success rate self-adaptation and neighborhood search, the papers reviewed in this section include two categories: papers related to ridesharing literature and papers relevant to self-adaptation and neighborhood search in an evolutionary computation approach. Papers related to ridesharing will be introduced first. Papers related to self-adaptation and neighborhood search in an evolutionary computation approach will be introduced next.

One of the sustainable development goals is to promote emerging paradigms to mitigate global warming by reducing the consumption of energies, greenhouse gas emissions and negative impact to the environment. With the global trend to achieve this goal, several transport modes such as ridesharing, car-sharing and bike-sharing have appeared in the transportation sector in the past two decades under the sharing economy. As one of the most important transport modes for shared mobility, ridesharing makes it possible for passengers and drivers with similar itineraries to share rides and enjoy cost savings. For a comprehensive survey of ridesharing literature, please refer to [6,7,18].

Although ridesharing is one of the transport modes with the most potential to achieve shared economy, there are still barriers and challenges for its acceptance by the general public. For example, the lack of trust in ridesharing is one factor that hinders users' acceptance of ridesharing [14]. Several studies have been done on the barriers to the acceptance of ridesharing. The acceptance of ridesharing mode is influenced by several monetary factors and non-monetary factors. Monetary factors for the acceptance of ridesharing are directly related to the financial benefits due to cost savings [12]. Non-monetary factors are directly related to the safety and comfortability of ridesharing such as trust, enjoyability and social awareness. For example, the trust issue in ridesharing has been studied in [14]. In particular, providing a monetary incentive is essential for the acceptance of ridesharing. In this study, we propose a scheme to provide a monetary incentive for ridesharing participants.

As cost savings is recognized as one of the most prominent benefits from ridesharing, the objective of the ridesharing problem considered in most studies is to maximize overall cost savings or minimize the overall travel costs while meeting the transportation requirements of riders and drivers [11]. However, individual ridesharing participants may not enjoy the benefits of cost savings even if overall cost savings have been maximized or the overall travel costs have been minimized. To make individual ridesharing participants enjoy the benefits of cost savings, the overall cost savings must be allocated to individual ridesharing participants properly such that the benefit of cost savings is sufficient for ridesharing participants to accept ridesharing.

In [12], a problem formulation has been proposed to maximize the overall rewarding ratio. However, there is no guarantee that the minimal rewarding rate can be guaranteed even if the overall cost savings is maximized [13]. In [13], a problem formulation and associated solution methods are proposed to ensure that the rewarding rate can be guaranteed. However, scalability of the algorithms was not studied. In this study, we will propose a new algorithm for the DGRP formulated in [13] to improve the performance and convergence rate to guarantee satisfaction of the rewarding rate for ridesharing participants.

As the DGRP is a typical integer programming problem in which the decision variables are binary, the complexity of the problem grows exponentially with the problem size. Exact optimization approaches are computationally feasible only for small instances due to the exponential growth of the solution space as the instances grow. Therefore, approximate optimization approaches will be adopted to solve the decision problem. In the past decades, a lot of evolutionary algorithms were proposed to find solutions for complex optimization problems. These include the Genetic Algorithm [19], Particle Swarm Optimization algorithm [20], Firefly algorithm [21] and metaheuristic algorithms such as the Differential Evolution algorithm [22]. In the literature, a wide variety of variants in the Genetic Algorithm, Particle Swarm Optimization algorithm, Firefly algorithm and Differential Evolution algorithm can be found in [23–26], respectively. Although these approaches may be applied to find solutions for optimization problems, their performances vary. The studies of [27,28] show several advantages of the PSO approach over the Genetic Algorithm. The previous study of [29] indicates that the Differential Evolution approach performs better than the Genetic Algorithm. Evolutionary computation approaches such as PSO or DE algorithms are well-known metaheuristic algorithms. A metaheuristic algorithm refers to a higher-level procedure that generates or selects a heuristic to find a good solution to an optimization problem. In [30–36], several adaptive Differential Evolution algorithms have been proposed to solve optimization problems.

The goal of this study is to propose a more effective solution algorithm to improve the performance of the DGRP. In this study, we will combine a Differential Evolution approach with a success rate self-adaptation mechanism to develop a solution algorithm for the DGRP. The characteristics of the DGRP are different from the problems addressed in [30–36] as the decision variables of the DGRP are discrete whereas the decision variables of the problems studied in [30–36] are continuous real values. In this paper, the self-adaptation mechanism of [37] and the concept of the neighborhood search of [38] are applied jointly to develop an effective problem solver. Note that the self-adaptation mechanism of [37] and the neighborhood search concept of [38] are originally proposed for a continuous solution space. This study will verify effectiveness of combining the self-adaptation mechanism and neighborhood search mechanism for problems with a large number of constraints and discrete decision variables.

The problem addressed in this paper is the DGRP, which was formulated in [13]. This paper is different from the previous work [13] in that the proposed success rate-based self-adaptive metaheuristic algorithm is different from the ten algorithms proposed in [13]. The contribution of this paper is to propose a novel self-adaptive algorithm to improve the performance and convergence rate of discount-guaranteed ridesharing systems. We verified the effectiveness of the self-adaptive algorithm by conducting experiments. The results indicated that the proposed method improves the performance of the solution and convergence rate for finding the solution. Although the algorithm proposed in this paper is designed for the DGRP, it can be applied to other optimization problems. For example, the work reported in [14] also applied a similar approach to another instance of a trust-based ridesharing problem.

3. The Formulation of the DGRP

In this section, we will present the formulation of the DGRP based on a combinatorial double auction mechanism [39]. The variables, parameters and symbols used in this paper are listed in Table 1. We first briefly introduce the combinatorial double auction model and then formulate the DGRP based on the combinatorial double auction model.

Table 1. Notation of symbols, variables, and parameters.

Variable	Meaning
P	Total passengers.
D	Total drivers.
p	Passenger index, where $p \in \{1,2,3,\ldots P\}$.
d	Driver index, where $d \in \{1,2,3,\ldots,D\}$.
K	The number of location indices for all passengers, i.e., K is equal to P.
k	Location index, $k \in \{1,2,\ldots,K\}$.
J_d	Total bids of driver $d \in \{1,2,\ldots,D\}$.
j	The j-th bid index of driver $d \in \{1,2,3,\ldots,D\}$ with $j \in \{1,2,\ldots,J_d\}$.
DB_{dj}	$DB_{dj} = (q^1_{dj1}, q^1_{dj2}, q^1_{dj3}, \ldots, q^1_{djP}, q^2_{dj1}, q^2_{dj2}, q^2_{dj3}, \ldots, q^2_{djP}, o_{dj}, c_{dj})$: driver d's j-th bid, where q_{djp}: the no. of seats allocated for passenger p, o_{dj}: the original cost of driver $d \in \{1,2,\ldots,D\}$ if he/she travels alone, c_{dj}: the bid's travel cost.
q^1_{djp}	No. of seats allocated at the pick-up location of passenger p, $q^1_{djp} = q_{djp}$.
q^2_{djp}	No. of seats released at the drop-off location of passenger p, $q^2_{djp} = q_{djp}$.
PB_p	$PB_p = (s^1_{p1}, s^1_{p2}, s^1_{p3}, \ldots, s^1_{pP}, s^2_{p1}, s^2_{p2}, s^2_{p3} \ldots, s^2_{pP}, f_p)$: passenger p's bid, where s_{pk}: the no. of seats requested by p at location k and f_p: the original cost of p without ridesharing.
s^1_{pk}	No. of seats requested at passenger p's pick-up location, $s^1_{pk} = \begin{cases} s_{pp} & if\ k = p \\ 0 & otherwise \end{cases}$.

Table 1. *Cont.*

Variable	Meaning
s^2_{pk}	No. of seats released at passenger p's drop-off location, $s^2_{pk} = \begin{cases} s_{pp} & if\ k = p \\ 0 & otherwise \end{cases}$.
x_{dj}	Decision variable for driver $d \in \{1,2,\ldots,D\}$: x_{dj} = 1 if DB_{dj} is a winning bid and x_{dj} = 0 otherwise.
y_p	Decision variable for passenger $p \in \{1,2,3,\ldots P\}$: y_p = 1 if PB_p is a winning bid and y_p = 0 otherwise.
r_D	Drivers' minimal expected cost savings discount.
r_P	Passengers' minimal expected cost savings discount.
$F(x,y)$	The objective function, $F(x,y) = \left(\sum\limits_{p=1}^{P} y_p\left(f_p\right)\right) + \left(\sum\limits_{d=1}^{D}\sum\limits_{j=1}^{J_d} x_{dj}o_{dj}\right) - \left(\sum\limits_{d=1}^{D}\sum\limits_{j=1}^{J_d} x_{dj}c_{dj}\right)$.
Γ_{dj}	The set of passengers on the ride of the bid DB_{dj} of driver $d \in \{1,2,\ldots,D\}$.
$F_{dj}(x,y)$	Cost savings of the bid DB_{dj} of driver $d \in \{1,2,\ldots,D\}$. $H_{dj}(x,y) = \left[\left(\sum\limits_{p\in\Gamma_{dj}} y_p f_p\right) + x_{dj}o_{dj} - \left(x_{dj}c_{dj}\right)\right]$.
cf_{pdj}	Travel cost for passenger $p \in \Gamma_{dj}$ on the ride of bid DB_{dj}.

3.1. An Auction Model for Ridesharing Systems

Just like buyers and sellers who trade goods in a traditional marketplace, the functions and operations of a ridesharing system are similar to a traditional marketplace. In a traditional marketplace, buyers purchase goods according to their need and sellers recommend goods based on the available items in stock. In a ridesharing system, individual passengers with transportation requirements are on the demand side. Individual drivers also have their transportation requirements and constraints. Individual drivers are on the supply side. The roles of passengers and drivers in a ridesharing system are similar to buyers and sellers in a traditional marketplace. Therefore, a ridesharing system can be modeled as a virtual "marketplace" in which potential passengers and drivers seek to find an opportunity for ridesharing. Auctions are a proper business model that can be applied to trade goods in a marketplace in which the price of goods is not fixed and is determined by buyers and sellers. They can also be applied to determine the passengers and drivers for ridesharing in ridesharing systems.

In the literature, a variety of auction models have been proposed and applied in different application scenarios. Depending on the number of buyers and sellers in an auction, auctions can be classified into two categories: single-side auctions and double auctions. There are two types of single-side auctions: (1) single seller and multiple buyers and (2) single buyer and multiple sellers. In a double auction, there are multiple buyers and multiple sellers. If there are multiple types of goods for trading in a double auction, buyers and sellers can purchase or sell a combination of goods in the auction. This type of double auction is called a combinatorial double auction.

For an auction scenario with multiple buyers and multiple sellers to trade multiple types of goods, although one can apply either multiple single-side auctions or one combinatorial double auction, the combinatorial double auction is more effective in terms of efficiency. Therefore, we apply the combinatorial double auction model to determine the passengers and drivers for ridesharing in ridesharing systems. There are three types of roles in a typical combinatorial double auction for trading goods: buyers, sellers and the auctioneer. In a ridesharing system modeled with a combinatorial double auction, there are three types of roles: passengers, drivers and the ridesharing information provider. The ridesharing information provider acts as the auctioneer and provides a ridesharing system to process the requests from the passengers and drivers.

3.2. A Formulation of the DGRP Based on Combinatorial Double Auctions

A passenger expresses his/her transportation requirements by sending a request to the ridesharing system provided by the ridesharing information provider. A driver expresses his/her transportation requirements by sending a request to the ridesharing system to indicate his/her transportation requirements and constraints. The ridesharing system must determine the passengers and drivers for ridesharing. In a combinatorial double auction model, buyers and sellers who place the winning bids are called winners. In a ridesharing system, each passenger and each driver on a shared ride determined by the ridesharing system are called winners.

The request submitted by a passenger takes the following form: $R_p = (Lo_p, Le_p, \omega_p^e, \omega_p^l, n_p)$, which includes the passenger p's start location, Lo_p, end location, Le_p, earliest departure time, ω_p^e, latest arrival time, ω_p^l, and requested seats, n_p, respectively. The request submitted by a driver takes the following form: $R_d = (Lo_d, Le_d, \omega_d^e, \omega_d^l, a_d, \overline{\tau}_d, \Gamma_d)$, which includes the driver's start location, Lo_d, end location, Le_d, earliest departure time, ω_d^e, latest arrival time, ω_d^l, available seats, a_d, and maximum detour ratio, $\overline{\tau}_d$. The earliest departure time and the latest arrival time in the request are used in the decision models of most papers on ridesharing. The earliest departure time and the latest arrival time are specified by the ridesharing participant sending the request. The ridesharing system will extract the information from the R_p of a passenger to form a bid $PB_p = (s_{p1}^1, s_{p2}^1, s_{p3}^1, \ldots, s_{pP}^1, s_{p1}^2, s_{p2}^2, s_{p3}^2 \ldots, s_{pP}^2, f_p)$, where s_{pk}^1 is the No. of seats requested at pick-up location k of passenger p, s_{pk}^2 is the No. of seats released at drop-off location k of passenger p and f_p is passenger p's original cost without ridesharing. The ridesharing system will extract the information from R_d of a driver to form a bid $DB_{dj} = (q_{dj1}^1, q_{dj2}^1, q_{dj3}^1, \ldots, q_{djP}^1, q_{dj1}^2, q_{dj2}^2, q_{dj3}^2, \ldots, q_{djP}^2, o_{dj}, c_{dj})$, where q_{djp}^1 is the No. of seats allocated at the pick-up location k of passenger p, q_{djp}^2 is the No. of seats released at the drop-off location k of passenger p, o_{dj} is the original cost of the driver when he/she travels alone and c_{dj} is the travel cost of the bid.

The DGRP to be formulated takes into account several factors: balance between demand and supply, the non-negativity of surplus, a maximum of one winning bid for each driver, minimal rewarding rate for drivers and minimal rewarding rate for passengers based on the bids submitted by passengers, $PB_p \forall p \in \{1, 2, 3, \ldots P\}$ and the bids submitted by DB_{dj} $\forall d \in \{1, 2, \ldots, D\}$, $j \in \{1, 2, \ldots, J_d\}$, submitted by drivers.

The surplus or total cost savings is $F(x, y) = \left(\sum_{p=1}^{P} y_p\left(f_p\right)\right) - \left(\sum_{d=1}^{D}\sum_{j=1}^{J_d} x_{dj}(c_{dj} - o_{dj})\right)$. The objective function is described in (1). Constraint (2) and (3) describe balance between demand and supply of seats in ridesharing vehicles. To benefit from ridesharing, the non-negativity of surplus (cost savings) described by Constraint (4) must be satisfied. A driver may submit multiple bids, a maximum of one bid can be a winning bid for each driver. This constraint is described by Constraint (5). To attract individual drivers to take part in ridesharing, Constraint (6) enforces the satisfaction of the minimal rewarding rate for drivers. To provide incentives for individual passengers to accept ridesharing, Constraint (7) enforces the satisfaction of the minimal rewarding rate for passengers. The constraint that all decision variables must be binary is described by Constraint (8).

Based on the objective function (1) and the constraints defined by Constraint (2) through (8), the DGRP is formulated as an integer programming problem as follows.

Problem Formulation of the DGRP

$$\max_{x,y} F(x, y) \tag{1}$$

$$\sum_{d=1}^{D}\sum_{j=1}^{J_d} x_{dj} q_{djk}^1 = y_p s_{pk}^1 \forall \mathrm{p} \in \{1, 2, \ldots, \mathrm{P}\}\ \forall \mathrm{k} \in \{1, 2, \ldots, \mathrm{P}\} \tag{2}$$

$$\sum_{d=1}^{D}\sum_{j=1}^{J_d} x_{dj} q_{djk}^2 = y_p s_{pk}^2 \forall \mathrm{p} \in \{1, 2, \ldots, \mathrm{P}\}\ \forall \mathrm{k} \in \{1, 2, \ldots, \mathrm{P}\} \tag{3}$$

$$\sum_{p=1}^{P} y_p f_p + \sum_{d=1}^{D}\sum_{j=1}^{J_d} x_{dj} o_{dj} \geq \sum_{d=1}^{D}\sum_{j=1}^{J_d} x_{dj} c_{dj} \tag{4}$$

$$\sum_{j=1}^{J_d} x_{dj} \leq 1 \forall d \in \{1, \ldots, D\} \tag{5}$$

$$x_{dj}\left(\frac{F_{dj}(x,y)}{\sum\limits_{p \in \Gamma_{dj}} y_p cf_{pdj} + x_{dj}c_{dj}} - r_D\right) \geq 0 \tag{6}$$

$$y_p\left(\frac{F_{dj}(x,y)}{\sum\limits_{p \in \Gamma_{dj}} y_p cf_{pdj} + x_{dj}c_{dj}} - r_P\right) \geq 0 \tag{7}$$

$$x_{dj} \in \{0,1\} \forall d \in \{1, \ldots, D\}\ \forall j \in \{1, \ldots, J_d\} \text{ and } y_p \in \{0,1\} \forall \mathrm{p} \in \{1,2, \ldots, \mathrm{P}\} \tag{8}$$

The determination of the ridesharing decisions is not solely based on price and locations, the model also considers the constraint that the minimal rewarding rate for drivers and passengers must be satisfied. As we focus on comparison with [13], we use the same model as the one used in [14]. Factors other than price and locations not considered in the model of this paper can be taken into consideration in the future.

4. A Self-Adaptive Meta-Heuristic Algorithm Based on Success Rate and Differential Evolution

The complexity of the DGRP is due to two characteristics: (1) discrete decision variables and (2) a large number of constraints. For these reasons, the development of an effective solution algorithm for the DGRP relies on a method to ensure values of the decision variables are discrete and a method to enforce the evolution processes to guide the candidate solutions in the population to move toward a feasible solution space. For the former, we use a function to systematically map the continuous values of decision variables to discrete values in the evolution processes. For the latter, a fitness function is used in this paper to provide a direction to improve solution quality by reducing the violation of constraints in the solution-finding processes. In this section, we first briefly describe the details of the methods to convert continuous values of decision variables to discrete values and the fitness function to guide the candidate solutions in the population to move toward feasible solution space, as mentioned. We then present the proposed algorithm.

4.1. The Conversion of Decision Variables and Fitness Function

We define a conversion function to ensure the values of the decision variables are discrete. The function *Convert2Binary* in (9) through (15) is used in our solution algorithm to map the continuous values of decision variables to discrete values in the evolution processes. This procedure makes it possible to adapt existing evolutionary algorithms that were originally proposed for problems with a continuous solution space to work for problems with a discrete solution space.

$$\text{Function } Convert\ 2\ Binary \tag{9}$$

$$\text{Input: } u \tag{10}$$

$$\text{Output: } \overline{u} \tag{11}$$

$$\text{Step 1: } v = \begin{cases} V_{\max}\ if\ u > V_{\max} \\ u\ \ if\ -V_{\max} \leq u \leq V_{\max} \\ -V_{\max}\ if\ u < -V_{\max} \end{cases} \tag{12}$$

$$\text{Step 2: } s(v) = \frac{1}{1+\exp^{-v}} \tag{13}$$

$$\text{Step 3}: \text{ Generate a random variable } rsid \text{ with uniform distribution } U(0,1)$$
$$\overline{u} = \begin{cases} 1 \; rsid < s(v) \\ 0 \; otherwise \end{cases} \tag{14}$$

$$\text{Step 4}: \text{ return } \overline{u} \tag{15}$$

To provide a direction for an evolutionary algorithm to improve solution quality by reducing the violation of constraints in the solution finding processes, we define the set of feasible solutions in the current population as S_f and use $S_{f\min} = \min_{(x,y)\in S_f} F(x,y)$ to denote the objective function value of the worst feasible solution in S_f. We introduce the following fitness function.

The fitness function $F_1(x,y)$ for the penalty method is defined in (16):

$$F_1(x,y) = \begin{cases} F(x,y) & if\ (x,y)\ is\ feasible \\ U(x,y) & otherwise \end{cases}, \tag{16}$$

where $U(x,y)$ is defined in (17).

$$\begin{aligned} U(x,y) = {} & S_{f\min} - \left(\sum_{p=1}^{P} \sum_{k=1}^{K} \left(\left| \sum_{d=1}^{D} \sum_{j=1}^{J_d} x_{dj} q_{djk}^{\backslash 1} - y_p s_{pk}^{1} \right| + \left| \sum_{d=1}^{D} \sum_{j=1}^{J_d} x_{dj} q_{djk}^{2} - y_p s_{pk}^{2} \right| \right) \right) \\ & + \min\left(\sum_{p=1}^{P} y_p f_p - \sum_{d=1}^{D} \sum_{j=1}^{J_d} x_{dj}(c_{dj} - o_{dj}), 0.0 \right) \\ & + \sum_{d=1}^{D} \sum_{j=1}^{J_d} \min\left(1 - \sum_{j=1}^{J_d} x_{dj}, 0.0\right) \\ & + \sum_{d=1}^{D} \sum_{j=1}^{J_d} x_{dj} \min\left(\left(\frac{F_{dj}(x,y)}{\sum_{p\in\Gamma_{dj}} y_p cf_{pdj} + x_{dj} c_{dj}} \right) - r_D, 0.0 \right) \\ & + \sum_{p=1}^{P} y_p \min\left(\left(\frac{F_{dj}(x,y)}{\sum_{p\in\Gamma_{dj}} y_p cf_{pdj} + x_{dj} c_{dj}} \right) - r_P, 0.0 \right) \end{aligned} \tag{17}$$

In (17), we define the penalty function $U(x,y)$ to penalize violation of constraints. The terms $\left| \sum_{d=1}^{D} \sum_{j=1}^{J_d} x_{dj} q_{djk}^{\backslash 1} - y_p s_{pk}^{1} \right|$ and $\left| \sum_{d=1}^{D} \sum_{j=1}^{J_d} x_{dj} q_{djk}^{2} - y_p s_{pk}^{2} \right|$ correspond to penalty due to the violation of Constraints (2) and (3), respectively. The term $\min\left(\sum_{p=1}^{P} y_p f_p - \sum_{d=1}^{D} \sum_{j=1}^{J_d} x_{dj}(c_{dj} - o_{dj}), 0.0 \right)$ corresponds to penalty due to the violation of Constraint (4). The term $\sum_{d=1}^{D} \sum_{j=1}^{J_d} \min\left(1 - \sum_{j=1}^{J_d} x_{dj}, 0.0\right)$ corresponds to penalty due to the violation of Constraint (5). The terms $\sum_{d=1}^{D} \sum_{j=1}^{J_d} x_{dj} \min\left(\left(\frac{F_{dj}(x,y)}{\sum_{p\in\Gamma_{dj}} y_p cf_{pdj} + x_{dj} c_{dj}} \right) - r_D, 0.0 \right)$ and $\sum_{p=1}^{P} y_p \min\left(\left(\frac{F_{dj}(x,y)}{\sum_{p\in\Gamma_{dj}} y_p cf_{pdj} + x_{dj} c_{dj}} \right) - r_P, 0.0 \right)$ correspond to penalties due to the violation of Constraints (6) and (7), respectively.

4.2. The Proposed Success Rate-Based Self-Adaptive Metaheuristic Algorithm

Based on the conversion function and the fitness function defined above, we introduce the proposed algorithm as follows. Instead of using one single mutation strategy, we use two different mutation strategies and adopt a self-adaptation mechanism to select the best strategy for improving the performance. The two different mutation strategies are DE-1 and DE-6, which are two well-known mutation strategies. Therefore, the self-adaptive metaheuristic algorithm is referred to as SaNSDE(DE1, DE6) or SaNSDE-1-6 in this paper for simplicity. The self-adaptation mechanism used by SaNSDE-1-6 keeps track of the number of times that a mutation strategy successfully improves the performance and calculates the success rate of each mutation strategy. A strategy selection index for a mutation strategy is calculated by dividing the success rate of the mutation strategy with the sum of success rate for all mutation strategies. The strategy selection index is used to select one mutation strategy used in the solution-finding processes.

Let N be the problem dimension. To describe a mutation strategy, we use $Z_{gbn} = (z_{gbn})$ to denote the value of the n-th dimension of the best individual in the population of the g-th generation. We use z_{gr_1n}, z_{gr_2n}, z_{gr_3n} and z_{gr_4n} to denote four individuals randomly selected from the current population. In this paper, we use the two strategies defined in (18) and (19) to design the proposed success rate-based self-adaptive metaheuristic algorithm. The n-th dimension of the mutant vector $v_{(g+1)in}$ of the i-th individual in the population of the $(g+1)$-th generation is calculated either by (18) or by (19), depending on the success rates of the two strategies. The flow chart of the success rate-based self-adaptive metaheuristic algorithm is shown in Figure 1.

$$v_{(g+1)in} = z_{gr_1n} + F_i(z_{gr_2n} - z_{gr_3n}) \tag{18}$$

$$v_{(g+1)in} = z_{gin} + F_i(z_{gbn} - z_{gin}) + F_i(z_{gr_1n} - z_{gr_2n}) + F_i(z_{gr_3n} - z_{gr_4n}) \tag{19}$$

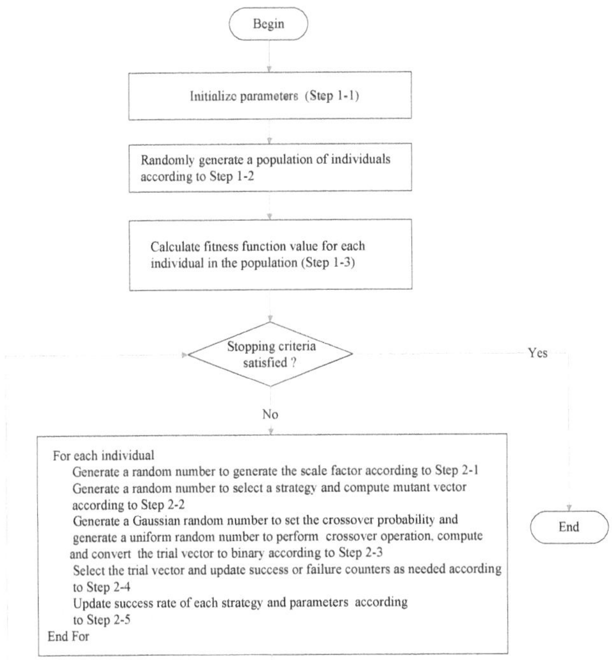

Figure 1. A flowchart of the proposed algorithm.

As we use two mutation strategies, a mutation strategy is referred to as s, where $s \in \{1,2\}$. In the proposed algorithm, the number of times that a mutation strategy s successfully improves the performance is stored in variable S_s. The number of times that a mutation strategy s fails to improve the performance is stored in variable U_s. The success rate of strategy s is $w_s = \frac{S_s}{S_s+U_s}$, where $s \in \{1,2\}$. The parameter f_p used to select the probability distribution to generate the scale factor and select the mutation strategy is calculated by $f_p = \frac{w_1}{w_1+w_2}$. A list L is used to store the crossover probability cr_i that successfully improves performance by executing the statement $L \leftarrow L \cup \{cr_i\}$. The list L is used to update the parameter cr by $cr = \frac{\sum_{k\in\{1,2,\ldots,|L|\}} L(k)}{|L|}$, which is used to generate the crossover probability cr_i in the next generation.

The discrete self-adaptive metaheuristic algorithm based on success rate and differential evolution is listed in Algorithm 1.

Algorithm 1: Discrete Self-Adaptive Metaheuristic Algorithm based on Success Rate and Differential Evolution

---

Step 1: Initialize the parameters and population of individuals
Step 1-1: Initial the parameters
    $cr = 0.5$
    $f_p = 0.5$
Step 1-2: Generate a population with NP individuals randomly
Step 2: Evolve solutions
    For $g = 1$ *to* G
      For $i = 1$ *to* NP
Step 2-1: Generate a uniform random number r from uniform distribution $U(0,1)$ ranging from 0 to 1

$$F_i = \begin{cases} r_1, \text{ where } r_1 \text{ is a Gaussian random number with } N(\mu, \sigma_1^2) & if\ r < f_p \\ r_2, \text{ where } r_2 \text{ is a uniform random number sampled from } U(0,1) & otherwise \end{cases}$$

Step 2-2: Generate a uniform random number r from uniform distribution $U(0,1)$ ranging from 0 to 1
    Calculate the mutant vector v_{gi} as follows.
    For $n \in 1, 2, \ldots, N$

$$v_{(g+1)in} = \begin{cases} z_{gr_1n} + F_i(z_{gr_2n} - z_{gr_3n}) & if\ r < f_p \\ v_{(g+1)in} = z_{gin} + F_i(z_{gbn} - z_{gin}) + F_i(z_{gr_1n} - z_{gr_2n}) + F_i(z_{gr_3n} - z_{gr_4n}) & otherwise \end{cases}$$

$$s = \begin{cases} 1 & if\ r < f_p \\ 2 & otherwise \end{cases}$$

    End For
Step 2-3: Generate a trial vector u_{gi}
    Generate a Gaussian random number cr_i with distribution $N(cr, \sigma_2^2)$
    For $l \in 1, 2, \ldots, N$
      Generate a uniform random number r from uniform distribution $U(0,1)$ ranging from 0 to 1

$$u_{gil} = \begin{cases} v_{gil} & if\ r < cr_i \\ z_{gil} & otherwise \end{cases}$$

      $\overline{u}_{gil} \leftarrow Convert2Binary(u_{gil})$
    End For
Step 2-4: Update the individual and success/failure counters
    If $H_1(\overline{u}_{gi}) \geq H_1(z_{gi})$
      $z_{(g+1)i} = u_{gi}$
      $L \leftarrow L \cup \{cr_i\}$
      $S_s = S_s + 1$
    Else
      $U_s = U_s + 1$
    End If
    End For
Step 2-5: Update the parameters as needed
    If $g > LP$
      $w_1 = \frac{S_1}{(S_1+U_1)}$
      $w_2 = \frac{S_2}{(S_2+U_2)}$
      $f_p = \frac{w_1}{w_1+w_2}$
      $cr = \frac{\sum_{k\in\{1,2,\ldots,|L|\}} L(k)}{|L|}$
    End If
  End For

---

5. Results

As the goal of this paper is to improve the performance of the quality of solutions for the DGRP and improve the convergence rate (the number of generations) for finding the best solutions, verification by the results of experiments is needed to demonstrate the advantage of the proposed algorithm. In this section, the results of experiments obtained by applying the algorithm developed in this paper will be analyzed. Our analysis focuses on two algorithmic properties: performance and convergence rate.

The evaluation process of the algorithms can be divided into five steps. The first step is to select the performance metrics for comparing different algorithms, the second step is to create instances for the DGRP, the third step is to set the parameters for different algorithms, the fourth step is to apply different algorithms to solve each instance of the DGRP and the fifth step is to calculate the performance metrics under consideration based on the results of experiments and compare all algorithms. For the first step, the performance metrics for comparing different algorithms include the average fitness function values, the average number of generations to find the best solutions and the average computation time to find the best solutions. For the second step, the locations of drivers and passengers are randomly generated based on a selected geographical area in Taichung City, which is located in the central part of Taiwan. The number of drivers and the number of passengers are increased gradually to generate instances of the DGRP with different size. For the third step, the parameters for PSO, NSDE, DE-1 and DE-3 are the same as the ones used in [13]. The parameters for SaNSDE-1-6 are specified later in this section. For the fourth step, we apply SaNSDE-1-6 ten times to solve each instance of the DGRP. As the results of applying PSO, NSDE, DE-1 and DE-3 to Case 1 through Case 10 are available in [13], we apply PSO, NSDE, DE-1 and DE-3 ten times to solve to Case 11 through Case 14. For the fifth step, we first calculate the average fitness function values, the average number of generations to find the best solutions and the average computation time to find the best solutions based on the results obtained. We then compare all algorithms based on the performance metrics mentioned above.

In [13], ten algorithms were developed to solve the DGRP. The study of [13] indicates that the NSDE, DE-1, DE-3 and PSO are the top four solvers among the ten algorithms for solving the DGRP in terms of performance and convergence rate (the number of generations to find the best solutions).

To illustrate effectiveness of the algorithm proposed, the experiments include Test Case 1–10 (available at [40]) used in [13] and Test Case 11–14 (available at [41]) to compare with the existing algorithms for the DGRP. To illustrate superiority of the algorithm proposed in terms of scalability with respect to problem size, we generated several test cases by increasing the problem size. We conducted these additional test cases by applying the algorithm proposed in this paper and the best four algorithms reported in [13]. We analyzed by comparing the results obtained by applying all of these algorithms to study the performance and convergence rate of these algorithms as problems grow.

As the effectiveness of evolutionary algorithms depends on the population size parameter, we conducted two series of experiments. The population size parameter of the first series of experiments is 30. The population size parameter of the second series of experiments is 50. The values of algorithmic parameters used by each algorithm are listed in Table 2. The number of generations parameter used by each algorithm is set to 1000 for Test Case 1 through Test Case 10. The number of generations parameter used by each algorithm is set to 50,000 for Test Case 11 through Test Case 14.

Experiments based on the parameters in Table 2 for $NP = 30$ were performed. The results were summarized in Tables 3 and 4 for $NP = 30$. Table 3 shows the average fitness function value and Table 4 shows the average number of iterations to find the best solutions.

Table 2. Parameters for different algorithms and test cases.

Algorithm	Parameters for Case 1 through Case 10	Parameters for Case 11 through Case 14
SaNSDE-1-6	POP = 30, Gen = 1000, LP = 1000	POP = 50, Gen = 50,000, LP = 1000
DE-1	POP = 30, Gen = 1000, $CR = 0.5$ F: a value arbitrarily selected from uniform (0, 2)	POP = 50, Gen = 50,000, $CR = 0.5$ F: a value arbitrarily selected from uniform (0, 2)
DE-3	POP = 30, Gen = 1000, $CR = 0.5$ F: a value arbitrarily selected from uniform (0, 2)	POP = 50, Gen = 50,000, $CR = 0.5$ F: a value arbitrarily selected from uniform (0, 2)
NSDE	POP = 30, Gen = 1000, $CR = 0.5$, $F_i = 0.5r_1 + 0.5$, where r_1 is a random value with Gaussian distribution $N(0,1)$.	POP = 50, Gen = 50,000, $CR = 0.5$, $F_i = 0.5r_1 + 0.5$, where r_1 is a random value with Gaussian distribution $N(0,1)$.
PSO	POP = 30, Gen = 1000, $c_1 = 0.4, c_2 = 0.6, \omega = 0.4$	POP = 50, Gen = 50,000, $c_1 = 0.4, c_2 = 0.6, \omega = 0.4$

Table 3. Fitness function values for discrete SaNSDE-1-6,DE-1, DE-3, NSDE and PSO algorithms with $NP = 30$; $r_D = r_P = 0.1$.

Case	D	P	SaNSDE-1-6	DE-1	DE-3	NSDE	PSO
1	3	10	32.998	32.998	32.998	32.998	32.998
2	5	11	63.615	63.615	63.615	63.615	63.615
3	5	12	41.715	41.715	41.715	41.715	41.2892
4	6	12	51.11	51.11	51.11	51.11	50.9085
5	7	13	30.063	30.063	30.063	30.063	28.4254
6	8	14	72.328	72.328	72.328	72.328	70.2629
7	9	15	89.03	89.03	89.03	89.03	80.8106
8	10	16	54.02	54.02	54.02	54.02	44.0023
9	11	17	74.05	74.05	74.05	74.05	49.356
10	12	18	50.9	50.0623	50.9	50.9	32.8349
11	20	20	112.906	112.906	112.906	112.906	97.7979
12	30	30	202.15	196.9089	200.1078	200.7964	141.6005
13	40	40	201.8256	190.1664	179.4244	186.6996	−1.5081
14	50	50	190.9436	137.1625	171.1756	161.3107	−3.9598

Table 4. Average number of generations for discrete SaNSDE-1-6,DE-1, DE-3, NSDE and PSO algorithms with $NP = 30$; $r_D = r_P = 0.1$.

Case	D	P	SaNSDE-1-6	DE-1	DE-3	NSDE	PSO
1	3	10	9.6	16.6	19.8	15.6	64.6
2	5	11	19.2	32.2	36.9	29.1	299.6
3	5	12	29.4	39.7	47.6	43.3	394.5
4	6	12	19.6	43.8	50.3	44.1	320.9
5	7	13	16.7	31.8	44.1	37.3	304.1
6	8	14	20	48.3	67.8	39.6	375.6
7	9	15	60.8	101.4	135	70.7	553.6
8	10	16	48.2	61.3	78.6	59.5	447.5
9	11	17	53.9	59.7	65	64.2	580.7
10	12	18	106.4	136.8	146.3	94.3	489.3
11	20	20	191	436.5	817.1	542.5	21,314.5
12	30	30	1392.5	5691.5	16,065.1	14,417.2	21,742.3
13	40	40	18,439.777	15,185.8	27,574.4	27,260.1	22,734.2
14	50	50	19,390.5	17,131.7	22,745.3	36,742.7	25,822.2

The results in Table 3 show that the top four algorithms are SaNSDE-1-6, NSDE, DE-1 and DE-3. For small test cases, including Case 1 through Case 11, the fitness function values obtained using SaNSDE-1-6, NSDE, DE-1 and DE-3 are the same. However, as the problem size grows, the average fitness function values obtained using SaNSDE-1-6 are significantly better than those obtained using NSDE, DE-1 and DE-3. For Case 12, the average fitness function value obtained using SaNSDE-1-6 is better than those obtained using NSDE, DE-1 and DE-3. The differences between the average fitness function value obtained using SaNSDE-1-6 and those obtained using NSDE, DE-1 and DE-3 are about 1% to 2%. For Case 12, the average fitness function value obtained using SaNSDE-1-6 is better than those obtained using NSDE, DE-1 and DE-3. For Case 13, the differences between the average fitness function value obtained using SaNSDE-1-6 and those obtained using NSDE, DE-1 and DE-3 are about 5% to 10%. For Case 14, the differences between the average fitness function value obtained using SaNSDE-1-6 and those obtained using NSDE, DE-1 and DE-3 are about 10% to 28%. In short, SaNSDE-1-6 outperforms NSDE, DE-1 and DE-3 in terms of scalability. To compare performance clearly, please refer to the bar chart shown in Figure 2 for the average fitness function values of Case 1 through Case 14.

In terms of convergence rate (the number of generations to find the best solutions), the results in Table 4 indicate that the average numbers of iterations for SaNSDE-1-6 to find the best solutions are significantly less than those for NSDE, DE-1 and DE-3 to find the best solutions for most test cases (with some exceptions). This indicates that SaNSDE-1-6 outperforms NSDE, DE-1 and DE-3 in terms of convergence rate. To compare the convergence rate clearly, please refer to the bar chart shown in Figure 3 for the average number of generations of Case 1 through Case 10 and please refer to the bar chart shown in Figure 4 for the average number of generations of Case 11 through Case 14.

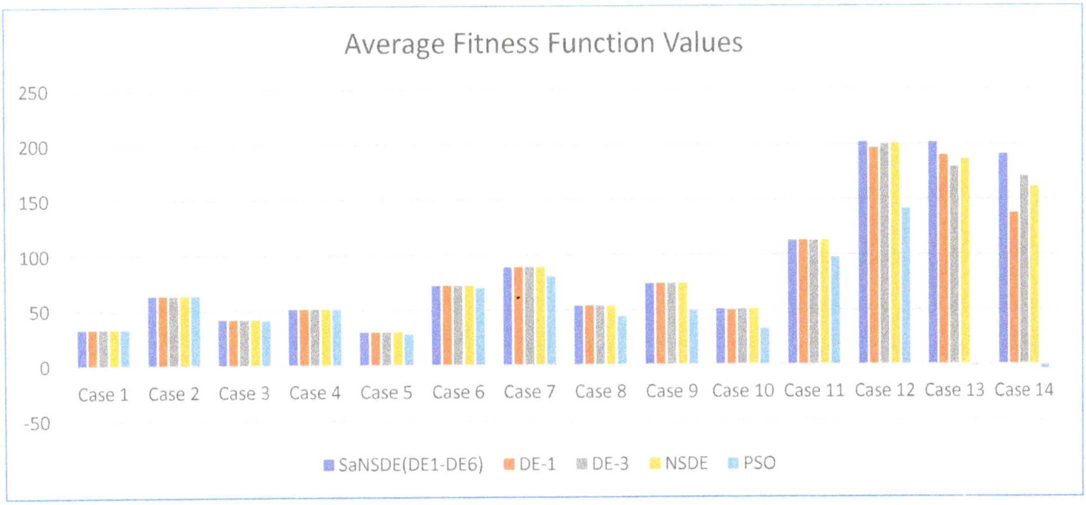

Figure 2. Average fitness function values for $r_D = r_P = 0.1$ with POP = 30.

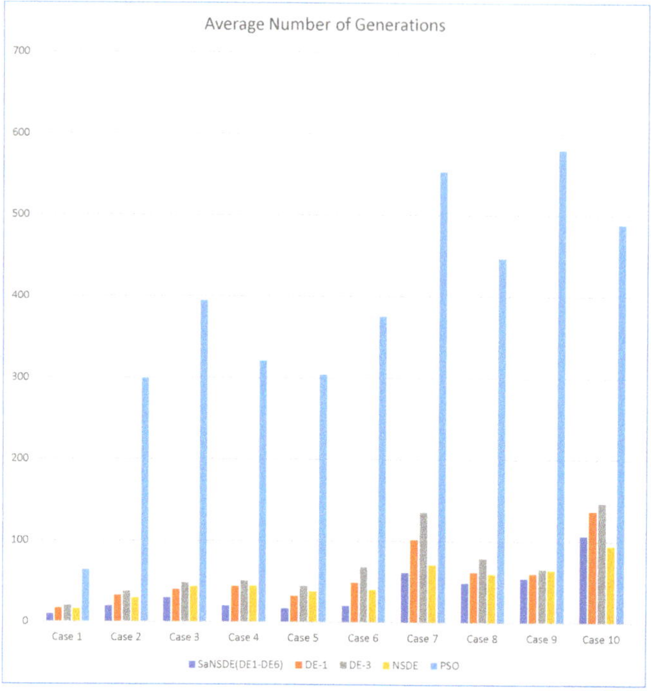

Figure 3. Average number of generations for Case 1 through Case 10 with $r_D = r_P = r = 0.1$ and POP = 30.

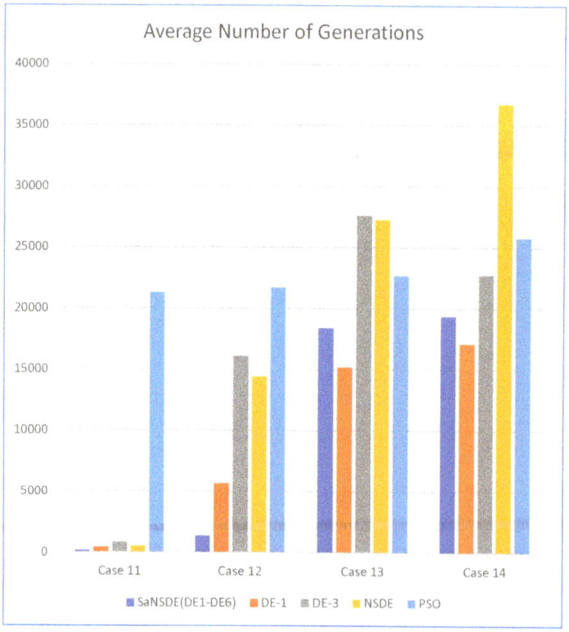

Figure 4. Average number of generations for Case 11 through Case 14 with $r_D = r_P = r = 0.1$ and POP = 30.

To verify the convergence rate for POP = 30, we show the results of several runs of Case 5, Case 11, Case 12, Case 13 and Case 14 in Figures 5–9, respectively.

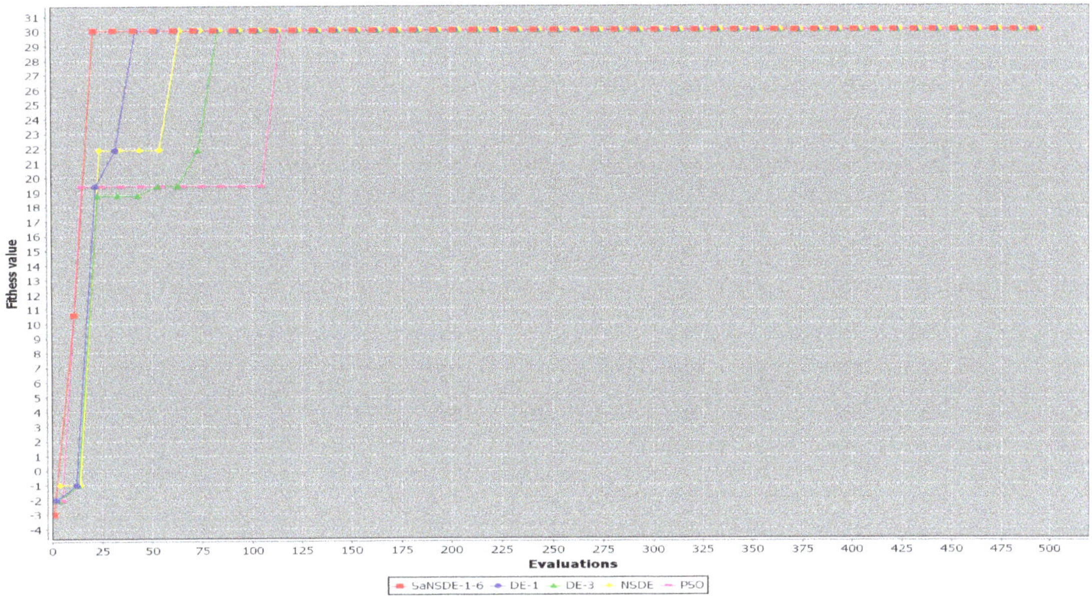

Figure 5. Convergence curves for a run of Case 5 for $r_D = r_P = r = 0.1$ with POP = 30.

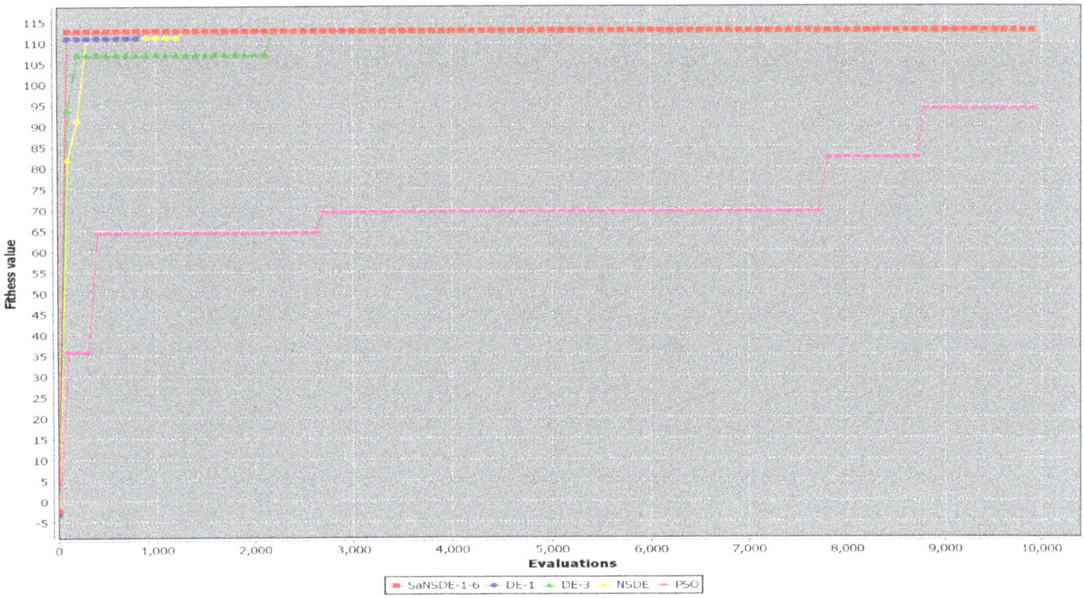

Figure 6. Convergence curves for a run of Case 11 for $r_D = r_P = r = 0.1$ with POP = 30.

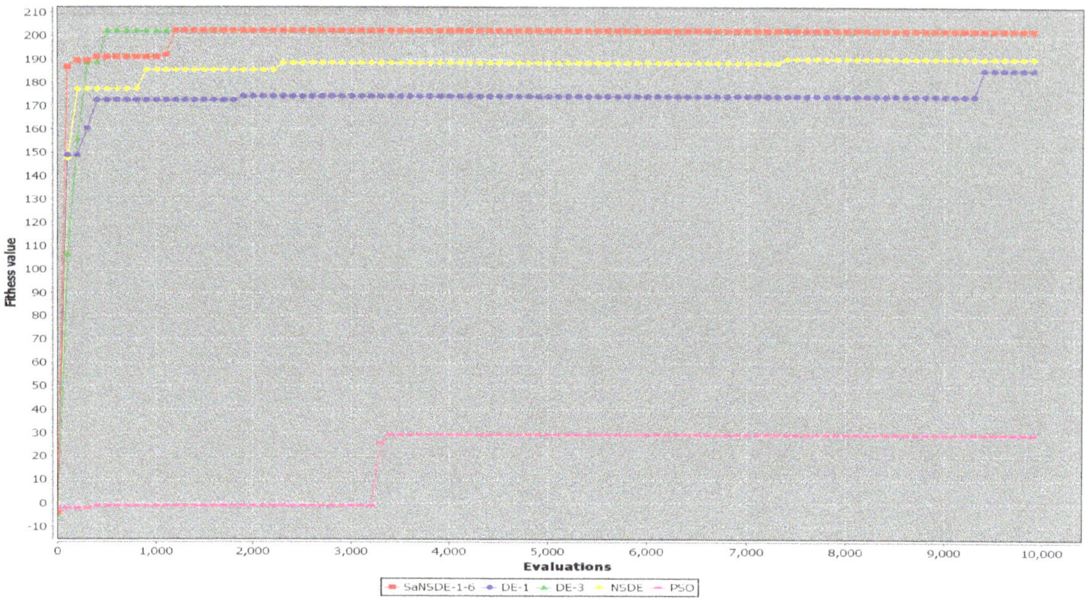

Figure 7. Convergence curves for a run of Case 12 for $r_D = r_P = r = 0.1$ with POP = 30.

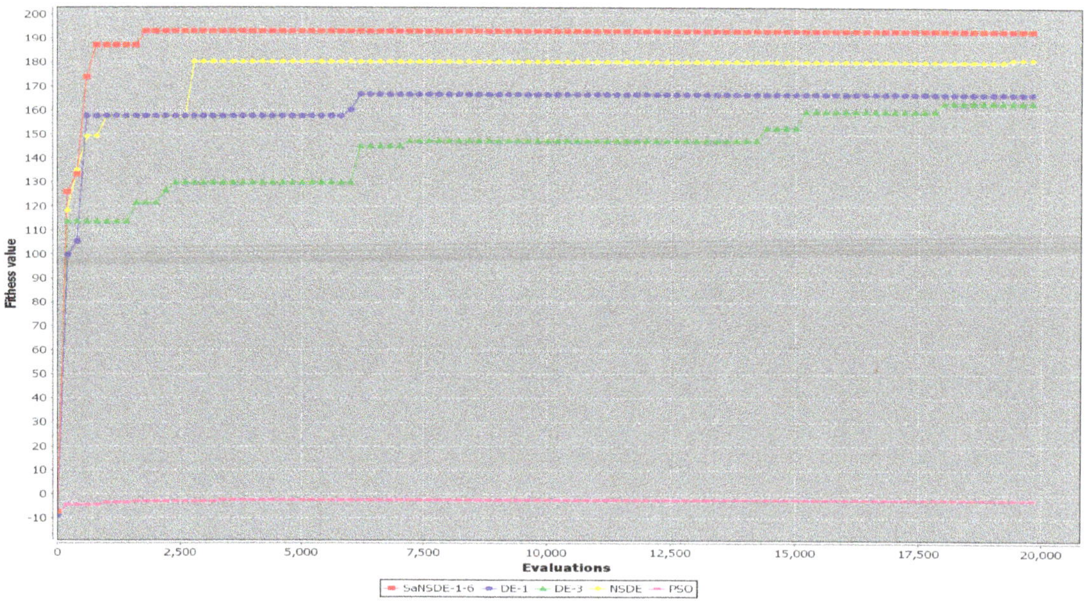

Figure 8. Convergence curves for a run of Case 13 for $r_D = r_P = r = 0.1$ with POP = 30.

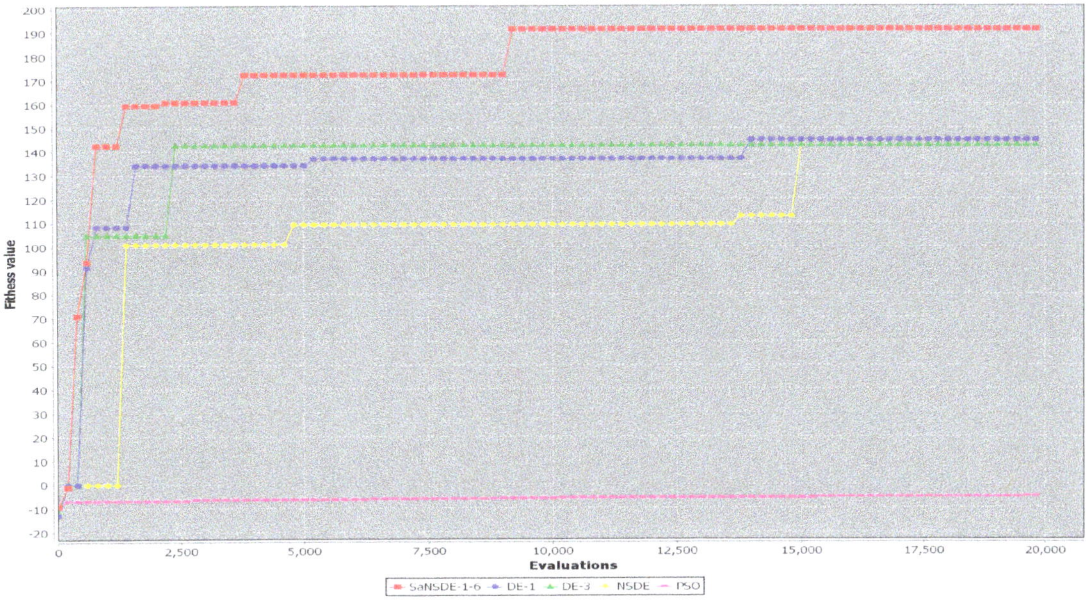

Figure 9. Convergence curves for a run of Case 14 for $r_D = r_P = r = 0.1$ with POP = 30.

The results presented above are based on a comparison of the average number of generations. For the comparison of computation time, the results in Table 5 indicate that the average computation time for SaNSDE-1-6 to find the best solutions is significantly less than that for PSO to find the best solutions for Case 1 through Case 9 and is greater than those of NSDE, DE-1 and DE-3 for Case 1 through Case 10. This indicates that SaNSDE-1-6 outperforms PSO in terms of computation time for Case 1 through Case 9 and NSDE, DE-1 and DE-3 outperform SaNSDE-1-6 in terms of computation time for Case 1 through Case 10. For Case 11, SaNSDE-1-6 outperforms PSO, NSDE, DE-1 and DE-3 in terms of computation time. For bigger cases, Case 12 through Case 14, PSO, NSDE, DE-1 and DE-3 outperform SaNSDE-1-6 in terms of computation time. As the experiments were done on the same platform as the one used in [13], which was an old laptop delivered in 2019 with Intel(R) Core(TM) i7 CPU, base clock speed of 2.6 GHz and16 GB of onboard memory, to compare different algorithms, the computation times of SaNSDE-1-6 are much longer for Case 12, Case 13 and Case 14. Obviously, a more powerful computer or a server class computer is required to apply the SaNSDE-1-6 algorithm.

Experiments based on the parameters in Table 2 for $NP = 50$ were performed. The results were summarized in Tables 6 and 7 for $NP = 50$. Table 6 shows the average fitness function values and Table 7 shows the average number of iterations to find the best solutions.

Table 5. Average computation time (in mini-second) for discrete SaNSDE-1-6,DE-1, DE-3, NSDE and PSO algorithms with $NP = 30$; $r_D = r_P = 0.1$.

Case	D	P	SaNSDE-1-6	DE-1	DE-3	NSDE	PSO
1	3	10	11.494	5.8907	6.7182	7.3339	15.1677
2	5	11	31.0079	21.2555	18.0998	19.4591	118.2929
3	5	12	45.7525	22.0871	25.3382	26.1856	130.6768
4	6	12	42.0781	25.937	25.8155	29.1062	113.5842
5	7	13	28.2401	18.4303	22.264	16.7606	100.2475
6	8	14	48.7842	25.7381	34.9751	28.5755	132.6016
7	9	15	139.7892	73.5523	87.1861	55.5141	264.7854
8	10	16	111.0234	45.3785	54.5258	51.003	200.8072
9	11	17	152.6207	50.241	50.4531	64.2084	286.7221
10	12	18	236.4897	108.65	106.4809	87.0697	199.9287
11	20	20	798.39717	1134.249	2142.048	1496.592	45,171.44
12	30	30	21,158.007	17,419.87	49,393.98	48,106.86	51,932.66
13	40	40	2,116,465.2	60,483.27	113,286.7	119,378.3	66,539.42
14	50	50	3,198,096.1	90,395.02	118,089.5	144,559.9	87,874.68

Table 6. Fitness function values for discrete SaNSDE-1-6,DE-1, DE-3, NSDE and PSO algorithms with $NP = 50$; $r_D = r_P = 0.1$.

Case	D	P	SaNSDE-1-6	DE-1	DE-3	NSDE	PSO
1	3	10	32.998	32.998	32.998	32.998	32.998
2	5	11	63.615	63.615	63.615	63.615	63.615
3	5	12	41.715	41.715	41.715	41.715	41.2892
4	6	12	51.11	51.11	51.11	51.11	51.11
5	7	13	30.063	30.063	30.063	30.063	30.063
6	8	14	72.328	72.328	72.328	72.328	69.9483
7	9	15	89.03	89.03	89.03	89.03	80.5986
8	10	16	54.02	54.02	54.02	54.02	46.8013
9	11	17	74.05	74.05	74.05	74.05	55.9356
10	12	18	50.9	50.9	50.9	50.9	31.1131
11	20	20	112.906	112.906	112.906	112.906	104.4808
12	30	30	202.15	201.8116	200.6549	201.8116	145.0514
13	40	40	202.4952	194.0284	190.4004	185.9914	−1.2835
14	50	50	192.343	190.4874	157.1781	162.1984	−3.6674

Table 7. Average number of generations for discrete SaNSDE-1-6,DE-1, DE-3, NSDE and PSO algorithms with $NP = 50$; $r_D = r_P = 0.1$.

Case	D	P	SaNSDE-1-6	DE-1	DE-3	NSDE	PSO
1	3	10	10.1	17.6	19.2	12.9	51.5
2	5	11	18.7	30.2	33	26.4	127.3
3	5	12	22.3	28.7	43.3	32.9	437.2
4	6	12	22.5	35.4	37.4	33.2	468.6
5	7	13	17.7	29.3	32.2	24.8	247.1
6	8	14	30	38	47.1	41.9	416.4
7	9	15	37.5	78.9	75.4	61.3	366.4
8	10	16	35.3	51.5	66.3	48.6	609.5
9	11	17	65.6	68.7	78.4	67	611.5
10	12	18	79.4	83.6	197.9	63.1	521.5
11	20	20	98.6	469.5	829.5	565.5	22,275.7
12	30	30	2216.7	8847.1	6494.7	26,379.5	20,738.9
13	40	40	15,231.4	14,944.6	18,551.7	21,166.5	30,168.9
14	50	50	16,730.2	28,597.9	32,025.9	25,665.8	33,480.8

The results in Table 6 show that the top four algorithms are SaNSDE-1-6, NSDE, DE-1 and DE-3. For small test cases, including Case 1 through Case 11, the fitness function values obtained using SaNSDE-1-6, NSDE, DE-1 and DE-3 are the same. However, as the problem size grows, the average fitness function values obtained via SaNSDE-1-6 are significantly better than those obtained via NSDE, DE-1 and DE-3. For Case 12, the average fitness function value obtained via SaNSDE-1-6 is better than those obtained by NSDE, DE-1 and DE-3. The differences between the average fitness function value obtained via SaNSDE-1-6 and those obtained via NSDE, DE-1 and DE-3 are about 0.1674% to 0.73959%. For Case 12, the average fitness function value obtained via SaNSDE-1-6 is better than those obtained via NSDE, DE-1 and DE-3. For Case 13, the differences between the average fitness function values obtained via SaNSDE-1-6 and those obtained via NSDE, DE-1 and DE-3 are about 4.00326% to 7.9796%. For Case 14, the differences between the average fitness function value obtained via SaNSDE-1-6 and those obtained via NSDE, DE-1 and DE-3 are about 3.1171% to 46.403%. In short, SaNSDE-1-6 outperforms NSDE, DE-1 and DE-3 in terms of scalability. To compare performance clearly, please refer to the bar chart shown in Figure 10 for the average fitness function values of Case 1 through Case 14.

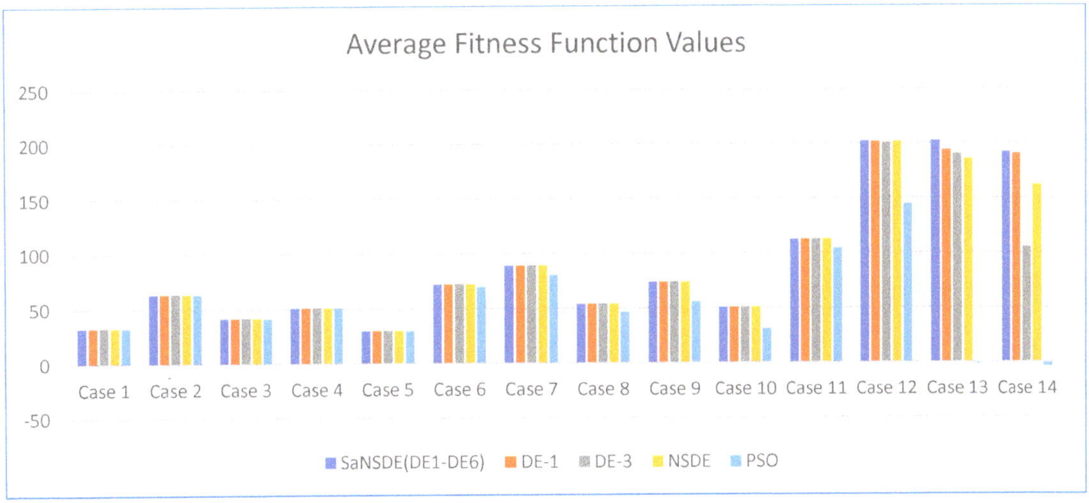

Figure 10. Average fitness function values for $r_D = r_P = r = 0.1$ with POP = 50.

In terms of convergence rate (the number of generations to find the best solutions), the results in Table 7 indicate that the average numbers of iterations for SaNSDE-1-6 to find the best solutions are significantly less than those for NSDE, DE-1 and DE-3 to find the best solutions for most test cases (with some exception). This indicates that SaNSDE-1-6 outperforms NSDE, DE-1 and DE-3 in convergence rate. To compare the convergence rate clearly, please refer to the bar chart shown in Figure 11 for the average number of generations of Case 1 through Case 10 and please refer to the bar chart shown in Figure 12 for the average number of generations of Case 11 through Case 14.

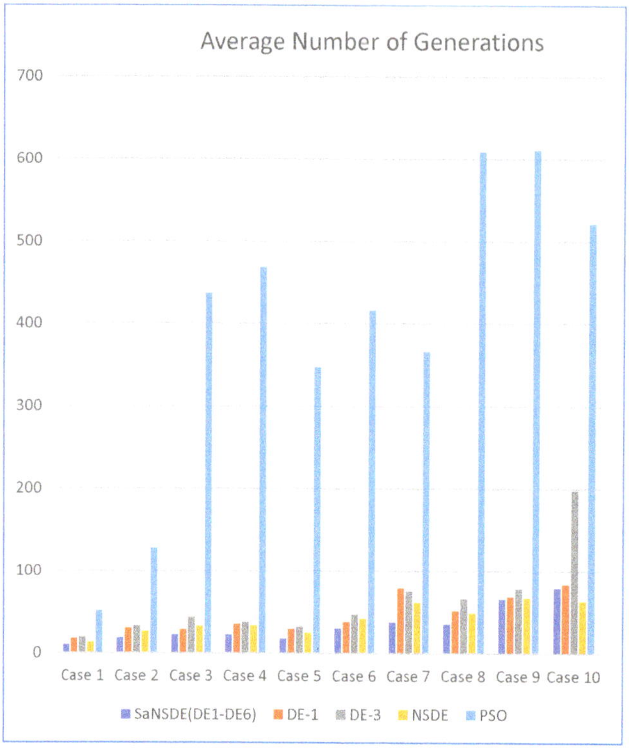

Figure 11. Average number of generations for Case 1 through Case 10 with $r_D = r_P = r = 0.1$ and POP = 50.

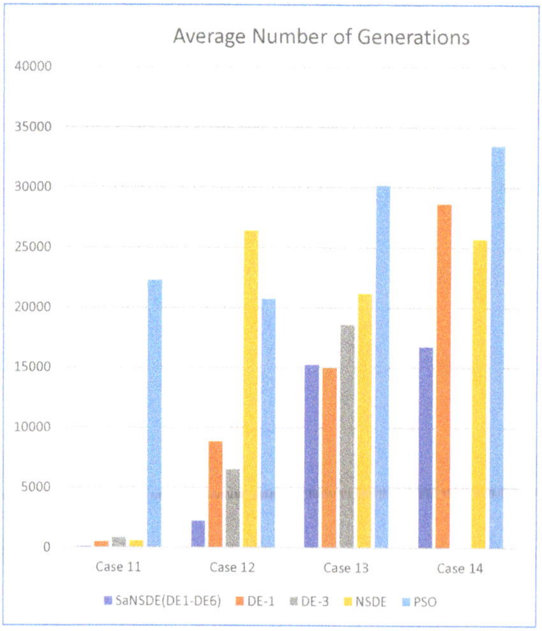

Figure 12. Average number of generations for Case 11 through Case 14 with $r_D = r_P = r = 0.1$ and POP = 50.

To verify the convergence rate for POP = 50, we show the results of several runs of Case 5, Case 11, Case 12, Case 13 and Case 14 in Figures 13–17, respectively.

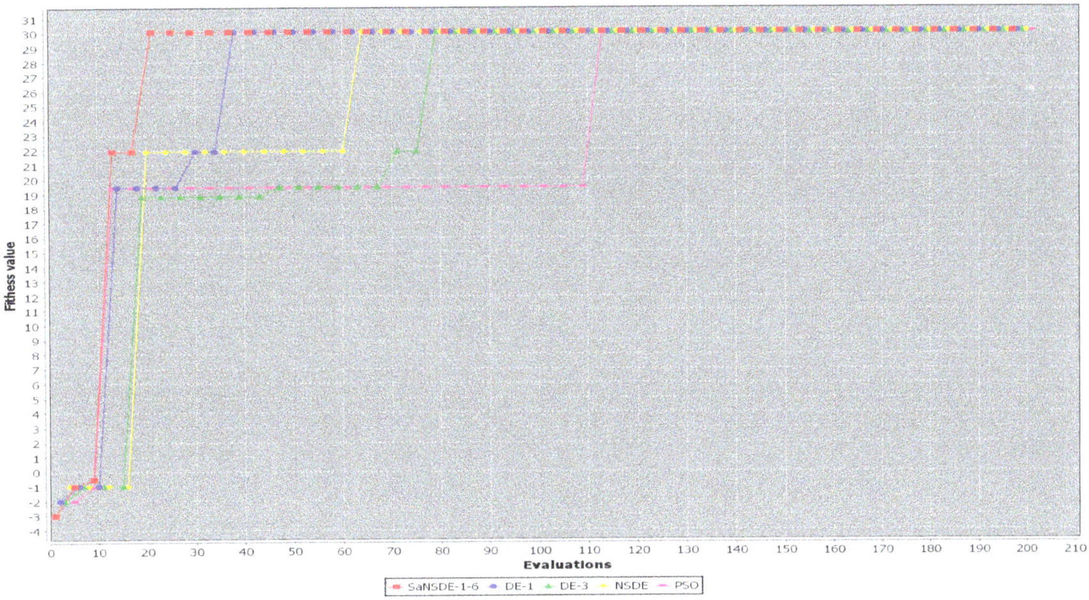

Figure 13. Convergence curves for a run of Case 5 for $r_D = r_P = r = 0.1$ with POP = 50.

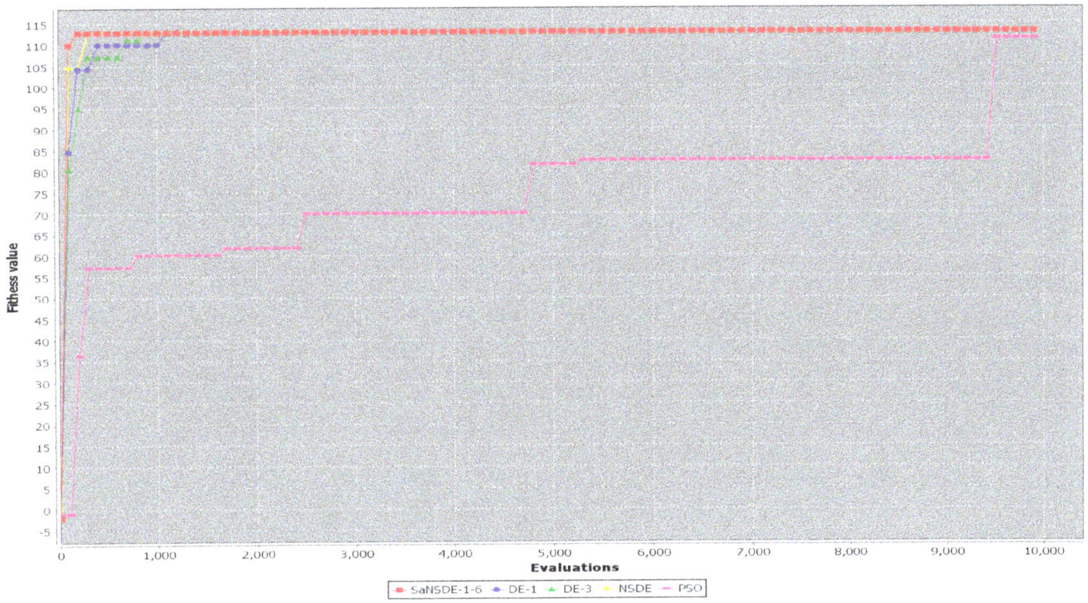

Figure 14. Convergence curves for a run of Case 11 for $r_D = r_P = r = 0.1$ with POP = 50.

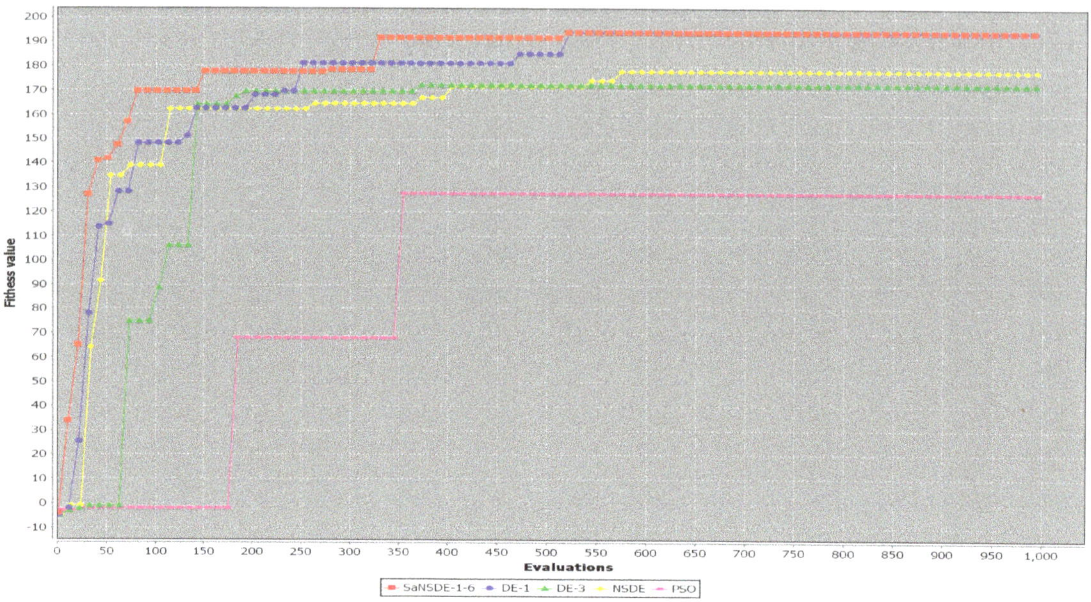

Figure 15. Convergence curves for a run of Case 12 for $r_D = r_P = r = 0.1$ with POP = 50.

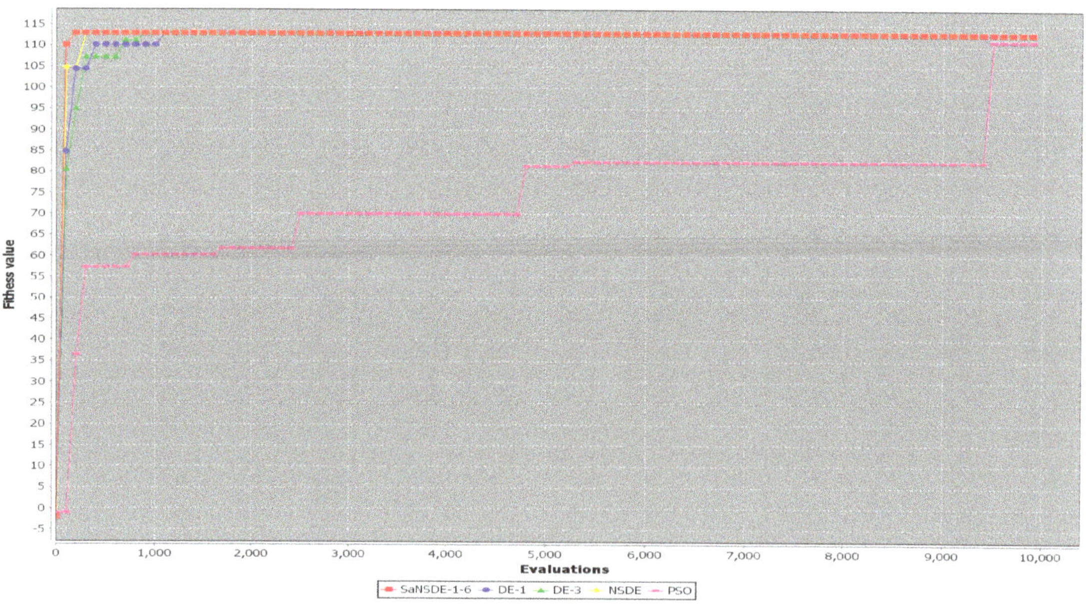

Figure 16. Convergence curves for a run of Case 13 for $r_D = r_P = r = 0.1$ with POP = 50.

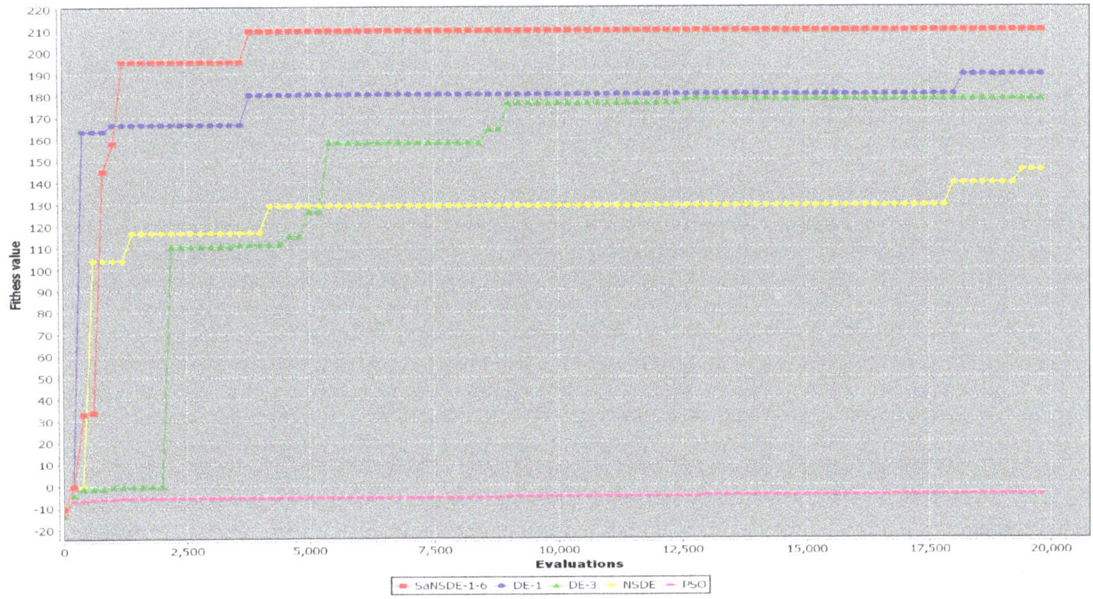

Figure 17. Convergence curves for a run of Case 14 for $r_D = r_P = r = 0.1$ with POP = 50.

The results presented above are based on comparison of average number of generations. The results in Table 8 indicate that the average computation time for SaNSDE-1-6 to find the best solutions is significantly less than that for PSO to find the best solutions for Case 1 through Case 10 and is greater than those of NSDE, DE-1 and DE-3 for Case 1 through Case 10. This indicates that SaNSDE-1-6 outperforms PSO in terms of computation time for Case 1 through Case 10 and NSDE, DE-1 and DE-3 outperform SaNSDE-1-6 in terms of computation time for Case 1 through Case 10. For Case 11 and Case 12, SaNSDE-1-6 outperforms PSO, NSDE, DE-1 and DE-3 in terms of computation time. For Case 13 through Case 14, PSO, NSDE, DE-1 and DE-3 outperform SaNSDE-1-6 in terms of computation time. As the experiments to compare the different algorithms were done on the same platform as the one used in [13], which was an old laptop delivered in 2019 with Intel(R) Core(TM) i7 CPU, base clock speed of 2.6 GHz and16 GB of onboard memory, the computation times of SaNSDE-1-6 are much longer for Case 12, Case 13 and Case 14. Obviously, a more powerful computer or a server class computer is required to apply the SaNSDE-1-6 algorithm.

Table 8. Average computation time (in mini-second) for discrete SaNSDE-1-6,DE-1, DE-3, NSDE and PSO algorithms with $NP = 50$; $r_D = r_P = 0.1$.

Case	D	P	SaNSDE-1-6	DE-1	DE-3	NSDE	PSO
1	3	10	10.2156	10.1452	10.1662	8.4324	18.3854
2	5	11	31.0283	24.8503	24.4498	23.7312	57.3819
3	5	12	42.8796	23.849	32.7592	31.2488	212.022
4	6	12	33.3958	31.4915	29.161	33.3068	205.7338
5	7	13	33.6951	25.0836	25.8515	25.2025	105.7623
6	8	14	57.5261	32.3747	37.1701	42.8416	213.5017
7	9	15	117.5981	83.4787	78.1544	75.29	205.419
8	10	16	84.5311	65.2448	71.6938	71.265	385.2561
9	11	17	105.9706	87.5031	93.5826	86.378	413.8388
10	12	18	148.0639	101.1059	232.0098	96.711	400.4348
11	20	20	679.74567	1395.58	2805.873	1912.745	51,907.85

Table 8. *Cont.*

Case	D	P	SaNSDE-1-6	DE-1	DE-3	NSDE	PSO
12	30	30	14,772.423	35,531.47	25,760.51	118,419.7	57,773.8
13	40	40	2,016,957.3	83,749.26	104,356	131,989.4	108,469.1
14	50	50	2,890,556.2	171,886.2	233,575.7	201,288.2	145,752.5

6. Discussion and Conclusions

In this paper, we applied the self-adaptation concept to develop an algorithm to improve the performance in finding solutions for the DGRP formulated in the previous study. The self-adaptation mechanism used in this paper attempts to identify a better strategy that can be selected in the future as the strategy for mutation with a higher probability. To identify a better strategy and the probability for serving as a mutation strategy in the future, the algorithm records the number of "success events" and the number of "failure events" in a learning period. The probability for serving as the mutation strategy is calculated based on the number of "success events" and the number of "failure events" in a learning period for each mutation strategy. A mutation strategy with a higher probability for serving as the mutation strategy will be selected with a higher probability. A mutation strategy with a lower probability for serving as the mutation strategy will be selected with a lower probability. In this way, the performance of the solution that is found can be improved more efficiently in terms of the average number of generations for most cases. However, due to the additional computation in each iteration, the computation time of SaNSDE-1-6 is much longer for big cases.

A mutation strategy with a higher probability for serving as the mutation strategy indicates that the ratio between the number of "success events" and the total number of "success events" and "failure events" is higher. It is expected that using a mutation strategy with a higher probability for serving as the mutation strategy tends to improve the performance of the solution that is found. The results presented in the previous section confirm that using a more effective mutation strategy with a higher probability for serving as the mutation strategy indeed improves the performance of the solution that is found significantly. The degree of improvement is case dependent. With $NP = 30$, for Case 12, the improvement achieved using SaNSDE-1-6 is about 1% to 2%. For Case 13, the improvement achieved using SaNSDE-1-6 is about 5% to 10%. For Case 14, the improvement achieved using SaNSDE-1-6 is about 10% to 28%. In short, SaNSDE-1-6 outperforms NSDE, DE 1 and DE-3 in terms of scalability. With $NP = 50$, for Case 12, the improvement achieved using SaNSDE-1-6 is about 0.1674% to 0.73959%. For Case 13, the improvement achieved using SaNSDE-1-6 is about 4.00326% to 7.9796%. For Case 14, the improvement achieved using SaNSDE-1-6 is about 3.1171% to 46.403%. In short, SaNSDE-1-6 outperforms NSDE, DE-1 and DE-3 in terms of scalability. The bigger the problem size, the more significant the improvement.

In the real world, when one person fails to solve a problem alone, it might be easier to solve the problem by asking another person for help and working together. The reason is that one may consult the other and/or help each other when taking actions or making decisions. This way to solve a problem effectively is commonly used in our daily life. The results of the experiments presented in this paper are consistent with the abovementioned phenomena in the real world. In our self-adaptation mechanism, there are two strategies involved in the solution-finding processes. The selection of one strategy in the solution-searching processes is based on the success probability learned from the learning period. To verify the effectiveness of the self-adaptation mechanism, we carried out experiments by applying several standard algorithms and our proposed algorithm. Two different population sizes were used to perform the experiments. We compared the effectiveness of several single strategy algorithms and the self-adaptation-based algorithm. Our results indicate that the proposed algorithm based on the self-adaptation mechanism improves the performance and convergence rate in terms of the average number of generations required for finding the solutions for most cases. Although our proposed algorithm outperforms all of the other four algorithms in terms of performance and convergence rate for most cases, the computation time of the proposed algorithm is much longer for several big cases due to the additional computation in each iteration. The results of this study have two implications. First, the performance in solving the DGRP with two strategies and a self-adaptation mechanism is better than with one strategy. Second, although the performance in solving the DGRP can be improved and the average number of generations required for finding the solution is reduced, the computation time of the proposed algorithm is much longer than all of the other four algorithms for bigger instances. This implies that either a more powerful computer or a proper divide-and–conquer strategy to divide a big instance of the DGRP into small ones must be used before applying the proposed algorithm. The computational experience showing that the proposed self-adaptive algorithm outperforms the other four algorithms for the test cases in this paper sparks an interesting research question: does the proposed self-adaptive algorithm outperform the other four algorithms? This research question requires further study in the comparative analysis of the proposed algorithm. A comparative analysis of the algorithms studied in this paper for specific performance indicators is a challenging future research direction. Studies of other performance evaluation indicators for the proposed algorithm are another interesting future research directions. The other interesting future research direction is to extend the success rate-based self-adaptive scheme proposed in this study to other evolutionary approaches.

Funding: This research was supported in part by the National Science and Technology Council, Taiwan, under Grant NSTC 111-2410-H-324-003.

Data Availability Statement: Data available in a publicly accessible repository described in the article.

Conflicts of Interest: The author declares no conflicts of interest.

References

1. Bruglieri, M.; Ciccarelli, D.; Colorni, A.; Luè, A. PoliUniPool: A carpooling system for universities. *Procedia-Soc. Behav. Sci.* **2011**, *20*, 558–567. [CrossRef]
2. Hwang, K.; Giuliano, G. The Determinants of Ridesharing: Literature Review. Working Paper UCTC No. 38, The University of California Transportation Center. 1990. Available online: https://escholarship.org/uc/item/3r91r3r4 (accessed on 29 November 2023).
3. Uber. Available online: https://www.uber.com (accessed on 29 November 2023).
4. Lyft. Available online: https://www.lyft.com (accessed on 29 November 2023).
5. BlaBlaCar. Available online: https://www.blablacar.com (accessed on 29 November 2023).
6. Agatz, N.; Erera, A.; Savelsbergh, M.; Wang, X. Optimization for dynamic ride-sharing: A review. *Eur. J. Oper. Res.* **2012**, *223*, 295–303. [CrossRef]
7. Furuhata, M.; Dessouky, M.; Ordóñez, F.; Brunet, M.; Wang, X.; Koenig, S. Ridesharing: The state-of-the-art and future direc-tions. *Transp. Res. Part B Methodol.* **2013**, *57*, 28–46. [CrossRef]
8. Mourad, A.; Puchinger, J.; Chu, C. A survey of models and algorithms for optimizing shared mobility. *Transp. Res. Part B Methodol.* **2019**, *123*, 323–346. [CrossRef]
9. Martins, L.C.; Torre, R.; Corlu, C.G.; Juan, A.A.; Masmoudi, M.A. Optimizing ride-sharing operations in smart sustainable cities: Challenges and the need for agile algorithms. *Comput. Ind. Eng.* **2021**, *153*, 107080. [CrossRef]
10. Ting, K.H.; Lee, L.S.; Pickl, S.; Seow, H.-V. Shared Mobility Problems: A Systematic Review on Types, Variants, Characteristics, and Solution Approaches. *Appl. Sci.* **2021**, *11*, 7996. [CrossRef]
11. Hsieh, F.S.; Zhan, F.; Guo, Y. A solution methodology for carpooling systems based on double auctions and cooperative coevolutionary particle swarms. *Appl. Intell.* **2019**, *49*, 741–763. [CrossRef]
12. Hsieh, F.S. A Comparative Study of Several Metaheuristic Algorithms to Optimize Monetary Incentive in Ridesharing Systems. *ISPRS Int. J. Geo-Inf.* **2020**, *9*, 590. [CrossRef]
13. Hsieh, F.-S. Development and Comparison of Ten Differential-Evolution and Particle Swarm-Optimization Based Algorithms for Discount-Guaranteed Ridesharing Systems. *Appl. Sci.* **2022**, *12*, 9544. [CrossRef]
14. Hsieh, F.S. Trust-Based Recommendation for Shared Mobility Systems Based on a Discrete Self-Adaptive Neighborhood Search Differential Evolution Algorithm. *Electronics* **2022**, *11*, 776. [CrossRef]
15. Hsieh, F.-S. A Comparison of Three Ridesharing Cost Savings Allocation Schemes Based on the Number of Acceptable Shared Rides. *Energies* **2021**, *14*, 6931. [CrossRef]

16. Hsieh, F.-S. Improving Acceptability of Cost Savings Allocation in Ridesharing Systems Based on Analysis of Proportional Methods. *Systems* **2023**, *11*, 187. [CrossRef]
17. Deb, K. An efficient constraint handling method for genetic algorithms. *Comput. Methods Appl. Mech. Eng.* **2000**, *186*, 311–338. [CrossRef]
18. Hyland, M.; Mahmassani, H.S. Operational benefits and challenges of shared-ride automated mobility-on-demand services. *Transp. Res. Part A Policy Pract.* **2020**, *134*, 251–270. [CrossRef]
19. Michalewicz, Z. *Genetic Algorithms + Data Structures = Evolution Programs*; Springer: New York, NY, USA, 1992.
20. Kennedy, J.; Eberhart, R.C. Particle swarm optimization. In Proceedings of the IEEE International Conference on Neural Networks, Perth, WA, Australia, 27 November–1 December 1995; pp. 1942–1948.
21. Yang, X.S. Firefly algorithms for multimodal optimization. In *Stochastic Algorithms: Foundations and Applications. SAGA 2009*; Lecture Notes in Computer Science; Springer: Berlin/Heidelberg, Germany, 2009; Volume 5792, pp. 169–178.
22. Storn, R.; Price, K. Differential evolution—A simple and efficient heuristic for global optimization over continuous spaces. *J. Global Optim.* **1997**, *11*, 341–359. [CrossRef]
23. Katoch, S.; Chauhan, S.S.; Kumar, V. A review on genetic algorithm: Past, present, and future. *Multimed. Tools Appl.* **2021**, *80*, 8091–8126. [CrossRef]
24. Shami, T.M.; El-Saleh, A.A.; Alswaitti, M.; Al-Tashi, Q.; Summakieh, M.A.; Mirjalili, S. Particle Swarm Optimization: A Comprehensive Survey. *IEEE Access* **2022**, *10*, 10031–10061. [CrossRef]
25. Li, J.; Wei, X.; Li, B.; Zeng, Z. A survey on firefly algorithms. *Neurocomputing* **2022**, *500*, 662–678. [CrossRef]
26. Ahmad, M.F.; Isa, N.A.M.; Lim, W.H.; Ang, K.M. Differential evolution: A recent review based on state-of-the-art works. *Alex. Eng. J.* **2022**, *61*, 3831–3872. [CrossRef]
27. Eberhart, R.C.; Shi, Y. Comparison between genetic algorithms and particle swarm optimization. In *Evolutionary Programming VII, Proceedings of the 7th International Conference, ep98, San Diego, CA, USA, 25–27 March 1998*; Porto, V.W., Saravanan, N., Waagen, D., Eiben, A.E., Eds.; Lecture Notes in Computer Science; Springer: Berlin/Heidelberg, Germany, 1998; Volume 1447, pp. 611–616.
28. Hassan, R.; Cohanim, B.; Weck, O.D. A comparison of particle swarm optimization and the genetic algorithm. In Proceedings of the 46th AIAA/ASME/ASCE/AHS/ASC Structures, Structural Dynamics and Materials Conference, Structures, Structural Dynamics, and Materials and Collocated Conferences, Austin, TX, USA, 18–21 April 2005. [CrossRef]
29. Tušar, T.; Filipič, B. Differential Evolution versus Genetic Algorithms in Multiobjective Optimization. In *Evolutionary Multi-Criterion Optimization*; Obayashi, S., Deb, K., Poloni, C., Hiroyasu, T., Murata, T., Eds.; Lecture Notes in Computer Science; Springer: Berlin/Heidelberg, Germany, 2007; Volume 4403, pp. 257–271.
30. Qin, A.K.; Suganthan, P.N. Self-adaptive Differential Evolution Algorithm for Numerical Optimization. *Proc. IEEE Congr. Evol. Comput.* **2005**, *2*, 1784–1791.
31. Omran, M.G.H.; Salman, A.; Engelbrecht, A.P. Self-adaptive differential evolution. *Proc. Comput. Intell. Secur. Lect. Notes Artif. Intell.* **2005**, *3801*, 192–199.
32. Huang, V.L.; Qin, A.K.; Suganthan, P.N. Self-adaptive differential evolution algorithm for constrained real-parameter optimization. In Proceedings of the 2006 IEEE International Conference on Evolutionary Computation, Vancouver, BC, Canada, 16–21 July 2006; pp. 324–331.
33. Qin, A.K.; Huang, V.L.; Suganthan, P.N. Differential evolution algorithm with strategy adaptation for global numerical optimization. *IEEE Trans. Evol. Comput.* **2009**, *13*, 398–417. [CrossRef]
34. Islam, S.M.; Das, S.; Ghosh, S.; Roy, S.; Suganthan, P.N. An adaptive differential evolution algorithm with novel mutation and crossover strategies for global numerical optimization. *IEEE Trans. Syst. Man Cybern. Part B Cybern.* **2011**, *42*, 482–500. [CrossRef]
35. Kumar, J.; Singh, A.K. Workload prediction in cloud using artificial neural network and adaptive differential evolution. *Future Generation Comput. Syst.* **2021**, *81*, 41–52. [CrossRef]
36. Rosic', M.B.; Simic´, M.I.; Pejovic´, P.V. An improved adaptive hybrid firefly differential evolution algorithm for passive target localization. *Soft. Comput.* **2021**, *25*, 5559–5585. [CrossRef]
37. Yang, Z.; Tang, K.; Yao, X. Self-adaptive differential evolution with neighborhood search. In Proceedings of the 2008 IEEE Congress on Evolutionary Computation, Hong Kong, China, 1–6 June 2008; pp. 1110–1116.
38. Yang, Z.; He, J.; Yao, X. Making a difference to differential evolution. In *Advances in Metaheuristics for Hard Optimization*; Michalewicz, Z., Siarry, P., Eds.; Springer: Berlin/Heidelberg, Germany, 2007; pp. 415–432.
39. Xia, M.; Stallaert, J.; Whinston, A.B. Solving the combinatorial double auction problem. *Eur. J. Oper. Res.* **2005**, *164*, 239–251. [CrossRef]
40. Data of Test Cases 1–10. Available online: https://drive.google.com/drive/folders/19Zj69lRsQP8z0uuiJOqfkHBegCvZE2Pe?usp=sharing (accessed on 11 August 2022).
41. Data of Test Cases 11–14. Available online: https://drive.google.com/drive/folders/1FxECvDt_5ZuXCuL0zNQUXza2Bg82G2Ds?usp=sharing (accessed on 8 July 2023).

algorithms

MDPI

Article

IWO-IGA—A Hybrid Whale Optimization Algorithm Featuring Improved Genetic Characteristics for Mapping Real-Time Applications onto 2D Network on Chip

Sharoon Saleem, Fawad Hussain * and Naveed Khan Baloch

Department of Computer Engineering, University of Engineering & Technology, Taxila 47050, Pakistan; sharoon.saleem@uettaxila.edu.pk (S.S.); naveed.khan@uettaxila.edu.pk (N.K.B.)
* Correspondence: fawad.hussain@uettaxila.edu.pk

Abstract: Network on Chip (NoC) has emerged as a potential substitute for the communication model in modern computer systems with extensive integration. Among the numerous design challenges, application mapping on the NoC system poses one of the most complex and demanding optimization problems. In this research, we propose a hybrid improved whale optimization algorithm with enhanced genetic properties (IWOA-IGA) to optimally map real-time applications onto the 2D NoC Platform. The IWOA-IGA is a novel approach combining an improved whale optimization algorithm with the ability of a refined genetic algorithm to optimally map application tasks. A comprehensive comparison is performed between the proposed method and other state-of-the-art algorithms through rigorous analysis. The evaluation consists of real-time applications, benchmarks, and a collection of arbitrarily scaled and procedurally generated large-task graphs. The proposed IWOA-IGA indicates an average improvement in power reduction, improved energy consumption, and latency over state-of-the-art algorithms. Performance based on the Convergence Factor, which assesses the algorithm's efficiency in achieving better convergence after running for a specific number of iterations over other efficiently developed techniques, is introduced in this research work. These results demonstrate the algorithm's superior convergence performance when applied to real-world and synthetic task graphs. Our research findings spotlight the superior performance of hybrid improved whale optimization integrated with enhanced GA features, emphasizing its potential for application mapping in NoC-based systems.

Keywords: whale optimization algorithm; genetics algorithm; network-on-chip; real-time; parameter control

1. Introduction

In the current era of multicore systems, core integration on System-on-Chip (SoC) devices has increased significantly due to ongoing research and development. However, this significant growth in the integration density of processing elements on System-on-Chip (SoC) devices raises significant performance and scalability concerns. The conventional bus-based architecture fails to satisfy the ever-increasing requirements for high-volume and high-speed communication imposed by the increasing number of cores. Therefore, looking for alternate approaches to overcome these constraints and raise the effectiveness of multicore systems is vital. Network-on-Chip (NoC) has surfaced as a viable solution [1,2] to meet the current communication needs of the very large scale integration (VLSI) paradigm at the deep nanoscale level. Network on Chip comprises integrated processor cores (IP), a network interface (NI), routers, and the links that connect them. In NOC, cores employ a packet-based switching technique for communication through the routers with the help of interconnection links. The NoC topology signifies the physical organization of the architecture, defining the arrangement of routers and cores. Various standard topologies have been designed and employed for NoC depending on the interconnected networks.

Out of these, the mesh topology is the most prominent [3,4], featuring shorter paths between interconnected cores and high bisection width. These features make it suitable for use in numerous application mapping techniques. Due to its built-in parallelism and concurrent communication features, NoC is a popular computing engine for future many-core real-time systems [5]. Of the many key concerns in NoC designs, such as router architecture, topologies, and routing algorithms, the NoC's IP mapping issue has gained substantial attention [6]. IP mapping is an NP-hard combinatorial optimization issue [7,8] that requires discovering the best way to map IPs onto a certain NoC architecture while attempting to meet predefined metrics [9]. The random placement of the IP core in NoC designs does not effectively influence the network's overall efficiency. Achieving high performance with low communication cost, latency, and throughput requires excellent or even optimal IP core mapping onto NoC systems.

Because a mapping problem with n number of IPs leads to n! possible solutions, obtaining optimal solutions through exhaustive enumeration of all permutations is not practical. Consequently, more efficient methods must be discovered to overcome this issue. As it is an NP-Hard problem, there are no exact approaches to finding the solution in polynomial time. Even small-scale cases may need significant computing time [10]. Therefore, heuristic approaches are a practical and effective strategy for finding high-quality solutions [6,11]. In addition, rapid growth in the amount of data represents a challenge to researchers and data scientists in analyzing and extracting relevant information. Swarm intelligence is increasingly used for many optimization tasks, including feature selection, a complex task essential for reducing data dimensionality in high-dimensional datasets [12]. Although various heuristic-based techniques have been employed to solve the NoC application mapping problem, they still solve the problem at the cost of certain factors, and all have certain drawbacks. The well known Ant Colony optimization method has disadvantages such as parameter sensitivity, sluggish convergence rate, and precociousness. Certain evolutionary algorithms suffer from longer computation times in certain cases, along with poor stability. Although heuristic searches are extensively employed to solve the IP mapping issue, they often have the drawback of quickly settling on a local optimum. However, efficient local and global search capabilities on the part of heuristic algorithms can help to achieve optimum results. Hence, an optimization algorithm that balances exploration and exploitation while avoiding local optima may solve application mapping problems for both small-scale and large-scale applications in an ideal manner. Metaheuristic algorithms use a mathematical model of social evolution to efficiently solve optimization problems by promoting high-level methods and local improvement strategies.

Based on the factors and characteristics of the algorithm, an improved version of a nature-inspired meta-heuristic mapping technique is presented to achieve better results under allowed bandwidth limitations. The modified whale optimization algorithm (IWOA) features strong exploration and exploitation capabilities and a competitive convergence rate for a given set of problems. IWOA is proposed as the first step to finding an optimal solution to NoC application mapping, followed by implementing a hybrid IWOA featuring Improved Genetic Mechanism (IWOA-IGA). The genetic-based technique incorporates crossover and mutation to optimize IWOA-based results for optimal NoC mapping. The integrated tweaked GA features assist the IWOA in achieving the optimal mapping with faster convergence, allowing it to avoid local optima with enhanced search ability. These capabilities of the proposed algorithm make it superior to conventional techniques.

To analyze the NoC efficiency metric, current research seeks to model power, latency, and communication cost. Successful mapping strategies generate optimum NoC mapping with low communication costs, fewer iterations, and fewer processing resources. X–Y routing simplifies mapping problems. In a regular mesh design, network performance is measured by the cost of communication:

$$C_T = \sum_{i=1}^{size(ACG)} \sum_{j=0}^{te} [CB_{(te_i)}] \times HopCount_{(te_i)} \quad (1)$$

where t_e represents a collection of edges within a core graph, $B_{(te_i)}$ represents the communication bandwidth on edge i, and the number of hops that separate two communicating cores, referred to as $HopCount_{(te_i)}$, is used to quantify delay and communication costs [13,14]. Thus, as suggested by Equation (1), lowering communication costs indirectly reduces latency and energy usage.

IWOA-IGA offers an effective trade-off between performance measures and faster mapping onto 2D NoCs. The hybrid mechanism of the suggested method optimizes mapping results, reducing communication, energy, power, and latency costs. The contributions of the planned research are as follows:

- Based on an improved whale framework incorporating GA characteristics, a novel and effective solution to NoC application mapping is proposed.
- Improved Initial Mapping for IWO and direction-based crossover and mutation ability are introduced based on a ranked selection-based method during GA evolution.
- The final mapping technique delivers superior performance in terms of total communication costs. The proposed IWOA-IGA improves power, energy, and communication costs compared to existing bio-inspired algorithms.

The rest of this paper is arranged in the following manner: Section 2 presents the literature review; Section 3 discusses the mathematical formulation and performance metrics; Section 4 presents the framework of the proposed IWOA algorithm; Section 5 shows the enhanced version of the WOA algorithm featuring the improved genetic mechanism; our computational findings and analyses are presented in Section 6; Finally, Section 7 contains concluding remarks and discusses future research directions.

2. Related Work

In the current section, we briefly look into the current approaches to application mapping challenges for NoC design from several perspectives, namely, minimizing communication energy, enhancing performance, and reducing computation time. Small-scale IP mapping issues can be formulated as integer programming problems [15,16] and resolved using the branch and bound approach [17,18]. However, neither approach can effectively address extensive IP mapping issues due to their lengthy computation times. The most prominent approach for large-scale IP mapping issues is heuristic search, which may be loosely split into two kinds: transformative heuristics and constructive heuristics [13]. Typically, heuristic approaches map cores onto routers using pre-established criteria. NMAP, an efficient NoC application mapping approach [19], maps the application tasks to the cores in three phases: initialization, minimum path computation, and pair-wise swapping until the optimum mapping solution is obtained. BMAP is a greedy binomial mapping method [20] in which a candidate solution is first proposed, followed by iterative improvement until the final mapping is achieved. To achieve more solutions, CastNet generates various alternative solutions for each core depending on the availability of adjacent free neighbors by using multiple tiles to work as initial tiles, leveraging the symmetric features of the mesh [21]. In CHMAP (constructive heuristic mapping approach) [22], the root is chosen as the core with the highest average communication traffic, then a maximum spanning tree is constructed from the communicative graph. CHMAP calculates the mapping order based on the degree and distance of the cores and then assigns each core to the most suitable router based on mapping criteria. CastNet and CHMAP do not employ further iterative-based improvements upon the initial solution to reach the best possible solution. Transformative algorithms often use evolutionary approaches to look for approximations. Examples of common evolutionary algorithms [6] for optimization encompass genetic algorithms (GAs), discrete particle swarm optimization (DPSO), and ant colony algorithms. A DPSO-based technique that employs a multi-stage PSO and certain deterministic particles for the initial population rather than random ones was reported in [14]. Optimized Mapping solutions for both 2D and 3D NoC were presented. A novel optimization technique grounded on the DPSO framework was introduced in [23], where the velocity update process included a perturbation particle and an elite particle introduced to facilitate the exploration of local optima. The algorithm allows particles to switch between elite and conventional pools, and uses a simplified local search of elite particles to find potential solutions. Multi-application mapping based on a reconfigurable NoC architecture was reported in [24], where a Mesh of Tree (MoT) topology was implemented. A two-step efficient Particle Swarm Optimization (PSO) approach was used to reduce the communication cost for a reconfigurable architecture. A contention-aware genetic algorithm-based application mapping solution was reported in [25], where both the spatial and temporal attributes of communication were considered in order to optimize performance by avoiding contention. To optimize the NoC in terms of communication cost, average latency, and energy, an efficient mapping technique was proposed [26] using the cuckoo search technique with Levy flight. First, a greedy algorithm was used to place the most communicative tasks together for an initial quality solution. Levy flying meta-heuristic cuckoo search was then used to optimize task placements for optimum mapping. In [27], the authors put forward the IHPSA optimization method, an enhanced hybrid PSO, and the simulated annealing technique. Enhanced Particle Swarm Optimization with SA was combined in the proposed IHPSA for application mapping. The K-means clustering machine learning method arranges tasks based on their communication bandwidth. The K-means clustering algorithm employed the elbow method to intelligently predict the number of clusters in extensive applications. Eventually. Heuristics were applied to achieve the optimal cost for real benchmark applications and synthetic instances. The authors of [28] presented RAMAN, a method inspired by Reinforcement Learning (RL), for 2D NoC-based application mapping. RAMAN is an enhanced Q-learning algorithm influenced by RL that aims to achieve mapping solutions with the lowest possible and optimized communication cost. The results of this method show tremendous potential in terms of lower complexity and cost while tackling application mapping problems. In [29], linear programming was employed to mathematically model the mapping problem in NoC. Various constraints related to communication capacity and power budget were incorporated. Finally, simmulated annealing integrated with GA was implemented to take into account and consider the constraints while finding the optimum solution. The run-time application mapping technique presented in [30] aimed to balance the load on the overall network when implementing NoC for a Cube–Tree Hybrid (CTH) topology. The authors exploited the low network diameter of this topology with its high scalability, resulting in reduced mapping overhead. A significant reduction in execution time was observed during experimentation. A hybrid task mapping HyDra was proposed in [31] that incorporates design time mapping and runtime reconfiguration with the aim of reducing communication energy. At design time, multiple mapping solutions are produced to reduce latency and energy. As the applications arrive at runtime, the design time mappings are either used according to the applications or reconfigured based on the requirements. While many such algorithms have been developed, they may suffer from suboptimal performance along with inefficient convergence and complexity. To tackle these challenges, the authors of [32] used Grey Wolf optimization (GWO), which starts with cluster-based initial mapping followed by a modified GWO heuristic algorithm with polynomial regression. The modified algorithm enhances runtime efficiency and optimizes the overall mapping quality. In [33], the authors discussed various real-time application mapping techniques for complex multicore platforms. They categorized these techniques by emphasizing their optimization goals, such as communication cost and energy consumption in NoC-based systems. Future challenges, trends, and simulation tools in this area were presented as well. Application mapping in both 2D and 3D remains a complex challenge that requires further development of efficient algorithms. To address the outstanding challenges, a neural mapping model with a reinforcement learning (RL) approach (NeurMap3D) was presented in [34] to develop application-specific 3D NoCs. In addition, a neural congestion-aware mechanism was presented to address application mapping issues via placement and mapping and to incorporate TSV placement and load balancing in NoC.

3. Mathematical Formulation/Performance Metrics

This section discusses NoC platform assessment criteria such as communication cost, latency, energy, and power.

3.1. Communication Cost Model

Communication Cost in NoC-based formulations is presented in Equation (2):

$$Cost = \sum_{i,j} [B_{t_i,t_j} \times N_{i,j}], \tag{2}$$

where B_{t_i,t_j} is the bandwidth between tile t_i and t_j and $N_{i,j}$ is the Manhattan distance. The NoC architecture's Manhattan distance between the source nodes (x_i, y_i) and destination nodes (x_j, y_j) is provided by

$$N_{i,j} = |x_i - x_j| + |y_i - y_j|. \tag{3}$$

3.2. Power Model

Power and energy are calculated based on the execution of a given traffic pattern. $Pw_{act,j}$ and $Pw_{inact,j}$ are component j's active and passive power at 1.0 V and 1.0 GHz, respectively. Let $alpha_{i,j}$ be the active reading of component j in router i; then, the average power Pw_{av} [35] is expressed by

$$Pw_{av} = \frac{1}{N} \sum_{i=1}^{N} \sum_{j=1}^{N_i} [\alpha_{i,j} \cdot Pw_{act,j} + (1 - \alpha_{i,j}) \cdot Pw_{inact,j}] \tag{4}$$

where $Pw_{act,j}$ is j's component power when active, $Pw_{inact,j}$ signifies component j's power while inactive, and $\alpha_{i,j}$ reflects active measurement of component j in the router i.

3.3. Latency Model

The average latency of the NoC is shown in Equation (5):

$$Lt_{av} = \frac{1}{N} \sum_{i=1}^{N} \frac{1}{N_i} \sum_{j=1}^{N_i} Lt_{i,j} \tag{5}$$

where $Lt_{i,j}$ is the latency of packet j from one tile to another, N represents the number of processors in the mesh, and N_i reflects the quantity of packets received by a processor after the warm-up period.

3.4. Energy Model

The energy model estimates the network router energy consumption, represented as follows:

$$E_B = E_s B + E_L B \tag{6}$$

where E_B consists of the switch energy $E_s B$ and link energy $E_L B$ of the NoC. It is this energy that is consumed to transfer one bit data between the source and the destination. The following equation computes the average network energy consumption, denoted as $EB_{(p_i, p_j)}$, used for the transfer of one bit of data between source (p_i) and destination (p_j):

$$E_{B(p_i, p_j)} = H_{count} E_{SB} + (H_{count} - 1) E_{LB} \tag{7}$$

where $H_c ount$ represents the Manhattan distance between the source nodes (a_i, a_j) and destination nodes (b_i, b_j).

$$H_{count} = |a_i - b_i| + |a_j - b_j| \tag{8}$$

Consequently, the network's total energy consumption (ET) is determined by the average network energy and the link bandwidth $BW_{(p_i,p_j)}$ from p_i to p_j.

$$E_T = \sum_{i,j}^{H_{count}} (E_{B(p_i,p_j)} \times BW_{(p_i,p_j)}) \tag{9}$$

Hence, the final equation takes the form

$$E_T = \sum_{i,j} \left[(H_{count} \times E_{SB} + (H_{count} - 1) \times E_{LB}) \times BW_{(p_i,p_j)} \right]. \tag{10}$$

For cost computation (CC), the following equation can be used:

$$CC = \sum_{i,j} (H_{count} \times BW_{(p_i,p_j)}). \tag{11}$$

Various mappings lead to distinct energy and communication cost values. The primary aim is to derive a mapping function for the NoC with minimal cost. In this research, we use communication cost as the main performance metric for distinct applications.

3.5. Mathematical Formulation Model

Application mapping for 2D NoC design, including preliminary knowledge and a mathematical model, is presented in this subsection. The following elements constitute the inputs for formulating the application mapping problem.

Definition 1. *A Communication Trace Graph (CTG) is defined as an undirected communication trace graph $G = (V, E, M)$ that is weighted and consists of a set of vertices or cores, with V, E representing the directed edge set and M referring to the volume of data and the connectivity between nodes in MS/s.*

$$V = \{c_1, c_2, c_3, \ldots, c_n\} \tag{12}$$

where $|V| = finite(n)$

$$E = \{e_{i,j} = (c_i, c_j) \in C \times C | (c_i, c_j) \in C, i \neq j\} \tag{13}$$

and

$$M : E \to M(c_i, c_j) = M_{ij} \tag{14}$$

where $|V| = finite(n)$.

Definition 2. *In the Topology Graph (TG) for a network, TG = (T, L, M) constitutes the tile collection T placed in the network topology, the collection of links for tiled pairs in T is referred to as L, and M specifies the data volume and link between the tiles in megabytes per second.*

$$T = \{t_1, t_2, t_3, \ldots, t_n\} \tag{15}$$

$$L = \{l_{i,j} = (t_i, t_j) \in TxC | (t_i, t_j) \in T, i \neq j\} \tag{16}$$

$$M : L \to M(t_i, t_j) = M_{ij} \tag{17}$$

Definition 3. *$NAG = (R, C)$: The NoC Architecture Graph corresponds to the routing path $(c_{i,j}) \in C)$ between any router pairs (r_i, r_j) in the network. The intermediate links traversing the path (r_i, r_j) are termed hop counts (Hops) from the router r_i to r_j. Data are transferred or received to and from the cores through the connected routers, and the channel $c_{i,j}$ acts as the physical link for data transfer between the cores. The routing channel has limited bandwidth $(B_{i,j})$ between the nodes.*

4. Proposed Framework of IWO Algorithm

This section presents a new metaheuristic IWOA for a 2D NoC-based multicore platform with strong local and global search capabilities. On the whole, the algorithm attains several mapping solutions (explores the promising search space) by emulating whale hunting behavior, eventually converging to the best possible solution.

4.1. Inspiration for NoC Application Mapping

The metaheuristic method described in [36] simulates hunting behavior using a random agent or the best search agent to track prey. Researchers have found that Humpback Whales pursue prey by creating distinctive bubbles in a circular pattern, and are the only species known to engage in bubble-net feeding [37,38]. The hunting mechanism of whales is modeled in the proposed work to carry out NoC application mapping optimization. The whales' hunting mechanism uses search agents to find the optimal position. As the search agent approaches the prey, the other agents update their locations to discover the best solution. Each search agent corresponds to the task mapping solution, which is assessed on the basis of communication cost. Based on the reference mapping solution, other mapping solutions update their locations (tasks to core mapping). The basic WOA algorithm employs random initial mapping solutions, which are improved over further WOA iteration. However, in this work we modify the WOA algorithm to generate an initial mapping solution using a set of criteria, providing the optimization algorithm with a head start. The algorithm then progressively improves the initial mapping through further iteration.

Modified WOA with Targeted Initial Mapping Generation Heuristic searches for IP mapping are frequently employed; however, they often become trapped in local optima. This shortcoming is primarily attributable to the first population/solution generation stage of heuristic searches. Most transformational heuristics generate a random starting population, and the search procedure can then be seen as a search of the entire search space. In some heuristic procedures, the initial population/solution is formed using predefined criteria, expert information, and/or system attributes. These criteria may not work for large or complex problems. If the initial population generation is not effective, incremental improvement towards finding an optimal solution will be limited. In swarm-based intelligence optimization techniques, the initial population quality has a significant impact on the algorithm's speed and accuracy [39,40]. In the proposed work, the WOA algorithm is modified in terms of initial mapping generation, and is then employed to enhance overall application mapping to NoC Cores. The initial population is produced using the good point set approach instead of the random initial population generation in the standard WOA. This method is an efficient approach that can aid in minimizing the number of attempts and reaching optimal solutions in fewer iterations. The solutions produced using the good point set sequence exhibit a better distribution of solutions than the sequences chosen by the general random method [41]. This approach is used to generate initial mapping solutions for mesh-based 2D NoC architectures. To generate a mapping sequence using the good point set method, a sequence of m points in s-dimensional space is generated and represented as discussed below: Generate points such that

$$r = r_1, r_2, r_3, \tag{18}$$

$$r_i = 2 \times Cos(2i/p), \tag{19}$$

where i can be set to any value for generating various mapping points, and let p be a prime which satisfies $p \geq 2s + 3$. Now, using the several mapping points generated, a mapping sequence is generated that is stored in $T_{(s-map)}$:

$$T_{(s-map)} = (r_1 \times k), (r_1 \times k), (r_1 \times k), \ldots .(r_m \times k) \tag{20}$$

where $r_1 - r_m$ are the total mapping points generated and $k = 1, 2, 3, \ldots$, n, where n is designated as the total number of cores for which mapping is generated. For NoCs having certain application tasks to be mapped onto s × s sized mesh architectures, we can take m as the number of generated tasks or points. As the population size can be set to any number, various individual $T_{(s-map)}$ mapping solutions are generated as an initial generation, e.g., for 16 tasks to be mapped onto a 4 × 4 Mesh 2D NoC platform, we can set m to a certain number and set *popSize* to another number for which diverse task mapping solutions are to be created. The algorithm of the good point set method is provided in Algorithm 1. The set of initial mappings generated using the good point set method can be considered as one of the pools of the initial mapping $initMapping_{gps}$. The good point set method does not use any key characteristics such as communication between tasks and the number of connected neighbors. Thus, application task graph expert knowledge can be utilized to generate initial solutions. This initial mapping can be placed in a second pool based on communication between the cores, termed $initMapping_{comm}$. The final initial mappings are selected from the top-ranked mappings of both pools, which participate in the optimization algorithm. The second pool of initial mappings maps application tasks to the NoC platform using the communication weights between core edges. A core is first selected from the core graph and that specific core's total communication bandwidth and average communication are noted.

Algorithm 1 Initial Mapping Algorithm

```
Initialization:
Input: popSize, m
stopCriteria: popSize
while stopCriteria do
    for i ← 1 to m do
        Calculate r_i using Equation (19) to generate mapping points considering value of
        P
        Create Tasks to core mappings T_(s−map) using Equation (20)
    end for
    initMapping_gps = T_(s−map)
end while
```

The total weight TW of the selected core is provided by

$$TW_i = \sum_{eije/E} tw_{ij}. \tag{21}$$

The average communication is expressed as follows:

$$TW_i = \sum_{eij\epsilon/E} tw_{ij} \times (1/N_{c_i}) \tag{22}$$

where TW_i is the total communication weight between the cores and $N_{(c_i)}$ is the number of available neighbor cores of core c_i. In this way, the total communication and average communication of each specific core are calculated and the neighbors of every core are noted along with communication-related details. The first task to be selected is the one with the highest total communication bandwidth, which is placed onto the NoC platform at any location, and its neighbors are mapped next. If multiple neighbors are available, then the neighbor with a higher communication weight with the mapped task is placed at the closest location to the NoC. Then, the other neighbors of the first mapped task are placed based on their communication weights. When all the neighbors of the mapped task have been placed, the neighbors of already-mapped (recently mapped) tasks are placed. The choice is based on the mapped task with the maximum weight; this core is selected and its neighbors mapped using the same communication weight criteria as above. All of the cores are mapped in this manner. This initial mapping procedure assumes that any job mapped to a core must not be mapped to any other NoC location or considered for any other placement. To serve this statement well in the implementation of this algorithm, a list of tasks that have been mapped is maintained in the list T_{mapped}, while the unallocated cores are maintained in $T_{un\text{-}allocated}$. While mapping neighbors of already mapped cores, these two lists are referenced to reach the appropriate decision. As the algorithm runs, various mappings are achieved based on placement of cores on the NoC platform in different orders. The communication cost for both pools is computed and the top-ranked elite mappings are taken from both initial mapping pools to generate an enhanced initial mapping solution. The reason for retaining and finally gathering mappings from both pools is to improve the overall diversity of the solution.

$$enhancedInitMapping = initMapping_{gps} + initMapping_{comm} \tag{23}$$

The best mapping solution is selected as the current best candidate solution. Optimization improves initial mapping solutions across numerous generations to provide final mappings for varied applications.

4.2. Mathematical Model of IWOA

This section provides the mathematical models of various hunting capabilities of the IWOA in terms of the NoC application mapping problem. Later, the IWOA algorithm using the specified features and characteristics is presented.

4.2.1. Encircling/Navigating the Hunt: Unveiling Optimal Application Mapping Strategies

The whales have a special feature of trying to identify and encircle the prey's location. Because the ideal location within the search domain is not known in advance, the algorithm implies that the current leading/best candidate mapping solution is near the optimal design. The algorithm runs for a specific number of generations and iterations; consequently, the other solutions endeavor to modify their positions with the optimal mapping solution. These equations serve to represent this:

$$D = |\vec{C}.\vec{X}^*(t) - \vec{X}(t)|, \tag{24}$$

$$\vec{X}(t+1) = \vec{X}^*(t) - \vec{A}\vec{D}, \tag{25}$$

where t represents the current iteration, $\vec{A}$ and $\vec{C}$ are quantities that represent coefficient vectors, $\vec{X}^*$ is supposed to be the most effective mapping solution found in the present generation, and $\vec{X}$ represents position vector. If a superior mapping solution exists, then $X\vec{X}^*$ should be updated after each generation or iteration. $\vec{A}$ and $\vec{C}$ are computed using the following equations:

$$\vec{A} = 2\vec{a}\vec{r} - \vec{a}, \tag{26}$$

$$\vec{C} = 2.\vec{r} \tag{27}$$

In both the exploration and exploitation phases, a diminishes linearly from 2 to 0 throughout the iterations, while r represents a random number between [0, 1]. Figure 1 demonstrates the information behind Equation (25) for the 2D NoC problem and shows how mapping solution 1 (with a certain communication cost, commCost1) is updated to a new mapping solution 2 with a better communication cost (commCost2), where commCost2 < commCost1. Equation (25) allows the algorithm to update the mapping solution within the vicinity of the current best solution, leading to an optimum solution over a certain number of iterations and population sizes. The same principle can be applied to an n-dimensional search space in 2D NoC, with the mapping search agents traversing in hypercubes around the current most promising mapping solution.

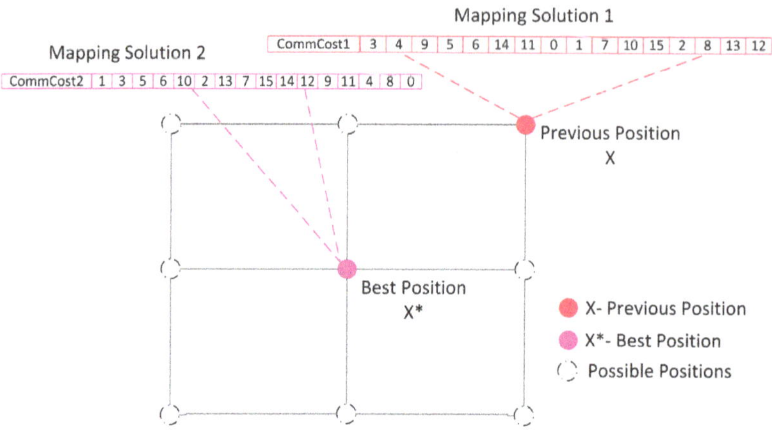

Figure 1. Task allocation and fitness updating process in IWOA for NoC platform.

As noted in the preceding section, the IWOA employs the bubble-net hunting tactic of the WOA integrated into NoC-based application mapping. This strengthens the algorithm's ability to exploit and explore, leading to better mapping solutions. The mathematical formulation of this procedure is as follows.

4.2.2. Exploitation Phase

To efficiently solve the application mapping problem, the IWOA uses the bubble net behavior of whales to achieve an optimum mapping solution within a faster convergence time.

1—Shrinking Encircling Mechanism: Exploitation for Mapping. In a 2D space, Figure 2 depicts the potential positions starting from (X) towards (X^*) that can be obtained by $0 \leq A \leq 1$. The shrinking encircling mechanism is modeled in a 2D NoC architecture such that X represents a mapping solution with a certain communication cost. In contrast, X^* is a better solution with a better communication cost than the earlier one. The mapping denoted as X tries to achieve a better solution, represented by X*, with a better communication cost. This mechanism runs for a certain number of iterations until final mapping is achieved. This mechanism to achieve optimal mapping uses Equation (26) by altering a value. In the equation, A represents a random value with a range of $-a$ to a, where a gradually diminishes from 2 to 0 over a specified number of iterations.

2—Spiral Strategies: Elevating NOC Application Mapping. This approach first approximates the deviation between the value of the objective function for the current mapping solution X and the optimal value that needs to be achieved. The spiral update position to achieve the optimal mapping is derived using the following equation:

$$\vec{X}(t+1) = \vec{D}'(t) \cdot e^{bl} \cdot cos(2\pi l) + \vec{X}^*(t) \tag{28}$$

where $\vec{D}'(t)$ is provided by

$$\vec{D}'(t) = |\vec{X}^*(t) - \vec{X}(t)|. \tag{29}$$

The equation shows the value of the ith mapping from the best mapping solution obtained thus far, where b determines the logarithmic spiral's shape while l is a random number. As the mapping is achieved using the shrinking circle and the spiral-shaped mechanism simultaneously, it is presumed that there is an equal 50% probability of selecting either the shrinking circular mechanism or the spiral model to update the mapping methods. The mathematical model takes the following form:

$$D_{it} = \begin{cases} \vec{X}^*(t) - \vec{X}(t) & \text{if } p < 0.5 \\ \vec{D}'(t) \cdot e^{bl} \cdot cos(2\pi l) + \vec{X}^*(t) & \text{if } p \geq 0.5 \end{cases} \tag{30}$$

where p represents a random value between 0 and 1. Figure 3 shows how various mapping solutions are created from X until reaching the final solution X^*.

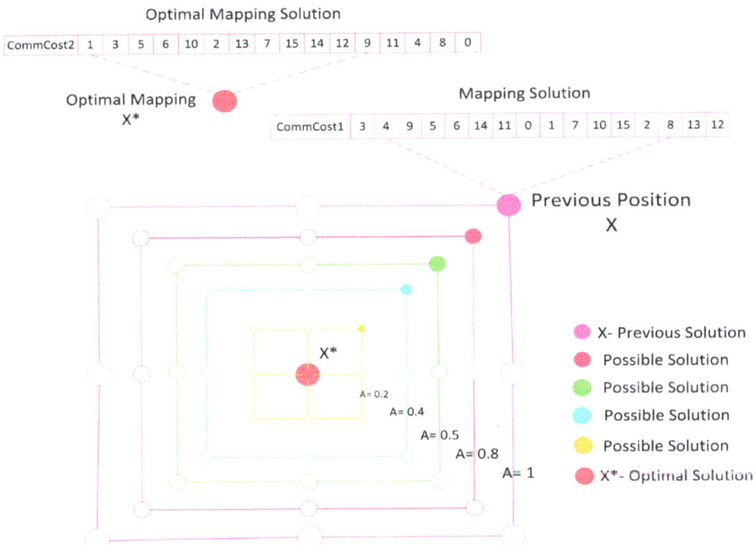

Figure 2. Shrinking encircling mechanism (X^* is the current best mapping solution).

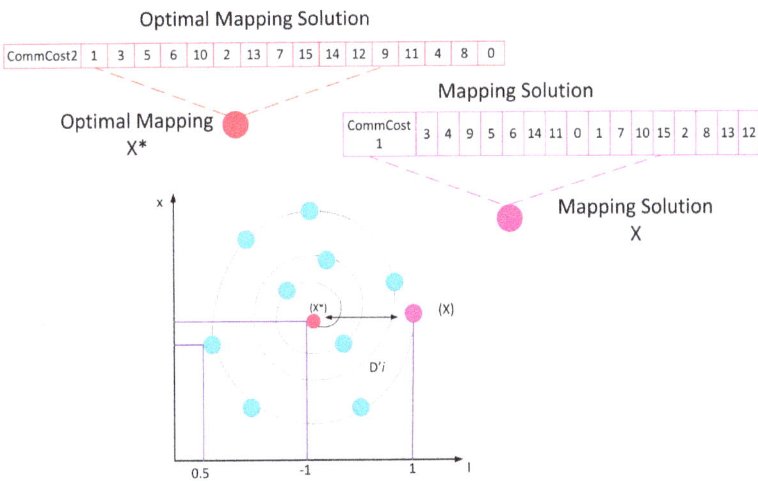

Figure 3. Spiral update position implemented in WOA (X^* is the current best solution).

4.2.3. Search for the Optimal Mapping Solution (Exploration Phase)

Thanks to its faster operation, the IWOA can use the strong exploration capabilities of the whale optimization algorithm to thoroughly explore in order to converge to the optimal solution. A variation pattern of vector A helps to find the best potential mapping with the lowest communication cost. As the mapping solutions are created by taking inspiration from the neighboring mapping solutions, we utilize A with certain randomized values between 1 and -1 in order to compel the search agents to distance themselves from the reference mapping solution and explore more of the overall search space. Compared to the exploitation phase, adjusting the placement of a search agent (mapping solution) does not take place during exploration based on the best solution. Instead, we choose a search agent (a random mapping solution) at random and perform updates based on it. This method, along with $\mid A \mid > 1$, emphasizes exploration and enables the IWOA to perform a global search. Its mathematical model is presented as follows:

$$\vec{D} = |\vec{C} \cdot \vec{X_{rand}} - \vec{X}|, \tag{31}$$

$$X(t+1) = \vec{X_{rand}} - \vec{A} - \vec{D}, \tag{32}$$

where X rand is a mapping solution picked at random from the current population. Several mapping solutions surrounding a specific solution with $A > 1$ are created.

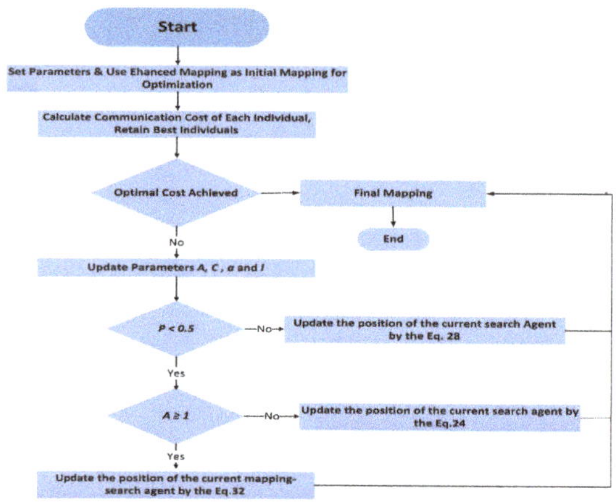

Figure 4. IWOA Flowchart.

4.3. Proposed IWOA

In this section, a new and improved metaheuristic IWOA is presented that has powerful local and global search functionality. Each search agent that constitutes the mapping solution is initialized using a certain approach between the minimum and maximum limit in the range. In the proposed technique, the fitness function is often a primary goal to be reduced. The proposed algorithm commences with a collection of solutions achieved using the enhanced initial mapping mechanism instead of the conventional random mappings. At each iteration, search agents that represent individual mapping solutions adjust their positions concerning either a randomly chosen search agent or the current optimal solution with the minimum communication cost. The number of iterations is predefined, and varies depending on the application. Each mapping solution S is in the solution space SS, where $S \epsilon SS$ represents its components such that $n > 0$ components, i.e., $S = S_1, S_2, \ldots . S_n$, where $i = 1, 2, 3, \ldots ., n$; here, n represents the scope of the optimization problem to be solved. To promote better exploration and exploitation, the value of parameter a is varied from 2 to 0. Random mapping is chosen in the proposed mapping technique when $\mid A \mid > 1$. In contrast, the selection of the best candidate mapping is carried out when $\mid A \mid < 1$ for modifying the mapping solutions of other search agents within the search space. This adaptive variation-based feature of the vector $\vec{A}$ permits the IWOA to transition seamlessly between exploration and exploitation while achieving better mapping results. After performing in-depth analysis and comparison, the appropriate values for the input parameters of the algorithm are determined. It should be noted that a parameter combination that works well for one situation may not work the same way for another. Algorithm 2 provides the pseudocode of the IWOA. In each iteration, the algorithm efficiently obtains a better mapping solution as compared to the previous iterations, depending upon the fitness function. The termination criteria is the achievement of the required fitness value, that is, when the mapping solution with the lowest communication cost has been obtained. The IWOA's method for solving the optimization issue is a global optimizer which includes exploration and exploitation capabilities. As the algorithm runs, the initial mappings are fine-tuned to adjust the placement of tasks to the cores, ensuring that better solutions are achieved in each iteration. The flowchat of IWOA is presented in Figure 4.

Algorithm 2 Improved Whale Optimization Mapping Algorithm

```
Initialization:
Input: enhancedInitMapping, taskGraph, itrWOA, itrIGA, a, C, l, A, p
stopCriteria: itrTotal, OptimalCommCost
// Find pop, v, fitness
initPopFitness(enhancedInitMapping, pop_size, meshSize)
// Find gbestpop, gbestfitness
getinitbest(fitness, pop) //X* = BestSearchAgent
while stopCriteria do
    for i ← 1 to itrWOA do
        for searchAgentj ← 1 to pop_size do
            Update parameters A, C, a, l, p
            A⃗ = 2a⃗.r⃗ − a⃗
            C⃗ = 2.r⃗
            if p < 0.5 then
                if |A| < 1 then
                    Update the current mapping solution using Equation (24), finally updating
                    pop[j]
                end if
                if |A| ≥ 1 then
                    Select a random mapping solution-Xrand
                    Update the position of the current mapping solution by the Equation (32),
                    finally updating pop[j]
                end if
            end if
            if p ≥ 0.5 then
                Update the position of the current mapping solution by the Equation (28)
                updating pop[j]
            end if
        end for
        // Check If a search agent-based solution exceeds the search space, modify it.
        for particlej ← 1 to pop_size do
            calCommCost(meshSize, pop[j], tashGraph)
        end for
        gbestfitness = mim(pop[j])
        gbestpop = min(pop)
    end for
    if gbestfitness == optimalCommCost then
        finalmapping ← gbestpop
        optimalCommCost ← gbestfitness
    end if
    itrIWOA = itrIWOA + 1
    update(gbestfitness, finalMapping
end while
```

5. IWOA-IGA—Enhanced WOA Algorithm Featuring Improved Genetic Mechanism

In this paper, a modified and improved WOA (IWOA) is presented to solve the application mapping problem in NoC. The incorporated IWOA algorithm has better global and local search ability; however, for certain scenarios, such as large-scale complex problems, it may undergo lower convergence accuracy and can fall into local optimum. To avoid such problems and scenarios, an improved GA is executed along with the IWOA to achieve optimal solutions. The GA uses a biological evolutionary mechanism, and, as a global-based search optimizer, it searches for the optimum solution in the search space for complex problems. GAs have proven to be robust and effective in exploring complex spaces. Hence, they can effectively solve a wide range of pattern recognition, artificial intelligence, resource allocation, and similar complex challenges. To achieve better mapping solutions, the proposed modified IWOA algorithm is integrated with an improved genetic algorithm incorporating improved crossover and mutation abilities. This further enhances the ability to find the optimum solution within the search space with better convergence towards the final mapping solution. In addition, according to the original whale optimization algorithm [30], in order to accurately mimic whale behavior and obtain improved optimization results, evolution-based characteristics must be added to the WOA algorithm. Instead of using conventional GA characteristics such as random selection of parents and choosing random points for crossover [42], the proposed technique uses modified crossover and mutation. This can help to avoid local optima and promotes faster convergence with better searchability and higher population diversity. Hence, to achieve optimal mapping solutions for the 2D NOC platform, the modified GA is directly integrated with the IWOA, influencing the overall algorithm's ability to generate high-quality individuals with reduced energy, latency, and power requirements.

5.1. Important Aspects of Genetic Algorithm

The genetic flowchart depicted in Figure 5 resembles biological evolution [9]. A traditional GA begins with an initial random population comprised of randomly chosen chromosomes that produce offspring via crossover and mutation. It continues to work iteratively until a predetermined count of iterations is completed or a termination criterion is satisfied. A pre-set fitness function determines the fitness of the chromosomes and the communication cost on the NoC platform.

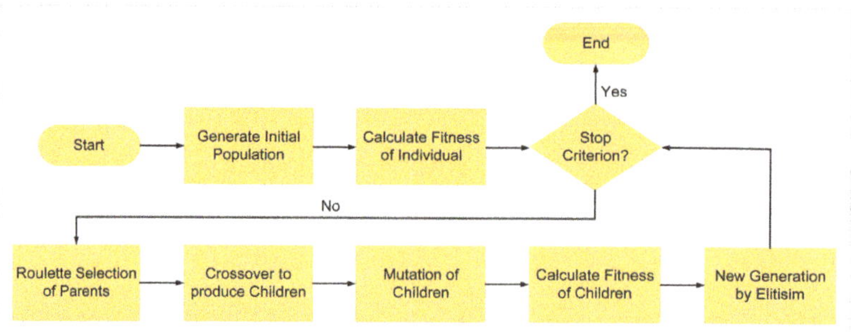

Figure 5. Genetic algorithm flowchart.

5.2. IGA Framework

This article focuses solely on a single objective, namely, use of an improved GA integrated with an improved WOA for solving application mapping through better search ability and population diversity. The population obtained from the IWOA is fed into the modified GA, which uses direction-based crossover features [43] and mutation ability to generate high-quality mapping solutions. Each chromosome in the GA signifies a mapping solution of the 2D NoC mapping problem. The subtasks imply the genes in the chromosomes. The various individual mapping solutions undergo through the genetic-based process for some number of iterations and generations to produce high-quality mapping solutions.

5.2.1. Improved Genetic Algorithm Formulation

The genetic operators exert different levels of impact on the algorithm, with selection identifying the most promising chromosomes for crossover to enhance the solutions. Crossover combines genetic data to improve population characteristics, while mutation introduces new genes to address the weaknesses of crossover.

5.2.2. Expert Initial Curation for NoC Application Mapping Excellence

The selection operator selects the individuals that participate in crossover and mutation; hence, their selection substantially impacts the entire GA process [44]. The proposed modified algorithm uses expert criteria-based selection to choose the initial mapping solutions to undergo crossover and mutation. This is contrary to the random selection used in the conventional GA. Initially, the mapping solutions achieved by the IWOA are ordered according to the objective function $F(X)$ (communication cost) and treated as an initial population for the IGA, represented as

$$X = \{X_1, X_2, \ldots\ldots X_n, \} \tag{33}$$

while the sorted population is provided by

$$X^s = \{X_1^s, X_2^s, \ldots\ldots X_n^s\} \tag{34}$$

which satisfies $F(X_1^s) \geq F(X_2^s) \geq, \ldots\ldots F(X_n^s)$. Each element in X and X^s shows individual mapping solutions. The population obtained by the IWOA is sorted and divided into four groups (X^1, X^2, X^3, X^4), which are paired with each other $((X^1, X^2), (X^1, X^3), (X^1, X^4), (X^2, X^3), (X^2, X^4), (X^3, X^4))$ to form $X^(map_a)$ and $X^(map_b)$. Crossover is performed using the elements of $X^{(map_a)}$ and $X^{(map_b)}$.

5.2.3. NoC Application Mapping: Directional Optimization Incorporating Crossover and Mutation

The GA's central process significantly affects the algorithm's ability to seek improved and optimal solutions [45]. The more optimal the objective function is for any individual solution, the closer the mapping combination will be to the optimal region. Hence, the proposed IWOA-IGA uses a direction-based crossover operator. The modified crossover operation has good ability for searching the overall search space, and can produce mapping solutions with a larger probability of enhancing the objective function, thereby speeding up the algorithm's convergence. As in the initial step of the GA, the population of mapping solutions is divided into two groups, $X^{(map_a)}$ and $X^{(map_b)}$. The communication cost of the individual solutions in $X^{(map_a)}$ is superior to those of the solutions in $X^{(map_b)}$; thus, $X^{(map_a)}$ is the leader in the direction of crossovers for producing high quality solutions. The mathematical equation that generates the mapping solutions with directional-based crossover is provided by

$$\begin{cases} X_i^* = X_i^{map_a} + r_{ij} \cdot \vec{D}_{ij} \\ \vec{D}_{ij} = X_i^{map_a} - X_i^{map_b} \quad i = 1, \ldots..3n/2, j = 1, \ldots.., m \end{cases}, \tag{35}$$

where $\vec{D}_{ij}$ represents the directional vector. The parameter r_{ij} is a uniformly distributed random number ranging from -1 to 1. Thus, in directional crossover based on grouping-wise solutions, each mapping pair develops into one mapping solution, eventually generating new individuals. Finally, it sorts the high-quality mapping solutions with the best communication cost, developing n individuals as offspring of the crossover.

To see the effect of the direction-based crossover operation on two of the paired mapping solutions, paired individuals are shown to occupy a certain space in the appropriate region. As r_{ij} takes on boundaries between 1 and -1, the paired mapping solutions will take various directions to reach the optimal communication cost value. The various directions for solution X_1 can be X_1 with D11, D12, D11, D12, as shown in Figure 6. The solution in the region is shown by the rectangle with its center at X_1 and X_2 as one vertex. Having now obtained various paired mapping solutions, they can be used to generate many such rectangular patterns; along with the selection process, direction-based crossover assists the algorithm in generating high-quality solutions in the direction of the optimal solution. Parents X_1 and X_2 can produce offspring that will tend to have better communication costs. The crossover operation recombines the parents' chromosomes, and the best offspring undergo mutation and continue on to produce superior results.

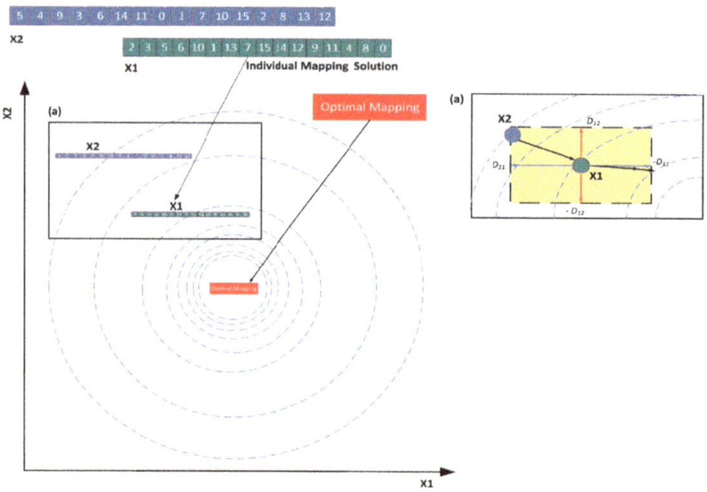

Figure 6. Direction-based crossover.

The mutation operation is applied to offspring to achieve more optimal results. Mutation allows for altering the placement of one or more mapped tasks of individuals in a population with a specified mutation probability, which can help to promote population diversity and prevent premature events. The communication cost of all the resulting mappings are computed after mutation, and the mutation is accepted if the communication cost of the mutated mapping is less than that of the original ones. To increase population diversity and prevent the GA version from falling into local optima, the mechanism of replacement operation is incorporated [43]. The basic goal of the replacement operation is to set aside the most elite K individuals from the population for every K generation. In traditional approaches, certain elite individual solutions are retained in parents. Nevertheless, the high-quality solutions created as a result of crossover may be affected by the mutation operation, and their quality may degrade to some extent. Thus, in this work we adopt a better approach to retain the most elite mapping solutions produced by the genetic operation. In this way, the number of elite mapping solutions can be maximized to enhance the population attributes.

5.3. IWOA-IGA: Modified IWO Algorithm Featuring IGA Characteristics

This section proposes a new enhanced hybrid IWOA-IGA application mapping technique for 2D NoCs. This algorithm combines a modified GA with improved whale features to create a powerful local and global search capability with faster convergence. The proposed hybrid approach effectively combines the exploratory advantages of the IWOA with improved search capability with the accelerated convergence and local optima avoidance features of the IGA. Their combination can maintain a high level of population diversity to achieve the global optimal application mapping solution. The initial mapping solution is generated using the enhanced initial mapping to give a head start to the optimization algorithm, enhancing its performance in the search space. To effectively integrate the IGA and IWOA, The IWOA incorporates the IGA into each of its K iterations until the ideal solution is found. The algorithm runs for a certain number of iterations and simulations. Each K iteration is predetermined and changes depending on the application. The fitness function holds significant importance within the realm of optimization mapping problems, as it represents the objective we seek to minimize in order to reach the optimal solution. If the best global solution remains unchanged for the last K number of iterations, this suggests that the algorithm might have become stuck at an optimal local value. The IWOA optimizes the mapping solution in certain generations, and the adaptive strategy drives a well balanced search operation. To enhance the rate of convergence and prevent the algorithm from becoming stuck in local optima, additional improved genetic features such as groupwise selection, direction-based crossover, and mutation characteristics are incorporated. Direction-based crossover allows the diverse mapping solutions in each generation to efficiently converge towards the optimum solution. The input parameters for the technique are chosen after a thorough assessment. Certain parameters are allowed to adapt according to the environment and the solutions acquired in each K iteration, allowing the algorithm to achieve better convergence and search features. The IWOA-IGA pseudocode and framework are presented in Algorithm 3 and Figure 7, respectively.

Parameter Settings for the Proposed Hybrid Algorithm

The proposed hybrid method necessitates the specification of a few fundamental parameters to determine the efficacy of group searching. Numerous simulations and satisfying results were used to pick the parameter values for the proposed algorithm in order to produce an effective solution. Well balanced exploration and exploitation ability is derived using adaptive parameter adjustment, which results in updating of the vector $\vec{A}$. The algorithm uses a few other parameters, such as r (a random vector between [0, 1]), and a linear decline from 2 to 0 is performed throughout iteration during both the exploration and exploitation phases. For the shrinking encircling mechanism, A takes a random value between $-a$ and a, where a gradually diminishes from 2 to 0 throughout iteration. Configuring A within $[-1, 1]$ can result in a new position for the mapping solution between the original position and the current best position. For the spiral mapping position, the parameter b is kept at a constant value to maintain the shape of the spiral, while l is varied between $[-1, 1]$ to help the algorithm achieve better mapping solutions.

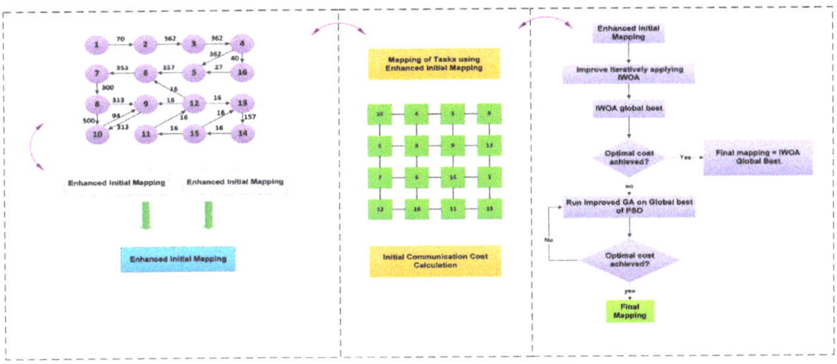

Figure 7. IWOA-IGA framework.

Algorithm 3 IWOA Integrated with IGA

```
Input: enhancedInitMapping, taskGraph, itrWOA, itrIGA, a, C, l, A, p
stopCriteria: itrTotal, OptimalCommCost
// Find pop, v, fitness and gbestpop, gbestfitness
initPopFitness(enhancedInitMapping, pop_size, meshSize)
getinitbest(fitness, pop) //X* = BestSearchAgent
while stopCriteria do
    for i ← 1 to itrWOA do
        for searchAgentj ← 1 to pop_size do
            Update parameters A, C, a, l, p
            A⃗ = 2a⃗.r⃗ − a⃗ , C⃗ = 2.r⃗
            if p < 0.5 then
                if |A| < 1 then
                    Update current solution using Equation (24), finally updating pop[j]
                end if
                if |A| ≥ 1 then
                    Choose random mapping solution-Xrand
                    Update current mapping solution by the Equation (32), updating pop[j]
                end if
            end if
            if p ≥ 0.5 then
                Update Current mapping solution by Equation (28), updating pop[j]
            end if
        end for
        for particlej ← 1 to pop_size do
            calCommCost(meshSize, pop[j], tashGraph)
        end for
        gbestfitness = mim(pop[j])
        gbestpop = min(pop)
    end for
    if (gbestfitness == optimalCommCost) then
        finalmapping ← gbestpop
        optimalCommCost ← gbestfitness
    end if
    //Apply IGA to global best solution
    itrIGA ← 0
    //Take Pop as initial population for IGA to Optimize
    //Find gbestpop, gbestfitness
    getinitbest(fitness, pop)
    currentbest ← gbestpop//X* = best_search_agent
    currentbestpop ← gbestfitness
    SetParametersr
    while itrIGA do
        //Execute Selection Process
        X^map_a & X^map_b
        //Apply Direction Based CrossOver
        Update X*_i and D⃗_ij using Equation (35)
        itrIWOA = itrIWOA + 1
        update(gbestfitness, finalMapping
        Keep Elite Solutions & then perform Mutation forming Offsprings
        if commCost_Offsprings > currentBestCost then
            currentBest ← gbest_offspring
            currentbestCost ← gbestfitness
        end if
        if gbestfitness == optimalCommCost then
            finalmapping ← gbestpop
            optimalCommCost ← gbestfitness
        end if
        itrIGA = itrIGA + 1
    end while
end while
```

The direction-based vector D allows the algorithm to converge more quickly towards the optimum. It entails a significant probability of producing high quality mapping solutions in the direction of the optimal communication cost. The parameter r_{ij} in the direction-based crossover mechanism is randomly generated. The parameter's population size and generations are set to 150 and 100, respectively. The algorithm first generates an initial mapping, which is iteratively improved for the population size and number of generations, eventually leading to an optimal mapping solution for a set of benchmarks.

6. Simulation Results

The simulation results of the IWOA and IWOA-IGA are examined in this part and compared with those of existing mapping methods.

6.1. Simulation Setup and Scenario

Experiments were conducted on available real-time benchmarks and synthetic task graphs to assess the algorithms. All benchmarks for real-world applications are limited to small-scale tasks, often demanding fewer than 32 cores; consequently, custom benchmarks were produced using task graph tools to evaluate the performance of our method on more challenging issues. TGFF instances can have a problem scale anywhere from 16 to 196, with small-scale (core count < 35), medium-scale (36 < core count < 70), and extensive-scale (core count > 70) problems all taken into account. The TGFF tool takes the heterogeneous communication characteristics of the cores into account and builds task graphs at random depending on this behaviour. The set of real-world benchmark instances we used were video object plan decoder (VOPD), 263decoder (263DEC), MPEG-4 decoder, 263encoder (263ENC), Mp3encoder (Mp3ENC), and MWD. All real-time benchmark applications we examined used the standard network size of 4 × 4. This network size is the same as in prior state-of-the-art architectures, allowing for a fair comparison. The VOPD application was divided into 16 subtasks, each of which could be allocated to a 4 × 4 mesh. Larger mesh sizes were used for the synthetic task graphs. Details of the real-time benchmark applications are listed in Table 1.

Table 1. Standard NoC benchmarks with 2D mesh sizes.

Benchmarks	Nodes	Edges	Mesh Size
PIP	8	8	3 × 3
MPEG-4	12	26	4 × 4
MWD	12	13	4 × 4
263encMP3dec	12	12	4 × 4
263decMP3dec	14	15	4 × 4
Mp3EncMp3Dec	13	14	4 × 4
VOPD	16	21	4 × 4
CAVLC	16	22	4 × 4
MMS	25	38	5 × 5

Modifications were implemented in the NoCTweak simulator to conduct a comparative analysis of different application mapping techniques [38]. ENoCTweak was used by applying the algorithm on various real-time and synthetic task graphs. NoCTweak is a SystemC-based open-source NoC simulator that offers a variety of performance characteristics, including energy, latency, communication cost, and throughput. In order to apply the proposed algorithm on both the synthetic and real-time applications, it was run on a computer system equipped with an Intel Core i3 platform, 8 GB of RAM, and a clock frequency of 1.6 Ghz. The simulation environment used to execute the benchmark applications on a 2D NoC architecture is detailed in Table 2. The proposed application mapping approach was implemented in Python to provide the optimal task mapping for the NoC. The resulting optimized mapping was executed on the NoC using the unified simulation framework known as ENoCTweak, which was used to simulate and analyze critical NoC system characteristics such as latency, throughput, energy, and power.

Table 2. Simulation setup details.

Configuration	Detail
Network Type	Mesh
Type of Platform	Embedded
Applications	VOPD, MP3encMp3dec, MPEG4, MWD, 263encMp3dec, 263decMp3dec
Mapping Algorithm	IWOA, IWO-IGA, SA, PSO
PacketDeliveryMode	Without ACK
SendingACKPolicy	Send ACK Optimally
Packet Distribution	Exponential
Fixed Packet Length	10 (flits) moment
FlitinInjectionRate	0.1 (flits/cycle/node)
Type of Router	Wormhole-Pipeline
Routing Algorithm	XY DIMENSION-ORDER
OutputChannelSelection	XY-ORDER
BufferSize	1 (flit)
Pipeline Type	8
StagesOfPipeline	4
InputVoltage	1 (V)
Operating Clock Frequency	100,000 (MHz)
Warm-Uptime	20,000 cycles

6.2. *Performance-Based Comparative Analysis*

As the main aim of this study was to enhance the communication cost, which represents the overall cost associated with running a particular application on NoC (depicted by Equation (1)), the communication cost for both the IWOA and IWOA-IGA were first evaluated based on real-world and TGFF-based graphs.

6.2.1. Performance Analysis Based on Standard Benchmark Instances

Both of the developed techniques were tested using real-world applications as well as TGFF-based graphs (presented in the next section). A comparison of the IWOA and IWOA-IGA with other advanced heuristic algorithms and with the exact mathematical solution was carried out. Table 3 depicts the performance comparison in terms of communication cost for various standard benchmark applications. When it comes to estimation of communication cost, ILP (Integer Linear Programming) is regarded as the optimal solution [3]. The results demonstrate that the proposed modified IWOA and its genetically integrated version obtained optimal results derived from real-world instances.

Table 3. Communication cost (bw $\times$ nh) comparison with real-world benchmark applications.

Algorithm	VOPD	CALVC	MMS	MPEG4	MWD	Mp3Enc	263enc	263dec	PIP
ILP	4119	-	-	3567	1120	17.021	230.407	19.823	-
ONMAP	4119	-	663,379	3567	-	-	-	-	640
GA	4141	-	-	3567	1321	17.133	230.69	19.911	-
SA	4125	-	-	3567	1451	-	-	-	-
BEMAP	4119	6701	664,636	3567	-	-	-	-	640
ACO	-	-	-	3670	-	17.231	-	-	-
PSO	4119	-	-	3567	1120	17.021	230.45	19.823	-
BA	4119	-	-	3567	1120	17.834	231.45	19.936	-
HDPSO	4119	-	-	3567	-	-	-	-	-
CSO	4119	6721	652,637	3567	-	-	-	-	640
SCSO	4119	-		3567	1122	17.021	230.407	19.823	-
DPSO	4119	-	688,297	3567	1120	17.021	-	19.823	640
RAMAN	4135	-	-	3774	1184	17.87	234.4	19.87	640
SFOA	4119	-	-	3567	1120	17.021	230.407	19.823	-
ACA	4119	6721	652,637	3567	1120	17.021	-	-	-
iHPSA	4119	-	-	3567	1120	17.021	230.407	19.823	-
IWOA	4119	6701	663,379	3567	1122	17.021	230.69	19.823	-
IWOA-IGA	4119	6701	663,379	3567	1122	17.021	230.69	19.82	-

6.2.2. Performance Analysis on TGFF Random Instances

This section presents the details of the generated TGFF random instances and their use in evaluating the algorithms with varying mesh sizes. For the sake of experimentation with large-scale applications, random synthetic task graphs were generated using the TGFF tool, which generates task graphs with 32, 64, and 128 cores. TGFF, known as Task Graphs For Free, is meticulously crafted to offer a standardized approach for generating arbitrary task graphs for experimentation purposes in research work. In our evaluation, we used mesh sizes of 6×6 for up to 32 core placements, 8×8 to accommodate 64 core graphs, and 12×12 for 128 core task graphs. The task graphs were configured to use a communication bandwidth of 50–600 MB/s between the interconnected tasks. As the tasks using the TGFF tools are randomly generated, there is no specific optimal value for the communication cost; however, there must be some criteria to evaluate the performance of the generated large-scale tasks. A few important considerations could include the number of generations for which the algorithms can execute and the communication cost value be achieved at that instance. Another consideration is the number of iterations for which the algorithm runs and the targeted communication cost to be achieved after a specified number of iterations. We observed that when the number of iterations was configured to a value between 180–200 and the IWAO-IGA was executed for the 200 iterations, the IWAO-IGA achieved the desired communication cost in fewer iterations than the IWOA for larger applications.

This is due to the integration of directional crossover and mutation features into the IWAO mechanism, allowing the algorithm to perform better when searching for optimal solutions in the search space and to attain the desired cost in a lesser number of iterations than to the simple IWOA. Moreover, the proposed algorithm attains the desired cost within a reduced number of iterations in comparison with other advanced and optimization algorithms, e.g., SA, GA, and PSO. A cost savings-based comparative analysis is presented in Tables 4 and 5 and in Figure 8. Across the tested benchmarks, the proposed IWOA reliably lowers the communication cost by an average of 41.31%, 46.61%, and 44.00% compared to PSO, SA, and GA, respectively, for 32 cores. In the case with 64 cores, the average improvement in cost is 38.70%, 31.94%, and 39.78%, respectively. The proposed algorithm consistently achieves a significant average reduction in communication cost when applied to scenarios involving 128 cores, with respective improvements of 17.37%, 20.99%, and 16.12% over PSO, SA, and GA. These outcomes highlight that the proposed algorithm outperforms applications with a moderate number of cores. However, as the nubmer of cores rises to surpass a certain threshold, the algorithm's performance becomes average compared to its efficiency with a lower number of cores. As the algorithm was further modified by adding an improved GA algorithm to ensure better results when larger applications are executed on it, the modified IWOA-IGA was tested again on larger TGFF-based applications implemented on 32, 64, and 128 cores. The IWOA-IGA minimizes communication costs effectively, with an average improvement of 45.66%, 50.56%, and 48.15%, respectively, for 32 cores. For the configuration with 64 cores, the average improvement in communication cost was 40.29%, 33.70%, and 41.33%, respectively. With 128 cores, the proposed technique produced respective average reductions in communication cost of 30.60%, 33.31%, and 26.67%. Table 5 shows the results.

Table 4. Percent communication cost savings with IWOA compared to other algorithms for extensive task graphs.

Task Graph	Over PSO	Over SA	Over GA
32 Cores	41.31%	46.661%	44.00%
64 Cores	38.70%	31.94%	39.78%
128 Cores	17.37%	20.99%	16.12%

Table 5. Percent communication cost savings with IWOA-IGA compared to other algorithms for extensive task graphs.

Task Graph	Over PSO	Over SA	Over GA
32 Cores	45.66%	50.56%	48.15%
64 Cores	40.29%	33.70%	41.33%
128 Cores	30.60%	33.31%	26.67%

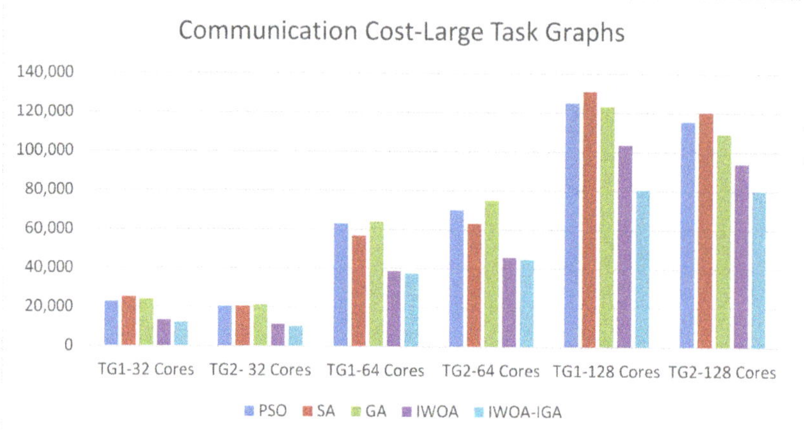

Figure 8. Comparison of communication costs for large synthetic task graphs.

6.2.3. Performance Comparison of IWOA and IWOA-IGA: Optimizing Power and Energy in 2D NoC Benchmark Applications.

To assess the proposed algorithm's power efficiency, we opted for the Orion Model [40,46], which analyzes the power and energy of the network and can be integrated into a simulation environment to compute the network's overall energy. The NoCtweak simulator was used for power estimation, which was conducted using a standard cell library and the ORION-2 model, allowing for accurate evaluation of power consumption at different CMOS nodes. This simulation tool derives both power and energy consumption. By utilizing post-layout power details sourced from 2D NoC components, these estimations are based on component behavior when executing traffic patterns or real-time programs with a given mapping. Regarding the power metrics, the proposed IWOA improves power reduction by an average of 21.93%, 30.20%, 22.07%, 26.92%, 15.84%, 8.70%, 7.78%, 1.9%, 7.66%, and 5.48% over the PSO, SA, GA, ACO, BA, SCSO, ILP, iHPSO, CSO, and SFO algorithms, respectively. We observed that IWOA integrated with GA further improves the power reduction by an average of 22.09%, 30.34%, 22.23%, 27.07%, 16.24%, 8.89%, 7.97%, 2.07%, 7.84%, and 5.63% over the PSO, SA, GA, ACO, BA, SCSO, ILP, iHPSO, CSO, and SFO algorithms, respectively. It is notable that there was an average power difference of 3–4% when the simulation was carried out using the IWOA-IGA instead of the IWOA with real-time benchmarks. This is because both algorithms have different features and abilities to achieve optimal mapping solutions. Overall, these results demonstrate that the proposed algorithms consume less power than existing bio-inspired algorithms. The comparison of projected power values through the proposed methods is visually depicted in Figure 9.

Figure 10 shows the energy consumption for the various methods. The mapping technique implemented in the proposed work outshines competing state-of-the-art techniques such as PSO, SA, GA, and iHPSA in terms of energy consumption. In comparison to PSO, SA, GA, and iHPSA, IWOA reduces energy consumption by 1.7%, 13.89%, 17.47%, and 0.2%. The improved version yields improvements of over 3.22% compared to PSO, 15.38% compared to SA, 27.58% compared to GA, and 1.64% compared to iHPSA.

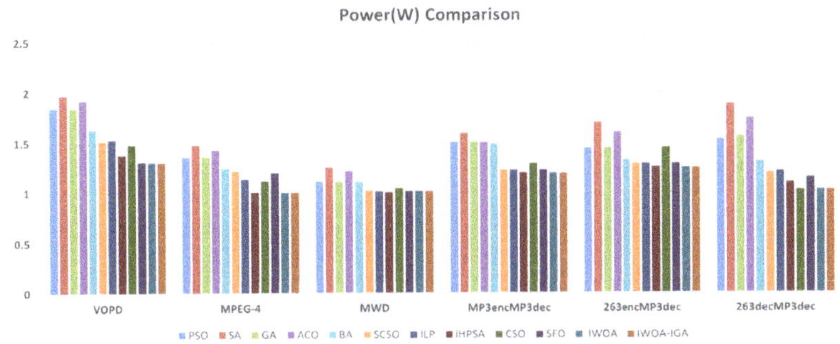

Figure 9. Power estimation for standard benchmarks.

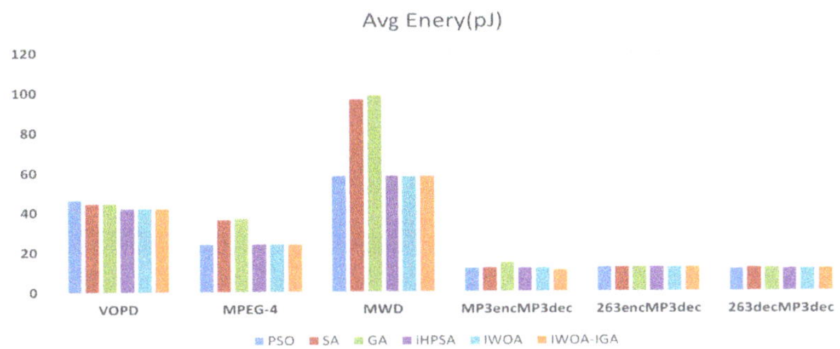

Figure 10. Energy estimation for standard benchmarks (normalized).

6.2.4. Performance Comparison of IWOA and IWOA-IGA: Optimizing Latency for 2D NoC Benchmark Applications

Network communication latency is a crucial factor to consider when evaluating mapping solutions. This subsection discusses the average latency derived with various approaches. Utilizing the NoC simulator [39], the proposed work was evaluated on a 4 × 4 mesh topology with an X–Y routing algorithm and wormhole-based routers. The evaluation showed that the IWO algorithm yielded latency improvements of 1.91% over PSO, 5.14% over SA, 5.7% over GA, and 1% over iHPSA. The extended IWOA-IGA version showed improvements of 2.87% over PSO, 6.08% over SA, 6.61% over GA, and 2% over iHPSA.

The comparison is illustrated in the Figure 11.

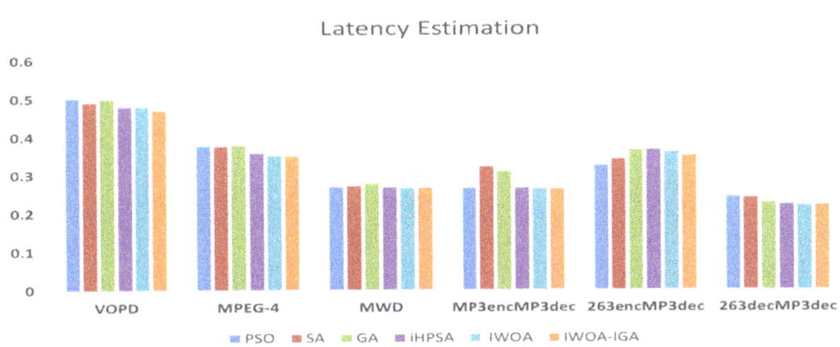

Figure 11. Average latency for standard benchmarks.

6.2.5. Convergence Evaluation Based on Convergence Factor

Convergence of any algorithm is one of the most important factors of an application mapping technique intended to acquire optimal mapping in a lesser number of iterations. Different algorithms running different applications may utilize different numbers of iterations and communication costs for the calculation of optimal mapping. Mathematically exact methods are able to solve small-scale problems (less than 20 cores) in around 7–8 h. While heuristic techniques are much faster, they may not offer optimal communication cost. They normally achieve near-optimal cost after the algorithm has run for a specific number of iterations. An algorithm that requires fewer iterations to achieve the lowest communication cost is considered to converge well, and is rated as highly suitable for the application mapping problem domain. Thus, one of the performance metrics considered in this paper is the convergence factor, which evaluates the algorithm's ability to converge effectively. Performance based on algorithm convergence ability is measured by calculating the average number of times the algorithm successfully converges, taking into account the number of hits achieved by the algorithm out of the total number of runs. To evaluate performance based on the convergence factor, the algorithms were tested 30 times while measuring the convergence factor for a range of real-time and synthetic task graphs of varying sizes. The proposed method was compared with PSO, SA, GA, and iHPSA based on the number of times the algorithms achieved the lowest communication cost over a certain number of runs. The equation for calculating the convergence factor is provided by

$$Perf_{cf} = Nav_c / NR_{total}, \tag{36}$$

where $Perf_{cf}$ is the performance based on the convergence factor, Nav_c is the average number of times the algorithm converged, and NR_{total} is the total number of times the algorithm was run. The comparison results for the real-world and synthetic applications are shown in Table 6, which shows that the algorithm converged several times when being run 30 times. The table shows Nav_c, the average number of times the algorithm converged for the specific application, which is calculated here based on half of the algorithm's executions R1–R15 (average halved) instead of 30, which is NR_{total}, the total number of times the algorithm was run for a specific application.

Table 6. Performance evaluation based on convergence.

Applications	Nav_c	NR_{total}	$Perf_{cf}$
PSO	10.71	15	7.14
SA	10.28	15	6.85
GA	11.14	15	7.42
IWOA	11.42	15	7.61
IWOA-IGA	12.42	15	8.2

The average value of convergence Nav_C calculated for PSO, SA, GA, IWOA, and IWOA-IGA against real and synthetic benchmarks was found to be 10.71, 10.28, 11.14, 11.42, and 12.42, respectively. As Nav_C is directly proportional to $Perf_{CF}$, a higher Nav_C value for an algorithm indicates better performance in terms of average convergence for a specified number of runs. The calculated $Perf_{CF}$ values for PSO, SA, GA, IWOA, and IWOA-IGA are 0.71, 0.68, 0.74, 0.76, and 0.82 respectively, which shows that the proposed algorithms converged several more times compared to the others when applied to the real and synthetic task graphs.

Figure 12 shows the graphical representation of the convergence factor results as derived using Nav_C. The graph shows the better values of Nav_C and performance based on the convergence factor achieved by the proposed algorithms in comparison to other competitive algorithms.

A statistical analysis of the results reveals that the IWOA-IGA hybrid algorithm significantly reduces communication costs across different core configurations. For 32 cores, the hybrid algorithm achieves an average reduction of 48.12% compared to the other optimization algorithms. Similarly, for 64 cores the reduction is 38.11%, while for 128 cores it is 30.53%. Regarding energy efficiency, the algorithm outperforms PSO by 3.22%, SA by 15.38%, GA by 27.58%, and iHPSA by 1.64%. Additionally, the IWOA-IGA exhibits superior convergence, with a $Perf_{CF}$ value of 0.82, indicating its effectiveness on both real and synthetic task graphs. The IWOA-IGA shows improved latency as well, achieving improvements of 2.87% over PSO, 6.08% over SA, 6.61% over GA, and 2% over iHPSA. In conclusion, the improved whale optimization algorithm integrated with specialized modified genetic algorithm properties presents superior performance in solving mapping problems within Network-on-Chip (NoC) architectures. Its effectiveness surpasses that of other state-of-the-art algorithms, establishing it as a viable and advantageous solution for addressing complex NoC challenges.

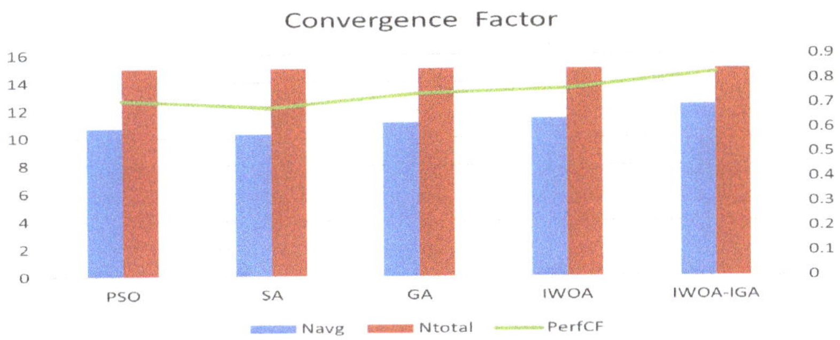

Figure 12. Performance based on convergence factor.

7. Conclusions

In this paper, we have presented a hybrid model incorporating an improved whale optimization algorithm with a modified enhanced version of genetic algorithm for the allocation of real-world applications onto 2D NoCs. The proposed algorithm incorporates enhanced initial mapping instead of random initial mapping to provide a head start to the optimization algorithm in achieving the global optimum solution. In the first step, the IWOA algorithm is introduced to address the application mapping challenge. To further improve the efficiency of the proposed algorithm, an enhanced and modified genetic algorithm which uses expert-based selection, direction-based cross-over, and mutation abilities is integrated with the IWOA to produce high-quality mapping solutions. This tweaked GA features helps the IWOA to achieve optimal mapping with faster convergence, allowing it to avoid local optima through its enhanced search capability. Extensive experimentation and analysis were performed with both real-time benchmarks and synthetic large-scale task graphs. The proposed IWOA-IGA shows significant improvements regarding communication cost, average power, energy, and latency over other competitive algorithms, demonstrating its high potential. In future work, the proposed algorithm can be employed to map real-time applications onto alternative NoC topologies.

Author Contributions: Conceptualization, S.S., F.H. and N.K.B.; methodology, S.S.; implementation, S.S.; validation, S.S., F.H. and N.K.B.; formal analysis, S.S.; investigation, S.S., F.H. and N.K.B.; data curation, S.S.; writing—original draft preparation, S.S.; writing—review and editing, F.H. and N.K.B.; supervision, F.H. and N.K.B.; project administration, N.K.B. All authors have read and agreed to the published version of the manuscript.

Funding: This research received no external funding.

Data Availability Statement: This research used publicly available data for experimentation and analysis purposes.

Conflicts of Interest: The authors declare no conflicts of interest.

References

1. Owens, J.D.; Dally, W.J.; Ho, R.; Jayasimha, D.; Keckler, S.W.; Peh, L.S. Research challenges for on-chip interconnection networks. *IEEE Micro* **2007**, *27*, 96–108. [CrossRef]
2. Kumar, S.; Jantsch, A.; Soininen, J.P.; Forsell, M.; Millberg, M.; Oberg, J.; Tiensyrja, K.; Hemani, A. A network on chip architecture and design methodology. In Proceedings of the IEEE Computer Society Annual Symposium on VLSI. New Paradigms for VLSI Systems Design, ISVLSI 2002, Pittsburgh, PA, USA, 25–26 April 2002; pp. 117–124.
3. Tosun, S.; Ozturk, O.; Ozen, M. An ILP formulation for application mapping onto network-on-chips. In Proceedings of the 2009 International Conference on Application of Information and Communication Technologies, Baku, Azerbaijan, 14–16 October 2009; pp. 1–5.
4. Ingle, V.V.; Gaikwad, M.A. Review of mesh topology of NoC architecture using source routing algorithms. *Int. J. Comput. Appl.* **2013**, *975*, 8887.
5. Han, J.J.; Lin, M.; Zhu, D.; Yang, L.T. Contention-aware energy management scheme for NoC-based multicore real-time systems. *IEEE Trans. Parallel Distrib. Syst.* **2014**, *26*, 691–701. [CrossRef]
6. Sahu, P.K.; Chattopadhyay, S. A survey on application mapping strategies for network-on-chip design. *J. Syst. Archit.* **2013**, *59*, 60–76. [CrossRef]
7. Pop, R.; Kumar, S. *A Survey of Techniques for Mapping and Scheduling Applications to Network on Chip Systems*; Research Report; School of Engineering, Jonkoping University: Jönköping, Sweden, 2004; Volume 4.
8. Sharma, P.K.; Biswas, S.; Mitra, P. Energy efficient heuristic application mapping for 2-D mesh-based network-on-chip. *Microprocess. Microsystems* **2019**, *64*, 88–100. [CrossRef]
9. Ogras, U.Y.; Hu, J.; Marculescu, R. Key research problems in NoC design: A holistic perspective. In Proceedings of the 3rd IEEE/ACM/IFIP International Conference on Hardware/Software Codesign and System Synthesis, Jersey City, NJ, USA, 19–21 September 2005; pp. 69–74.
10. Tosun, S.; Ozturk, O.; Ozkan, E.; Ozen, M. Application mapping algorithms for mesh-based network-on-chip architectures. *J. Supercomput.* **2015**, *71*, 995–1017. [CrossRef]
11. Wang, X.; Liu, H.; Yu, Z.; Shen, K. A novel two-phase heuristic for application mapping onto mesh-based Network-on-Chip. *IEICE Electron. Express* **2016**, *13*, 20151097. [CrossRef]
12. Brezočnik, L.; Fister Jr, I.; Podgorelec, V. Swarm intelligence algorithms for feature selection: A review. *Appl. Sci.* **2018**, *8*, 1521. [CrossRef]
13. Amin, W.; Hussain, F.; Anjum, S.; Khan, S.; Baloch, N.K.; Nain, Z.; Kim, S.W. Performance evaluation of application mapping approaches for network-on-chip designs. *IEEE Access* **2020**, *8*, 63607–63631. [CrossRef]
14. Sahu, P.K.; Shah, T.; Manna, K.; Chattopadhyay, S. Application mapping onto mesh-based network-on-chip using discrete particle swarm optimization. *IEEE Trans. Very Large Scale Integr. (VLSI) Syst.* **2013**, *22*, 300–312. [CrossRef]
15. Tosun, S. Cluster-based application mapping method for network-on-chip. *Adv. Eng. Softw.* **2011**, *42*, 868–874. [CrossRef]
16. Khajekarimi, E.; Hashemi, M.R. Energy-aware ILP formulation for application mapping on NoC based MPSoCs. In Proceedings of the 2013 21st Iranian Conference on Electrical Engineering (ICEE), Mashhad, Iran, 14–16 May 2013; pp. 1–5.
17. Khan, S.; Anjum, S.; Gulzari, U.A.; Afzal, M.K.; Umer, T.; Ishmanov, F. An efficient algorithm for mapping real time embedded applications on NoC architecture. *IEEE Access* **2018**, *6*, 16324–16335. [CrossRef]
18. Liu, L.; Wu, C.; Deng, C.; Yin, S.; Wu, Q.; Han, J.; Wei, S. A flexible energy-and reliability-aware application mapping for NoC-based reconfigurable architectures. *IEEE Trans. Very Large Scale Integr. (VLSI) Syst.* **2014**, *23*, 2566–2580. [CrossRef]
19. Murali, S.; De Micheli, G. Bandwidth-constrained mapping of cores onto NoC architectures. In Proceedings of the Design, Automation and Test in Europe Conference and Exhibition, Paris, France, 16–20 February 2004; Volume 2, pp. 896–901.
20. Shen, W.T.; Chao, C.H.; Lien, Y.K.; Wu, A.Y. A new binomial mapping and optimization algorithm for reduced-complexity mesh-based on-chip network. In Proceedings of the First International Symposium on Networks-on-Chip (NOCS'07), Princeton, NJ, USA, 7–9 May 2007; pp. 317–322.
21. Tosun, S. New heuristic algorithms for energy aware application mapping and routing on mesh-based NoCs. *J. Syst. Archit.* **2011**, *57*, 69–78. [CrossRef]
22. Cheng, C.H.; Chen, W.M. Application mapping onto mesh-based network-on-chip using constructive heuristic algorithms. *J. Supercomput.* **2016**, *72*, 4365–4378. [CrossRef]
23. Wang, X.; Choi, T.M.; Yue, X.; Zhang, M.; Du, W. An effective optimization algorithm for application mapping in network-on-chip designs. *IEEE Trans. Ind. Electron.* **2019**, *67*, 5798–5809. [CrossRef]
24. Upadhyay, M.; Shah, M.; Bhanu, P.V.; Soumya, J.; Cenkeramaddi, L.R. Multi-application based network-on-chip design for mesh-of-tree topology using global mapping and reconfigurable architecture. In Proceedings of the 2019 32nd international conference on VLSI Design and 2019 18th International Conference on Embedded Systems (VLSID), Delhi, India, 5–9 January 2019; pp. 527–528.
25. Yan, R.; Zhou, Y.; Cai, A.; Li, C.; Yan, Y.; Yin, M. Contention-aware mapping and scheduling optimization for NoC-based MPSoCs. In Proceedings of the International Conference on Automated Planning and Scheduling, Nancy, France, 26–30 October 2020; Volume 30, pp. 305–313.
26. Mohiz, M.J.; Baloch, N.K.; Hussain, F.; Saleem, S.; Zikria, Y.B.; Yu, H. Application mapping using cuckoo search optimization with Lévy flight for NoC-based system. *IEEE Access* **2021**, *9*, 141778–141789. [CrossRef]

27. Amin, W.; Hussain, F.; Anjum, S. iHPSA: An improved bio-inspired hybrid optimization algorithm for task mapping in Network on Chip. *Microprocess. Microsystems* **2022**, *90*, 104493. [CrossRef]
28. Choudhary, J.; Soumya, J.; Cenkeramaddi, L.R. Raman: Reinforcement learning inspired algorithm for mapping applications onto mesh network-on-chip. In Proceedings of the 2021 ACM/IEEE International Workshop on System Level Interconnect Prediction (SLIP), Munich, Germany, 4 November 2021; pp. 52–58.
29. Reza, M.F.; McCloud, Z. Heuristics-enabled high-performance application mapping in network-on-chip based multicore systems. In Proceedings of the 2023 IEEE International Conference on Omni-layer Intelligent Systems (COINS), Berlin, Germany, 23–25 July 2023; pp. 1–6.
30. Bose, A.; Ghosal, P. The CTH Network: An NoC Platform for Scalable and Energy Efficient Application Mapping Solution. *IEEE Trans. Nanotechnol.* **2023**, *22*, 58–69. [CrossRef]
31. Amin, W.; Hussain, F.; Anjum, S.; Saleem, S.; Ahmad, W.; Hussain, M. HyDra: Hybrid Task Mapping Application Framework for NOC-based MPSoCs. *IEEE Access* **2023**, *11*, 52309–52326. [CrossRef]
32. Amin, W.; Hussain, F.; Anjum, S.; Saleem, S.; Baloch, N.K.; Zikria, Y.B.; Yu, H. Efficient application mapping approach based on grey wolf optimization for network on chip. *J. Netw. Comput. Appl.* **2023**, *219*, 103729. [CrossRef]
33. Saleem, S.; Hussain, F.; Amin, W.; Ahmed, R.; Zikria, Y.B.; Ishmanov, F.; Yu, H. A Survey on Dynamic Application Mapping Approaches for Real-Time Network-on-Chip-Based Platforms. *IEEE Access* **2023**, *11*, 122694–122721. [CrossRef]
34. Ramesh, S.; Manna, K.; Gogineni, V.C.; Chattopadhyay, S.; Mahapatra, S. Congestion-Aware Vertical Link Placement and Application Mapping Onto Three-Dimensional Network-On-Chip Architectures. *IEEE Trans.-Comput.-Aided Des. Integr. Circuits Syst.* **2024**, 1. [CrossRef]
35. Tran, A.T.; Baas, B. *NoCTweak: A Highly Parameterizable Simulator for Early Exploration of Performance and Energy of Networks On-Chip*; Technical Report ECE-VCL-2012-2; VLSI Computation Lab, ECE Department, University of California: Davis, CA, USA, 2012.
36. Mirjalili, S.; Lewis, A. The whale optimization algorithm. *Adv. Eng. Softw.* **2016**, *95*, 51–67. [CrossRef]
37. Watkins, W.A.; Schevill, W.E. Aerial observation of feeding behavior in four baleen whales: Eubalaena glacialis, Balaenoptera borealis, Megaptera novaeangliae, and Balaenoptera physalus. *J. Mammal.* **1979**, *60*, 155–163. [CrossRef]
38. Goldbogen, J.A.; Friedlaender, A.S.; Calambokidis, J.; Mckenna, M.F.; Simon, M.; Nowacek, D.P. Integrative approaches to the study of baleen whale diving behavior, feeding performance, and foraging ecology. *BioScience* **2013**, *63*, 90–100. [CrossRef]
39. Chen, Q.; Huang, W.; Peng, Y.; Huang, Y. A reinforcement learning-based framework for solving the IP mapping problem. *IEEE Trans. Very Large Scale Integr. (VLSI) Syst.* **2021**, *29*, 1638–1651. [CrossRef]
40. Haupt, R.L.; Haupt, S.E. *Practical Genetic Algorithms*; John Wiley & Sons: Hoboken, NJ, USA, 2004.
41. Ning, G.Y.; Cao, D.Q. Improved whale optimization algorithm for solving constrained optimization problems. *Discret. Dyn. Nat. Soc.* **2021**, *2021*, 8832251. [CrossRef]
42. Tei, Y.Z.; Marsono, M.N.; Shaikh-Husin, N.; Hau, Y.W. Network partitioning and GA heuristic crossover for NoC application mapping. In Proceedings of the 2013 IEEE International Symposium on Circuits and Systems (ISCAS), Beijing, China, 19–23 May 2013; pp. 1228–1231.
43. Song, Y.; Wang, F.; Chen, X. An improved genetic algorithm for numerical function optimization. *Appl. Intell.* **2019**, *49*, 1880–1902. [CrossRef]
44. Ismkhan, H. Black box optimization using evolutionary algorithm with novel selection and replacement strategies based on similarity between solutions. *Appl. Soft Comput.* **2018**, *64*, 260–271. [CrossRef]
45. Elhoseny, M.; Tharwat, A.; Hassanien, A.E. Bezier curve based path planning in a dynamic field using modified genetic algorithm. *J. Comput. Sci.* **2018**, *25*, 339–350. [CrossRef]
46. Wang, H.S.; Zhu, X.; Peh, L.S.; Malik, S. Orion: A power-performance simulator for interconnection networks. In Proceedings of the 35th Annual IEEE/ACM International Symposium on Microarchitecture, 2002 (MICRO-35), Istanbul, Turkey, 18–22 November 2002; pp. 294–305.

algorithms

MDPI

Article

Multi-Objective Majority–Minority Cellular Automata Algorithm for Global and Engineering Design Optimization

Juan Carlos Seck-Tuoh-Mora *,†, Ulises Hernandez-Hurtado †, Joselito Medina-Marín, Norberto Hernández-Romero and Liliana Lizárraga-Mendiola

Academic Area of Engineering and Architecture, Institute of Basic Sciences and Engineering, Autonomous University of the State of Hidalgo, Pachuca 42184, Hidalgo, Mexico; ulixexido@comunidad.unam.mx (U.H.-H.); jmedina@uaeh.edu.mx (J.M.-M.); nhromero@uaeh.edu.mx (N.H.-R.); mendiola@uaeh.edu.mx (L.L.-M.)

* Correspondence: jseck@uaeh.edu.mx

† These authors contributed equally to this work.

Abstract: When dealing with complex models in real situations, many optimization problems require the use of more than one objective function to adequately represent the relevant characteristics of the system under consideration. Multi-objective optimization algorithms that can deal with several objective functions are necessary in order to obtain reasonable results within an adequate processing time. This paper presents the multi-objective version of a recent metaheuristic algorithm that optimizes a single objective function, known as the Majority–minority Cellular Automata Algorithm (MmCAA), inspired by cellular automata operations. The algorithm presented here is known as the Multi-objective Majority–minority Cellular Automata Algorithm (MOMmCAA). The MOMmCAA adds repository management and multi-objective search space density control to complement the performance of the MmCAA and make it capable of optimizing multi-objective problems. To evaluate the performance of the MOMmCAA, results on benchmark test sets (DTLZ, quadratic, and CEC-2020) and real-world engineering design problems were compared against other multi-objective algorithms recognized for their performance (MOLAPO, GS, MOPSO, NSGA-II, and MNMA). The results obtained in this work show that the MOMmCA achieves comparable performance with the other metaheuristic methods, demonstrating its competitiveness for use in multi-objective problems. The MOMmCAA was implemented in MATLAB and its source code can be consulted in GitHub.

Keywords: majority–minority cellular automata algorithm (MmCAA); multi-objective optimization; metaheuristic; cellular automata; real-world engineering problems

1. Introduction

Many real-world problems in engineering and other research areas require multi-objective optimization, where it is necessary to find a set of solutions in the search space that are well-distributed along the Pareto-optimal front. Generally, in this type of problem the computation of possible solutions must consider the existence of two or more conflicting objective functions. A multi-objective optimization problem can be defined as follows:

$$\begin{array}{c}\min \boldsymbol{h}(\boldsymbol{z}) = (h_1(\boldsymbol{z}), h_2(\boldsymbol{z}), \ldots, h_m(\boldsymbol{z})) \\ \text{where } \boldsymbol{z} \in Q, m \geq 2\end{array} \quad (1)$$

where each $h_i(\boldsymbol{z})$ is a real-valued scalar function, $\boldsymbol{h}(\boldsymbol{z})$ is the set of objective or cost functions that produce an m-dimensional vector in the $\mathbb{R}^m$-objective space when evaluated, $\boldsymbol{z} = (z_1, z_2, \ldots, z_n)$ is an n-dimensional vector in the search space $\mathbb{R}^n$, and $Q \subseteq \mathbb{R}^n$ is the set of all feasible solutions of Equation (1).

In this type of problem, identifying the set of best possible solutions can in many cases be highly complicated or impossible. Rather than looking for globally optimal solutions to a multi-objective problem, it is possible to instead seek to compute satisfactory solutions that can be obtained in adequate time, mainly to find the Pareto optimal (PS) set of solutions. The mapping from the PS to the objective space is the Pareto front (*PF*). Examples of classical optimization techniques adapted to multi-objective problems are the Weighted Sum Method [1], e-constraint method [2], goal programming [3], and lexicographic ordering [4] among others.

Multi-objective optimization evolutionary algorithms (MOEAs) are highly suitable for solving problems involving multiple objectives because they can generate a set of *PF*-approximate solutions in a single run [5]. In recent decades, many multi-objective optimization algorithms have been derived from classical single-objective metaheuristics, showing efficiency and effectiveness on various complex problems. Single-objective metaheuristics are optimization techniques that focus on finding a single optimal solution. Common examples include genetic algorithms (GA) [6], particle swarm optimization (PSO) [7], ant colony optimization (ACO) [8], and simulated annealing (SA) [9].

Several techniques based on single-objective metaheuristics have been developed to address multi-objective problems. Examples of the most outstanding options include the Non-Dominated Sorting Genetic Algorithm (NSGA-II), which extends GAs to handle multiple objectives using a non-dominated sorting scheme and population diversity [10]; the Multi-objective Evolutionary Algorithm Based on Decomposition (MOEA/D), which splits a multi-objective problem into several single-objective problems [11]; SPEA2, which uses a list of non-dominated solutions and assigns strengths to each solution to guide the search process [12]; Multi-objective PSO (MOPSO), which is an extension of PSO that allows it to handle multiple objectives while maintaining an archive of non-dominated solutions [13]; Multi-Objective Simulated Annealing (MOSA), which extends simulated annealing to handle multiple targets in order to maintain a balance between exploration and exploitation [14]; the Multi-objective Ant Lion Optimizer (MOALO), which expands the Ant Lion Optimizer by applying a repository to store non-dominated solutions in the Pareto set [15]; the Multi-objective Multi-Verse Optimizer (MOMVO), which builds on the MVO to compare and optimize test and practical problems [16]; and the Multi-objective ACO (MOACO), which adapts the ACO for multi-objective optimization using multiple populations to improve computational efficiency [17,18]. MOACO has also been employed to address multi-objective problems in airline crew turnover [19] and to improve the supply chain configurations [20].

Another work that is an extension of a recent metaheuristic algorithm is the Multi-objective Salp Swarm Algorithm (MSSA) [21]. Inspired by the swarming behavior of sea salps, the MSSA splits the population into a leader and followers, using an external file to store the non-dominated solutions. This approach shows adequate convergence and coverage of the PS. The interplay of several PSO algorithms for simultaneous optimization of single objectives in a multi-objective problem (MPMO) is described in [22] using multiple populations. In discrete problems, the MPMOGA is a new algorithm inspired by the GA which uses multiple populations to solve problems with multiple objectives. It was used in [23] to address the job-shop scheduling problems, obtaining satisfactory results. Multi-Objective Heat Transfer Search (MOHTS) based on Heat Transfer Search was proposed in [24] to optimize structural multi-objective problems, obtaining better results compared with other algorithms based on ant colonies and symbolic organisms. In related research, a multi-objective version of the symbolic organism algorithm was presented in [25] for optimal reinforcement design.

Multi-Objective Teaching–Learning-based Optimization (MOTLBO) is an extension of the Teaching–Learning-based Optimization (TLBO) algorithm. This approach uses two main phases: the teacher phase, where solutions are improved based on the best individual, and the learner phase, where solutions are optimized through knowledge sharing between individuals [26]. Multi-Objective Thermal Exchange Optimization (MOTEO), inspired by the principles of thermodynamics and heat exchange, seeks to solve optimization problems by considering multiple criteria simultaneously [27]. Multi-objective Plasma Generation

Optimization (MOPGO) is inspired by the generation and behavior of plasmas, where charged particles interact and move towards lower energy states [28]. the Multi-Objective Crystal Structure Algorithm (MOCSA) is motivated by the formation and organization of crystalline structures; solutions resemble atoms that organize themselves into configurations that minimize the energy of the system [29]. The Multi-Objective Forest Optimization Algorithm (MOFOA) follows the dynamics and ecology of forests; solutions resemble trees competing for resources, allowing the solutions (trees) to evolve and adapt in a competitive and cooperative environment [30]. The Competitive Mechanism Integrated Multi-Objective Whale Optimization Algorithm with Differential Evolution (CMI-MOWOA-DE) combines whale social behavior and differential evolution; the solutions simulate whale movement and hunting strategy, while differential evolution introduces variation and diversification to achieve a balance between multiple conflicting objectives. The Multi-Objective Harris Hawks Optimizer (MOHHO) is an extension of the Harris Hawks Optimizer (HHO) algorithm, which evokes the cooperative hunting strategies of Harris Hawks [31]. The Marine Predators Algorithm (MPA) emulates the behavior of species such as sharks and dolphins for multi-objective optimization, using search tactics and solution space exploitation to optimize multiple objectives [32]. The Multi-Objective Sine–Cosine Algorithm (MOSCA) is a variant of the Sine–Cosine Algorithm (SCA) that uses sine and cosine functions to guide the exploration and exploitation of the search space, dynamically adjusting the positions of candidate solutions to maintain diversity and ensure convergence to the FP [33]. The Multi-objective Atomic Orbital Search (MAOS) algorithm is based on the concept of atomic orbitals from quantum chemistry; the potential solutions are treated as electrons in different orbitals, and the search process resembles the movement of these electrons to reach lower energy configurations [34]. In the branch-and-bound framework for continuous global multi-objective optimization, the search space is recursively divided into smaller subregions, then lower and upper bounds are computed for the objective functions in these subregions. Subregions that cannot contain optimal solutions are discarded, which reduces the overall search space. Multi-Objective Differential Evolution (MODE) uses a population of candidate solutions that evolve through mutation operators. The multi-objective optimization method based on adaptive parameter harmony search algorithm simulates the improvisation process of musicians searching for the best harmony, dynamically adapting its parameters using memory and tuning operators to explore new solutions and preserve the best ones [35]. The Guided Population Archive Whale Optimization Algorithm (GPA-WOA) is a variant of the Whale Optimization Algorithm (WOA) that simulates the hunting behavior of humpback whales and uses guides or benchmark solutions to direct the search and improve convergence to the FP; a population file is dynamically updated to preserve diversity and ensure that solutions are optimal and well-distributed [36]. The quantum-inspired Decomposition-based Quantum Salp Swarm Algorithm (DQSSA) combines quantum mechanical principles with the swarming behavior of salps to divide multi-objective problems into more tractable subproblems, allowing a set of well-distributed optimal solutions to be found at the FP [37].

The above works are just a sample of the many single-objective algorithms that have recently been extended in various ways to deal with multi-objective problems. Single-objective optimization algorithms inspired by cellular automata are practical and have competitive results on these types of problems compared to more recent metaheuristics.

For instance, Cellular Particle Swarm Optimization (CPSO) is a variant of the classical PSO algorithm that organizes particles into a cellular lattice structure in which each particle only interacts with its nearest neighbors, thereby improving exploration and reducing the probability of premature convergence [38]. Island Cellular Model Differential Evolution combines the principles of Differential Evolution (DE) with a distributed population structure; a cellular scheme divides the population into subpopulations, which promotes genetic diversity, reduces premature convergence, and enhances exploration capability [39]. The Continuous-State Cellular Automata Algorithm (CCAA) is inspired by cellular automata but adapted to work with continuous rather than discrete variables. In this algorithm, individuals (or smart-cells) are organized in a spatial grid, each cell updates its state (candidate solution) based on the solutions of its local neighbors. Continuous states allow for finer exploration of the search space, while the restricted neighborhood structure favors a balance between local exploitation and global exploration [40]. The Cellular Learning Automata and Reinforcement Learning (CLARL) approach combines the principles of learning automata and reinforcement learning. In this method, learning automata are organized in a cellular mesh, where each automaton represents a potential solution and adapts its behavior through local interactions and reward-based feedback. This scheme allows for learning strategies that improve the system's dynamic adaptability [41]. The Reversible Elementary Cellular Automata Algorithm (RECAA) uses reversible rules, meaning that the system can return to previous states without losing information. In this algorithm, each potential solution follows simple local rules to update its state but with the property of reversibility, enabling a more controlled and efficient search space exploration [42].

However, only a few works have applied the concept of cellular automata for general multi-objective optimization. One of the most representative examples is the Cellular Ant Algorithm (CAA) for multi-objective optimization, which combines the ant colony structure with a cellular mesh in which ants only interact with their close neighbors. This mechanism simultaneously optimizes several objective functions, achieving balanced solutions to complex problems with multiple criteria [43]. Multi-objective Cellular Automata Optimization is another approach that applies cellular automata. Potential solutions are cells in a network that evolve based on local rules and interaction with their neighbors. This approach seeks to reach a balance by facilitating the identification of solutions [44]. Cellular Multi-objective Particle Swarm Optimization (CMPSO) is a variant of PSO in which particles are arranged in a cellular structure and only interact with their close neighbors. This promotes solution diversity by limiting global influences and encourages better exploration, and it is beneficial in applications that require simultaneous optimization of several criteria [45]. Cellular Teaching–Learning-Based Optimization (CTLBO) is a teaching-=learning-based approach to optimization. In this method, solutions are organized in a cellular structure, where each cell represents an individual who learns from its neighbors and a virtual teacher who guides the process. This approach enhances the algorithm's ability to adapt to dynamic changes for multiple objectives that may vary over time [46].

Following this trend, a recent single-objective optimization algorithm is the Majority-=minority Cellular Automata Algorithm (MmCAA), which was tested on several test problems in multiple dimensions and for various applications in engineering, obtaining satisfactory results against other well-recognized algorithms [47].

This paper presents a multi-objective version of this algorithm called MOMmCAA. This algorithm is inspired by the local behavior of cellular automata, particularly the majority and minority rules, which are intermixed and able to generate complex behaviors in order to perform the tasks of exploration and exploitation in the search space.

The problem to be addressed in this work is the optimization of multi-objective problems using a modification of the MmCAA to obtain an adequate approximation of its PS. Although multi-objective algorithms are continuously proposed in the specialized literature, they have yet to fully exploit the advantages offered by the different cellular automata rules, such as the diversities and richness of their dynamic behaviors and their easy implementation. Thus, this work aims to test and demonstrate the feasibility of modifying the MmCAA to deal with multi-objective problems in a manner comparable to current well-recognized algorithms for performing this task. The manuscript's originality lies in the fact that it is the first to propose an algorithm for multi-objective optimization inspired by cellular automata using majority and minority rules, and is complemented by managing a repository to control the density of solutions in the FP.

To test the performance of the MOMmCAA, we used the DLTZ benchmark, ten quadratic problems, and ten CEC2020 problems. The proposed algorithm was also tested on two practical engineering problems, obtaining satisfactory results. In these cases, five other algorithms were also considered for comparison: Multi-Objective Lightning Attachment Procedure Optimization (MOLAPO) [48], Grid Search (GS) [49], Multi-Objective Particle Swarm Optimization (MOPSO) [13], the Non-dominated Sorting Genetic Algorithm (NSGA-II) [10], and the Multi-objective Nelder–Mead Algorithm (MNMA) [50].

Non-parametric Wilcoxon statistical tests were performed to show the statistical significance of the experiments. The results indicate that the proposed algorithm ranks among the best with respect to the other methods used in this work.

The rest of this article is organized as follows. Section 2 presents the details of the Multi-Objective Majority–minority Cellular Automata Algorithm (MOMmCAA); Section 3 presents the results of our experiments on various test benches (DTLZ benchmark, ten quadratic problems, and ten CEC2020 problems), providing a statistical comparison of the MOMmCAA through the Wilcoxon test that relates it to other multi-objective algorithms recognized for their performance; Section 4 describes the application of the MOMmCAA to two practical engineering problems (design of a four-bar truss and a disk brake); finally, Section 5 provides the paper's conclusions.

2. The Proposed Multi-Objective Majority–Minority Cellular Automata Algorithm (MOMmCAA)

This section briefly explains the concept of cellular automata, the general characteristics of the Majority–minority Cellular Automata Algorithm, and the multi-objective implementation of this algorithm. The concept of hypercubes is used to delimit a repository for managing the PS solutions generated by the algorithm.

2.1. Basic Concepts of Cellular Automata with Majority Rule

Cellular Automata (CA) are dynamic systems of cells that initially take a value from a finite set of possible states. The dynamics of CA proceed in discrete steps, making CA discrete systems in time and space. At each step, a cell considers its current state and that of its close neighbors in order to update its state at the next time step. In this way, a mapping from blocks of states to individual states is called an evolution rule. CA can generate chaotic and complex global behaviors depending on the evolution rule that defines their local mapping. Because of this, they have been widely investigated and applied in various engineering and computational problems [51,52].

One of the rules of evolution extensively studied in recent work is the majority rule. In this rule, each cell takes its new state as the most common state in its local neighborhood. The dynamics of this rule are characterized by patchy patterns that stabilize as the system's evolution progresses. Its counterpart is the minority rule, which takes the least common element of each neighborhood to update the state of a cell in the next generation. The evolution of the minority rule is characterized by the fact that it does not tend quickly to a fixed or periodic state, as a minority state tends to become the majority and vice versa, resulting in oscillating global dynamics.

Figure 1 shows various dynamic behaviors of the majority rule and minority rule along with the application of the majority rule with probability in cellular automata of two states and various neighborhood sizes. In these examples, 500 cells and 250 evolutions were used. Evolution generates a periodic pattern for the original majority rule, while the minority rule generates a chaotic pattern of heterogeneous triangular shapes. When the rules are alternated probabilistically, the result is the formation of complex patterns in which non-periodic structures are combined with a stable background.

Figure 1. Examples of cellular automata with two states and a neighborhood size of 3, applying majority, minority, and majority with probability evolution rules.

This combination of majority and minority rules was used as inspiration to define the Majority–minority Cellular Automata Algorithm (MmCAA) for single-objective optimization [47]. The inspiration behind the MmCAA is to emulate the dynamic behavior of applying majority and minority rules in cellular automata.

The MmCAA starts by generating a random population S of n_S smart-cells, where each smart-cell is represented as $\boldsymbol{s} \in \mathbb{R}^n$. The dynamics of each $\boldsymbol{s}$ is defined by a set of rules, with one of them chosen randomly to improve the position of the smart-cell. The rules take information from other smart-cells to generate new neighbors, from which the best one is selected to upgrade the smart-cell position. With this mechanism, the positions of all smart-cells in the population are improved and the system evolves iteratively during the optimization process. The rules used by the MmCAA for smart-cell evolution are as follows.

The majority (minority) rule applied to a single smart-cell is described in Algorithm 1. The input is a smart-cell $\boldsymbol{s}_i$ for $1 \leq i \leq n_S$ and a weight parameter *prop_w* that defines a limit on the change in the values of $\boldsymbol{s}_i$. The rule takes the differences $\boldsymbol{cm}$ between the values in $\boldsymbol{s}_i$ and the most repeated element *el* in the smart-cell, and *rand* generates a random value between 0 and 1. A new solution $\boldsymbol{evol}$ is formed by taking the differences between the original smart-cell and $\boldsymbol{cm}$ randomly weighted between 0 and *prop_w*. This rule helps to bring the values of $\boldsymbol{s}_i$ closer to the most repeated value *el*.

Algorithm 1: Majority (minority) rule for a single smart-cell [47]

Result: New smart-cell $\boldsymbol{evol}$
Input: $\boldsymbol{s}_i$, $prop\_w$;
el = most repeated element in $\boldsymbol{s}_i$;
forall k *in* $length(\boldsymbol{s}_i)$ **do**
  $\boldsymbol{cm}(k) = \boldsymbol{s}_i(k) - el$;
end
$\boldsymbol{cm} = \boldsymbol{cm} * prop\_w * rand$;
$\boldsymbol{evol} = \boldsymbol{s}_i - \boldsymbol{cm}$;

The majority (minority) rule applied to a neighboring smart-cell is described in Algorithm 2. The most repeated value of a neighboring smart-cell $\boldsymbol{s}_j$ is taken as the change factor, depending on the average weight of the neighbor cost $\mu(\boldsymbol{h}(\boldsymbol{s}_j))$. If the cost $\mu(\boldsymbol{h}(\boldsymbol{s}_i))$ is larger than $\mu(\boldsymbol{h}(\boldsymbol{s}_j))$, then the weight *pon* is large and there is a higher probability of changing $\boldsymbol{s}_i$ by taking a random ratio between $-prop\_w$ and $prop\_w$ of the most repeated element in $\boldsymbol{s}_j$ and modifying each randomly selected position in $\boldsymbol{s}_i$.

Algorithm 2: Majority (minority) rule with one neighbor [47]

```
Result: New smart-cell evol
Input: s_i, μ(h(s_i)), s_j, μ(h(s_j)) prop_w;
evol = s_i;
sm = μ(h(s_i)) + μ(h(s_j));
pon = 1 − (μ(h(s_j))/sm);
el = most repeated element in neighbor s_j;
r = (rand * prop_w) − (prop_w/2);
forall k in length(evol) do
    if rand <= pon then
        evol(k) = evol(k) + (r * el);
    end
end
```

The rule for rounding values in a smart-cell (Algorithm 3) consists of rounding off n_r to the most significant decimal values of the selected elements in $\boldsymbol{s}_i$ [42]. The least significant decimal digit is the n_r-th digit to the right of the decimal point. The least significant digit remains unchanged if the first non-significant digit is less than 5; otherwise, the least significant digit is incremented by 1. This rule is applicable to find proper parameters for optimization problems.

Algorithm 3: Rounding rule [42]

```
Result: New smart-cell evol
Input: s_i, f(s_i), f(b_S), n_r;
evol = s_i;
sum = f(s_i) + f(b_S);
pon = 1 − (f(s_i)/sum);
forall k in length(evol) do
    if rand <= pon then
        evol(k) = round(evol(k), n_r);
    end
end
```

The adaptation of the majority and minority rules considers the elements of each solution that are repeated to a greater or lesser degree in the same smart-cell or in one or two additional smart-cells to obtain a new position. The randomness in choosing evolution rules and neighbors allows for access to the information in the rest of the population, thereby generating large and small changes in the position of a smart-cell, which favors the exploration and exploitation phases to escape from local optima and avoid the stagnation of the solutions. The MmCAA is presented in Figure 2. In (A), each smart-cell checks its neighbors using different majority and minority rules. These rules produce new solutions (B) and the best solution in the neighborhood is selected to update the smart-cell (C). The pseudo-code of MmCAA is described in (D).

2.2. Multi-Objective Majority–Minority Cellular Automata Algorithm (MOMmCAA)

The MmCAA was devised to solve single-objective optimization problems; therefore, it needs to be modified to deal with multi-objective problems. This results in the creation of the proposed multi-objective variant, MOMmCAA.

Taking as a basis the MmCAA inspired by the combination of majority and minority cellular automata and the handling of a solution repository on the Pareto front of the MOPSO algorithm, the MOMmCAA uses the following mechanisms for multi-objective optimization.

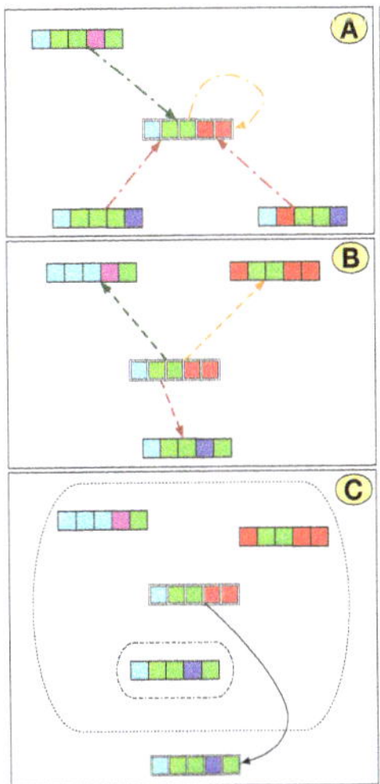

Algorithm Majority-minority cellular automata algorithm (MmCAA) **D**

Result: Best smart-cell $\boldsymbol{b}_S$ and fitness value $f(\boldsymbol{b}_S)$
Calculate random population S of n_S smart-cells;
Evaluate S in f;
forall $i = 2$ *to total iteration number* **do**
  Take the best elitist smart-cells in a new population;
  forall *No elitist solutions* j **do**
    Take $\boldsymbol{s}_j$ and other two random solutions $\boldsymbol{s}_{r_1}$, $\boldsymbol{s}_{r_2}$ from S for rules with extra parameters;
    forall *Neighbors* k **do**
      Take a random rule R ;
      Obtain $\boldsymbol{evol}_k = R(\boldsymbol{s}_j, \text{additional parameters})$;
      Review that the values of $\boldsymbol{evol}_k$ are between desired limits;
      Calculate $f(\boldsymbol{evol}_k)$;
    end
    Take the best solution from the k neighbors with a minimum cost $f(\boldsymbol{evol})$;
    if $rand < 0.25$ *or* $f(\boldsymbol{s}_j) > f(\boldsymbol{evol})$ **then**
      $\boldsymbol{s}_j = \boldsymbol{evol}$;
    end
  end
end
Return the best smart-cell $\boldsymbol{b}_S$ and $f(\boldsymbol{b}_S)$;

Figure 2. Majority–minority Cellular Automata Algorithm (MmCAA) [42].

Updating of each smart-cell: Each smart-cell solution generates a new set of neighboring solutions by taking its information or the information from one or two neighboring smart-cells (depending on the evolution rule that is randomly selected). Certain rules favor exploration by taking information from other smart-cells, while others favor exploitation by only taking information contained in the same smart-cell. From this set of neighbors, the one that is not dominated by the rest is used to update the smart-cell if it dominates it.

Repository with non-dominated solutions: The non-dominated smart-cells are stored in a repository, which also serves as a file to take neighbors when applying the different evolution rules that define the MOMmCAA optimization process. In order for a smart-cell to enter the repository, either it must dominate another solution or it must be the case that no other solution in the repository dominates it. The repository has a limited capacity; if a new smart-cell enters the repository, then the smart-cell that is dominated or the one that is in a region of the objective space with high density is deleted.

Hypercube density management in the objective space: Taking inspiration from the MOPSO mechanism, solutions in the PS are ranked depending on the density of the hypercube in which they are found in the solution space. If any other solution in the repository does not dominate a new solution and in turn is in a hypercube of lower density, then a solution that is in the hypercube with higher density is removed from the repository. This allows the repository to contain a better diversity of solutions. If a new solution is found in a new hypercube, then the boundaries of the solution space are expanded and the hypercube densities are recalculated. Figure 3 shows the handling of smart-cell selection in the repository using hypercubes; in (A), a new smart-cell replaces another smart-cell in a hypercube if it dominates it. In (B), if the new smart-cell falls into a higher-density hypercube and the repository is complete, then one of the smart-cells is randomly removed. In (C), if the new smart-cell falls into a less dense hypercube and the repository is complete, then one smart-cell is randomly removed from the denser hypercube. In (D), the hypercube boundaries are updated if the new smart-cell falls outside the current hypercube's boundaries.

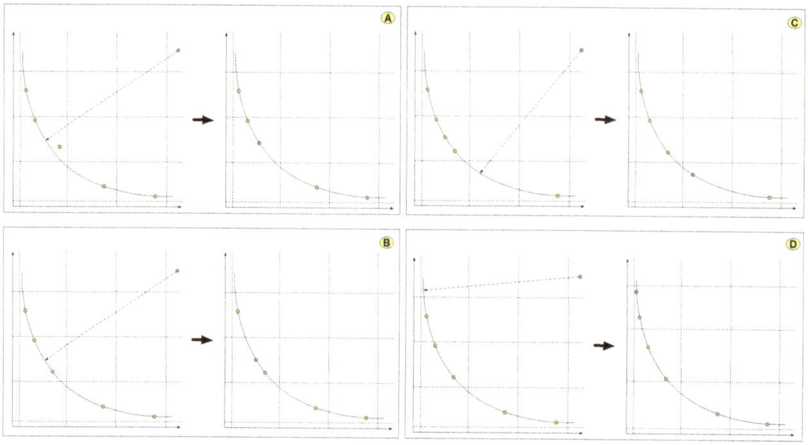

Figure 3. PS repository management using *PF* hypercubes.

The pseudocode of the MOMmCAA is presented in Algorithm 4.

Algorithm 4: Multi-Objective Majority–minority Cellular Automata Algorithm (MOMmCAA)

Result: Repository $\boldsymbol{B}_S$ of best smart-cells approximating PS
Generate random population S of n_S smart-cells;
Evaluate S in $\boldsymbol{h}$;
Initialize repository $\boldsymbol{B}_S$ of non-dominated smart-cells;
Calculate hypercube with n_{int} intervals and capacity cap;
forall $i = 2$ *to* n_{it} **do**
  Keep the best n_{el} smart-cells in a new population;
  forall $j = n_{el} + 1$ *to* n_S **do**
    Take $\boldsymbol{s}_j$ and other two random smart-cells $\boldsymbol{s}_{r_1}$, $\boldsymbol{s}_{r_2}$ from S for rules requiring extra solutions;
    forall $k = 1$ *to* n_{ne} **do**
      Choose a random rule R ;
      Obtain $\boldsymbol{evol}_k = R(\boldsymbol{s}_j,$ additional parameters of the rule$)$;
      Check that the n_d values of $\boldsymbol{evol}_k$ are between lb and ub and correct if necessary;
      Calculate $\boldsymbol{h}(\boldsymbol{evol}_k)$;
    end
    Choose the neighbor $\boldsymbol{evol}$ that dominates $\boldsymbol{s}_j$ and is in a hypercube with equal or lower density than $\boldsymbol{s}_j$ from the k generated neighbors. In other case, conserve $\boldsymbol{s}_j$;
    If $\boldsymbol{s}_j$ has been improved, update $\boldsymbol{B}_S$;
    If $\boldsymbol{B}_S$ has been improved, update the hypercube;
  end
end
Return the repository $\boldsymbol{B}_S$ with approximated PS;

2.3. *Computational Complexity of the Proposed Algorithm*

The computational time and space complexity of the MOMmCAA depends on the number of smart-cells N, number of objective functions to be evaluated m, management of the repository (where cap non-dominated solutions are stored), and total number of iterations n_{it}.

- **Smart-cell evaluation**: Each iteration evaluates all smart-cells, with a cost of $O(Nm)$.
- **Repository management**: The repository stores the non-dominated solutions. The cost of the Pareto dominance ordering mechanism can be high because the ordering of the non-dominated solutions has a complexity of $O(cap^2)$. Methods with similar

strategies attempt to reduce this cost by limiting the repository size; however, it is still costly, especially as the number of non-dominated solutions grows.

- **Iterations of the algorithm**: The total number of iterations impacts the complexity, as evaluations and repository management are performed at each step.

The overall complexity of the MOMmCAA can be approximated as $O(n_{it}(Nm + cap^2))$. In summary, the time complexity of the MOMmCAA is quadratic in terms of the repository size of non-dominated solutions. The space complexity is linear with respect to the population of smart-cells, the number of objectives, and the number of iterations. One of the advantages of using adaptive hypercubes is that the computational cost is better than when using a niche strategy such as the one used by NSGA-II [53]. The only case where both strategies have the same quadratic complexity is when the hypercubes are updated at each generation [54]. Thus, the complexity of the MOMmCAA is similar to that of MOPSO.

3. Computational Experiments Comparing MOMmCAA to Other Algorithms

MOLAPO, GS, MOPSO, NSGA-II, MOMVO, and MNMA were compared to MOMmCAA in order to identify the best performance in calculating Pareto-optimal solutions. The initial parameters of all described algorithms are summarized in Table 1. Each experiment employed 50 PS solutions and a maximum of 1000 iterations. The proposed algorithm was tested in 29 diverse case studies, including 27 unconstrained and constrained mathematical problems and two real-world engineering design problems.

Table 1. Parameters of the algorithms used for comparison: MOLAPO, GS, MOPSO, NSGA-II, MNMA, and MOMmCAA.

Algorithm	Parameters
MOLAPO	$FE = 2, cap = 50$
GS	$numparts = 20, cap = 50$
MOPSO	$C1 = C2 = 2, w = wmax - t \times (wmax - wmin)/(tmax)$, $wmax = 0.9, wmin = 0.4, cap = 50$
NSGA-II	$pcross = 0.7, ncross = 2 * round(pcross * 100/2), pmut = 0.4$, $nmut = round(pmut * 100), \mu = 0.02, Sigma = 0.1 * (vmax - vmin)$
MNMA	$\delta^e = 2, \delta^{oc} = 0.5, \delta^{ic} = -0.5, \gamma = 5$
MOMmCAA	$cap = 50, n_S = 11, n_{ne} = 3, n_{int} = 5, prop\_w = 5, 2 \leq n_r \leq 5$

The original Matlab implementations of these algorithms were taken directly from the web addresses indicated in the reference articles. The Matlab code of the MOMmCAA can be downloaded from Github using the link https://github.com/juanseck/MOMmCAA (accessed on 2 September 2024). The MOMmCAA and other algorithms were executed in Matlab 2015a on a PC with a 3.1 GHz Intel Xeon CPU with 64 GB of RAM using the macOS Sonoma operating system. Thirty independent runs were made for each algorithm on every benchmark function. A number of different metrics were used to compare the results of the algorithms, as described below.

Hypervolume (*HV*): The diverseness in the search space through the hypervolume metric was first introduced by Ulrich et al. to escalate the diversity in both decision space and objective space [55]. The *HV* of a set of solutions measures the size of the portion of the objective space dominated by those solutions as a group. In general, *HV* is favored because it captures both the closeness of the solutions to the optimal set and (to some extent) the distribution of solutions across the objective space in a single scalar. The *HV* value measures both convergence and diversity, and can be calculated using the equation

$$HV = \cup_s Z(s) \mid s \in PF, \tag{2}$$

where $Z(s)$ refers to the hypercube bounded by a solution s in the obtained PF. A larger HV value indicates a better approximation of the PF.

Contribution (C): The Contribution metric counts the number of PS points used in the combined solution of all algorithms. This metric is an extension of the Purity metric [56]

For two approximation Pareto sets A and B, where $B \subset A$, the C metric assigns A a higher measure than B.

For $O \geq 2$ MOEAs applied to a problem, let R_i be the non-dominated solutions obtained by the i-th MOEA for $1 \leq i \leq O$. The union of all these sets is $R = \cup_{i=1}^{O} R_i$. The set R^* of non-dominated solutions is calculated from R. Let r_i^* be the number of non-dominated sets in R^* obtained by the i-th MOEA:

$$r_i^* = \{s \mid s \in R_i \text{ and } s \in R^*\}. \tag{3}$$

Thus, the C_i metric of the i-th MOEA is defined as

$$C_i = \frac{|r_i^*|}{|R_i|}. \tag{4}$$

The C_i value may lie between $[0, 1]$, with a value nearer to 1 indicating better performance.

Epsilon Indicator (EI):

The Epsilon Indicator was defined in [57]. It measures the minimum value of the scalar ϵ required to make the Pareto front (PF) dominated by the approximation set S. Epsilon values fall within the range of $[1, \infty)$.

$$I_\epsilon(PF, S) = \inf_{\epsilon}\{\epsilon \mid \forall \mathbf{s} \in S, \exists \mathbf{b} \in PF \text{ such that } \mathbf{b} \leq \epsilon \mathbf{a}\} \tag{5}$$

In this case, the output of the epsilon indicator function is $1/\epsilon$; a value in the $(0, 1]$ range with a value near 1 is a close fit with the solution set.

3.1. Benchmark Instances

A total of 27 benchmark instances with complicated characteristics were used to compare the performance of the proposed MOMmCAA: DLTZ1-DLTZ7, ten quadratic problems, and ten CEC2020 test instances. These problems exhibit various characteristics, such as a convex, concave, mixed, disconnected, or degenerated PFs and a multimodal, biased, deceptive, and nonlinear variable PS.

For each instance, the compared algorithms are ranked according to the performance metrics, with the ranks shown in square brackets. The mean rank (MR) for each algorithm for each instance is also presented in the tables. As a result of the Wilcoxon rank sum test at a 5% significance level, a result labeled + denotes that the compared algorithm outperforms the MOMmCAA; in contrast, − means that the MOMmCAA has a better performance than the compared algorithm, while ≈ means that there is no statistically significant difference between MOMmCAA and the compared algorithm. The data in orange in every table show the best mean metric values yielded by the algorithms for each instance over 30 independent runs.

3.2. DLTZ Instances

Tables 2–4 present the results of the metric values obtained by algorithms.

As shown in Table 2, the MOMmCAA obtains significantly better HV values than MOLAPO, GS, MOPSO, NSGA-II, and MNMA for four, four, one, five, and seven out of the seven instances, respectively. Regarding the overall mean rankings, the MOMmCAA obtains the second optimal mean rank value, below MOPSO, followed by GS, NSGA-II, MOLAPO, and MNMA. The MOMmCAA has poor performance on the DLTZ1 and DLTZ3 test instances. In summary, the MOMmCAA is superior to the other four MOEAs on this metric. Table 3 shows that the MOMmCAA achieves seven, seven, six, six, and seven better C metric values than MOLAPO, GS, MOPSO, NSGA-II, and MNMA, respectively. This indicates the quality of the solutions obtained by the MOMmCAA. Table 4 summarizes the overall performance of six algorithms in terms of EI metric values. The MOMmCAA yields significantly better EI values than MOLAPO, GS, MOPSO, NSGA-II, and MNMA for six, four, one, seven, and five out of the seven instances, respectively. Overall, the EI statistics are similar to those for HV. Figure 4 plots the representative PFs obtained by the six comparison MOEAs. In summary, the MOMmCAA shows competitive performance on the DLTZ benchmark.

Table 2. Statistics (mean (std. dev.)) of Hypervolume metric values of the final populations obtained by MOLAPO, GS, MOPSO, NSGA-II, MNMA, and MOMmCAA over 30 independent runs on the DLTZ benchmark. The ranking of each algorithm is in braces.

Fn	MOLAPO	GS	MOPSO	NSGA-II	MNMA	MOMmCAA
DLTZ1	0.9942 (0.0019) [3]+	**0.9999 (0.0035)** [1]+	0.9962 (0.0157) [2]+	0.8081 (0.0372) [5]−	0.9579 (0.0121) [6]−	0.9867 (0.0028) [4]
DLTZ2	0.9960 (0.0039) [4]≈	0.9709 (0.0026) [6]−	**0.9985 (0.0016)** [1]+	0.9969 (0.0030) [2]≈	0.9850 (0.0057) [5]−	0.9969 (0.0033) [3]
DLTZ3	0.9905 (0.0045) [3]+	**0.9989 (0.0022)** [1]+	0.9973 (0.0032) [2]+	0.6326 (0.0529) [6]−	0.7055 (0.0608) [5]−	0.9746 (0.0058) [4]
DLTZ4	0.8115 (0.0480) [3]−	0.1597 (0.0039) [6]−	**0.9916 (0.0346)** [1]+	0.6762 (0.1710) [4]−	0.6434 (0.1716) [5]−	0.8940 (0.0588) [2]
DLTZ5	0.0733 (0.0260) [6]−	0.3479 (0.0171) [3]−	0.9413 (0.0511) [2]−	0.1733 (0.0260) [4]−	0.1106 (0.0056) [5]−	**0.9949 (0.0103)** [1]
DLTZ6	0.9900 (0.0025) [5]−	0.9973 (0.0008) [3]≈	**0.9997 (0.0005)** [1]≈	0.979 (0.0057) [6]−	0.9944 (0.0146) [4]−	0.9984 (0.0008) [2]
DLTZ7	0.8767 (0.0306) [5]−	0.8489 (0.0643) [6]−	0.9895 (0.0074) [2]≈	**0.9898 (0.0008)** [1]≈	0.9389 (0.0952) [4]−	0.9894 (0.0330) [3]
Mean rank	4.14	3.71	1.57	4.00	4.85	2.71
+/−/≈	2/4/1	2/4/1	4/1/2	0/5/2	0/7/0	—

Table 3. Statistics (mean (std. dev.)) of Contribution metric values of the final populations obtained by MOLAPO, GS, MOPSO, NSGA-II, MNMA, and MOMmCAA over 30 independent runs on the DLTZ benchmark.

Fn	MOLAPO	GS	MOPSO	NGSGA-II	MNMA	MOMmCAA
DLTZ1	0.815 (2.5×10^{-16}) [2]−	0.797 (1.8×10^{-16}) [5]−	0.814 (2.6×10^{-16}) [3]−	0.8095 (0.0124) [4]−	0.5406 (0.0061) [6]−	**0.999 (1.5×10^{-16})** [1]
DLTZ2	0.6088 (0.0461) [5]−	0.7600 (0.0644) [3]−	**0.9908 (0.0404)** [1]+	0.6421 (0.1655) [4]−	0.1969 (0.0361) [6]−	0.9248 (0.0116) [2]
DLTZ3	0.815 (6.8×10^{-17}) [2]−	0.797 (2.5×10^{-16}) [5]−	0.814 (6.8×10^{-17}) [3]−	0.8120 (0.0087) [4]−	0.4612 (0.0301) [6]−	**0.997 (2.3×10^{-17})** [1]
DLTZ4	0.810 (6.9×10^{-17}) [3]−	0.797 (1.5×10^{-16}) [5]−	0.814 (6.9×10^{-17}) [2]−	0.804 (7.9×10^{-17}) [4]−	0.6655 (0.0641) [6]−	**0.966 (1.9×10^{-17})** [1]
DLTZ5	0.0011 (0.0056) [6]−	0.1416 (0.0179) [4]−	0.8995 (0.1141) [2]−	0.1619 (0.2573) [3]−	0.0071 (0.0025) [5]−	**0.9135 (0.1269)** [1]
DLTZ6	0.0793 (0.0001) [6]−	0.1285 (0.0010) [5]−	0.3526 (0.0168) [3]−	**0.9791 (0.0003)** [1]+	0.2968 (0.0908) [4]−	0.5203 (0.0135) [2]
DLTZ7	0.1476 (0.0034) [5]−	0.0976 (0.0205) [6]−	0.7690 (0.0170) [3]−	0.7783 (0.0327) [2]−	0.7655 (0.0684) [4]−	**0.9489 (0.0001)** [1]
Mean rank	4.14	4.71	2.42	3.14	5.28	1.28
+/−/≈	0/7/0	0/7/0	1/6/0	1/6/0	0/7/0	—

Table 4. Statistics (mean (std. dev.)) of Epsilon Indicator metric values of the final populations obtained by MOLAPO, GS, MOPSO, NSGA-II, MNMA, and MOMmCAA over 30 independent runs on the DLTZ benchmark.

Fn	MOLAPO	GS	MOPSO	NGSGA-II	MNMA	MOMmCAA
DLTZ1	0.9925 (0.0064) [3]≈	0.9963 (0.0032) [2]+	**0.9994 (0.0015)** [1]+	0.8791 (0.0395) [6]−	0.9655 (0.0140) [5]−	0.9924 (0.0053) [4]
DLTZ2	0.9936 (0.0060) [3]−	0.9731 (0.0089) [6]−	**0.9985 (0.0015)** [1]≈	0.9931 (0.0070) [4]−	0.9819 (0.1230) [5]−	0.9984 (0.0059) [2]
DLTZ3	0.9555 (0.0241) [4]−	0.9702 (0.0249) [2]+	**0.9995 (0.0020)** [1]+	0.6953 (0.0780) [6]−	0.7779 (0.0963) [5]−	0.9689 (0.0217) [3]
DLTZ4	0.8053 (0.1128) [4]−	0.9047 (0.0235) [2]+	**0.9978 (0.0069)** [1]+	0.7507 (0.1543) [5]−	0.7380 (0.1501) [6]−	0.8434 (0.1174) [3]
DLTZ5	0.8790 (0.0086) [5]−	0.737 (2.9×10^{-10}) [6]−	0.911 (3.3×10^{-10}) [4]−	0.979 (8.6×10^{-10}) [2]−	0.921 (3.4×10^{-10}) [3]−	**0.999 (4.9×10^{-11})** [1]
DLTZ6	0.8967 (0.0500) [5]−	0.9276 (0.0021) [4]−	**0.9991 (0.0002)** [1]+	0.4459 (0.1339) [6]−	0.9687 (0.0611) [2]≈	0.9684 (0.0228) [3]
DLTZ7	0.8996 (0.0315) [6]−	0.9427 (0.0038) [5]−	**0.9957 (0.0072)** [1]+	0.9429 (0.0944) [4]−	0.9689 (0.0142) [3]≈	0.9694 (0.0146) [2]
Mean rank	4.28	3.85	1.43	4.71	4.14	2.57
+/−/≈	0/6/1	3/4/0	5/1/1	0/7/0	0/5/2	—

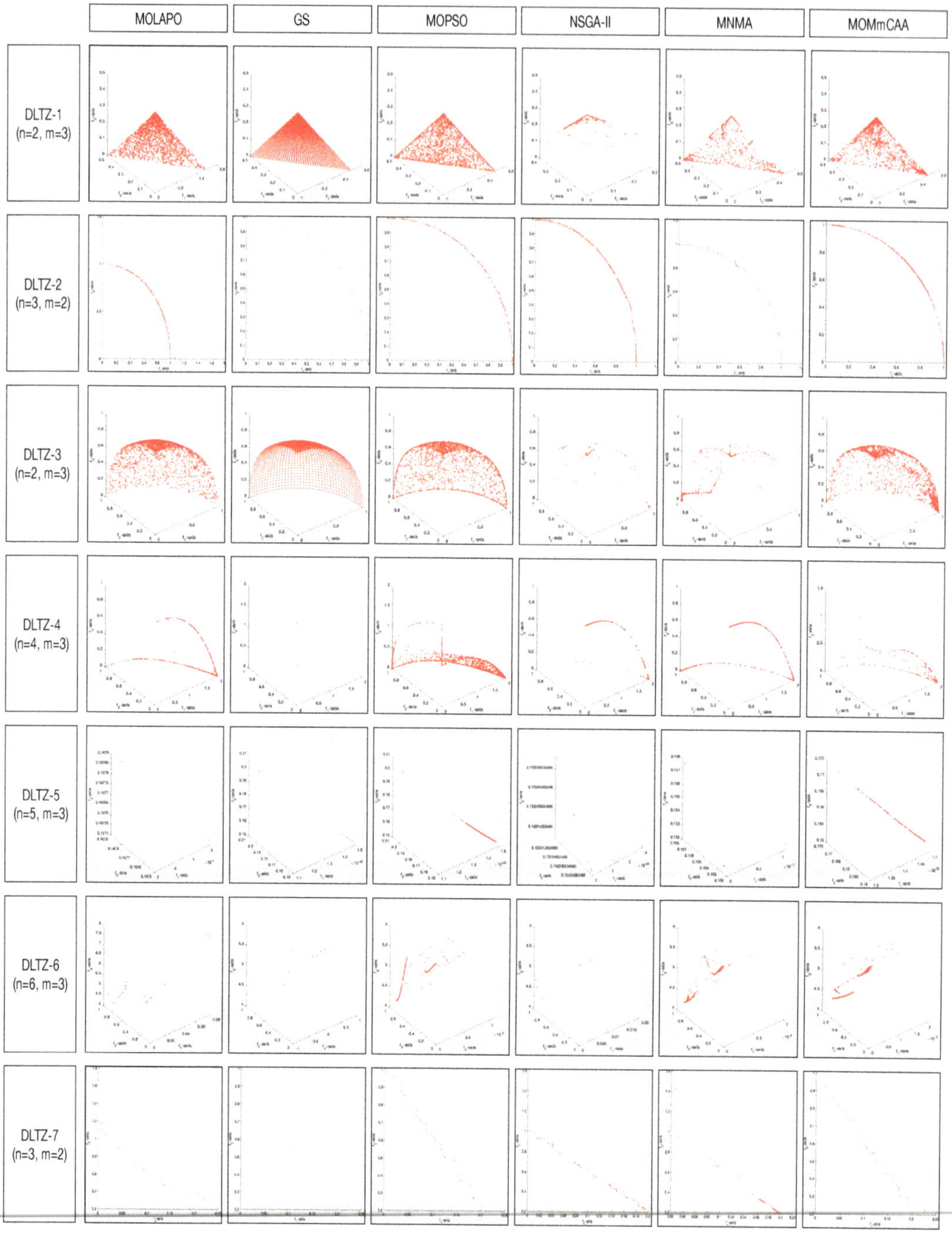

Figure 4. Representative *PF*s obtained by the seven MOEAs on the DLTZ benchmark.

3.3. Quadratic Instances

The Quadratics test set is a randomly generated test set described in [50]. The objective functions are all of the form $\frac{1}{2}x^T Ax + bx + c$, where the components A, b, and c are random numbers in the range $[-1, 1]$. A is not a symmetric matrix, and the test set is non-convex. Tables 5–7 expose the results of the metric values obtained by the algorithms over the ten quadratic problems.

Table 5 shows that the MOMmCAA obtains significantly better HV values than MOLAPO, GS, MOPSO, NSGA-II, and MNMA for six, seven, two, eight, and seven out of the eight instances, respectively. Regarding the overall mean rankings, the MOMmCAA obtains the second optimal mean rank value after MOPSO, followed by the other algorithms. The MOMmCAA demonstrates performance that is significantly equivalent to MOPSO regarding the other three instances. Table 6 shows that the MOMmCAA achieves ten, ten, five, eight, and seven better C metric values than MOLAPO, GS, MOPSO, NSGA-II, and MNMA, respectively. This indicates the quality of the solutions obtained by the MOMmCAA. Table 7 depicts the overall performance of the six algorithms in terms of their EI metric values. The MOMmCAA yields significantly better EI values than MOLAPO, GS, MOPSO, NSGA-II, and MNMA for nine, seven, six, eight, and nine out of the ten instances, respectively. Figure 5 shows representative Pareto fronts (PFs) obtained by the six comparison MOEAs. In summary, the MOMmCAA shows competitive performance on the Quadratic benchmark.

3.4. CEC2020 Instances

Tables 8–10 present the results of the metric values obtained by the algorithms on ten benchmark CEC2020 problems.

Table 8 shows that the MOMmCAA obtains significantly better HV values than MOLAPO, GS, MOPSO, NSGA-II, and MNMA for eight, nine, four, eight, and seven out of the ten instances, respectively. In the case of MOPSO, there are six results with no significant difference. Concerning the overall mean rankings, the MOMmCAA obtains the optimal mean rank value. The MOMmCAA demonstrates poor performance on the MMF-2 and MMF-7 test instances. In summary, the MOMmCAA is superior to all the other MOEAs in terms of this metric. Table 9 shows that the MOMmCAA achieves ten, ten, seven, eight, and six better C metric values than MOLAPO, GS, MOPSO, NSGA-II, and MNMA, respectively. In the case of MNMA, there are four results with the worst significant difference. Table 10 summarizes the overall performance of the six algorithms in terms of their EI metric values. The MOMmCAA yields significantly better EI values than MOLAPO, GS, MOPSO, NSGA-II, and MNMA for ten, ten, six, ten, and eight out of the ten instances, respectively. Figure 6 depicts the representative Pareto fronts (PFs) obtained by the seven comparison MOEAs. In summary, the MOMmCAA shows competitive behavior on the CEC2020 benchmark.

Table 5. Statistics (mean (std. dev.)) of Hypervolume metric values of the final populations obtained by MOLAPO, GS, MOPSO, NSGA-II, MNMA, and MOMmCAA over 30 independent runs on the Quadratic test suite.

Fn	MOLAPO	GS	MOPSO	NSGA-II	MNMA	MOMmCAA
Quad-1	0.7832 (0.0465) [5]−	0.9192 (0.2361) [3]−	**0.9999 (0.0105)** [1]+	0.9008 (0.0359) [4]−	0.7501 (0.1797) [6]−	0.9730 (0.0100) [2]
Quad-2	0.7556 (0.0364) [6]−	0.7834 (0.2361) [5]−	**0.9942 (0.0986)** [1]+	0.8271 (0.0389) [4]−	0.9267 (0.1588) [3]−	0.9668 (0.0118) [2]
Quad-3	0.9828 (0.0038) [5]−	0.9744 (0.0040) [6]−	0.9901 (0.0034) [4]−	**0.9983 (0.0022)** [1]+	0.9954 (0.0041) [2]≈	0.9953 (0.0050) [3]
Quad-4	**0.9988 (0.0017)** [1]≈	0.9957 (0.0046) [3]≈	0.9911 (0.0032) [4]−	0.9836 (0.0037) [5]−	0.9751 (0.0048) [6]−	0.9969 (0.0029) [2]
Quad-5	0.9409 (0.0108) [4]≈	0.9232 (0.0164) [5]−	**0.9999 (0.0044)** [1]+	0.9932 (0.0037) [2]+	0.8947 (0.0141) [6]−	0.9687 (0.0059) [3]
Quad-6	0.9834 (0.0050) [4]≈	0.9674 (0.0082) [6]−	0.9876 (0.0072) [3]≈	0.9759 (0.0114) [5]−	**0.9993 (0.0015)** [1]≈	0.9965 (0.0038) [2]
Quad-7	0.4328 (0.2074) [6]−	0.4650 (0.0808) [5]−	**0.9880 (0.0653)** [1]+	0.7888 (0.0595) [3]−	0.5682 (0.1853) [4]−	0.9374 (0.0143) [2]
Quad-8	0.9929 (0.0042) [4]≈	**0.9983 (0.0028)** [1]≈	0.9930 (0.0041) [3]≈	0.9649 (0.0151) [5]−	0.9622 (0.0204) [6]−	0.9947 (0.0058) [2]
Quad-9	0.9692 (0.0092) [4]−	0.9788 (0.0061) [3]−	**0.9977 (0.0010)** [1]≈	0.9010 (0.0203) [6]−	0.9077 (0.0257) [5]−	0.9973 (0.0058) [2]
Quad-10	0.8677 (0.0117) [4]−	0.9744 (0.0046) [2]+	**0.9814 (0.0055)** [1]+	0.7300 (0.0677) [6]−	0.8431 (0.0594) [5]−	0.9210 (0.0101) [3]
Mean rank	4.30	3.90	2.00	4.10	4.40	2.30
+/−/≈	0/6/4	1/7/2	5/2/3	2/8/0	0/8/2	—

Table 6. Statistics (mean (std. dev.)) of Contribution metric values of the final populations obtained by MOLAPO, GS, MOPSO, NSGA-II, MNMA, and MOMmCAA over 30 independent runs on the Quadratic test suite.

Fn	MOLAPO	GS	MOPSO	NGSGA-II	MNMA	MOMmCAA
Quad-1	0.7548 (0.0227) [4]−	0.8559 (0.3276) [3]−	**0.9938 (0.0237)** [1]+	0.6563 (0.0716) [5]−	0.6304 (0.0708) [6]−	0.9158 (0.0698) [2]
Quad-2	0.6150 (0.0140) [6]−	0.7628 (0.1674) [4]−	**0.9189 (0.1903)** [1]≈	0.7564 (0.2903) [5]−	0.8209 (0.0154) [3]−	0.9181 (0.1407) [2]
Quad-3	0.5834 (0.0181) [5]−	0.5745 (0.0118) [6]−	0.6594 (0.0327) [4]−	**0.9827 (0.0828)** [1]+	0.8591 (0.1329) [2]+	0.7498 (0.0357) [3]
Quad-4	0.9591 (0.1329) [2]−	0.9498 (0.0357) [3]−	0.8594 (0.0227) [4]−	0.6745 (0.0818) [6]−	0.6834 (0.0181) [5]−	**0.9827 (0.0823)** [1]
Quad-5	0.8493 (0.0419) [4]−	0.6621 (0.0211) [5]−	**0.9848 (0.0452)** [1]+	0.9398 (0.1482) [2]+	0.6382 (0.0162) [6]−	0.8862 (0.0741) [3]
Quad-6	0.8290 (0.0486) [3]−	0.6627 (0.0202) [6]−	0.7066 (0.0408) [4]−	0.6643 (0.0172) [5]−	**0.9965 (0.0130)** [1]+	0.9176 (0.0761) [2]
Quad-7	0.5266 (0.0516) [5]−	0.4667 (0.0821) [6]−	**0.9711 (0.0453)** [1]+	0.8471 (0.2180) [3]−	0.7424 (0.0153) [4]−	0.9264 (0.0303) [2]
Quad-8	0.7772 (0.0072) [4]−	0.8870 (0.0048) [2]−	0.8073 (0.0092) [3]−	0.6046 (0.0122) [5]−	0.5718 (0.0162) [6]−	**0.9951 (0.0031)** [1]
Quad-9	0.5452 (0.0215) [5]−	0.4369 (0.0118) [6]−	0.8706 (0.0423) [2]−	0.6800 (0.0718) [4]−	0.8547 (0.0887) [3]−	**0.9990 (0.0051)** [1]
Quad-10	0.5170 (0.0055) [6]−	0.6607 (0.0112) [5]−	0.9051 (0.0803) [2]+	0.7536 (0.1651) [4]−	**0.9722 (0.0694)** [1]+	0.0585 (0.8018) [3]
Mean rank	4.40	4.60	2.30	4.00	3.70	2.00
+/−/≈	0/10/0	0/10/0	4/5/1	2/8/0	3/7/0	—

Table 7. Statistics (mean (std. dev.)) of Epsilon Indicator metric values of the final populations obtained by MOLAPO, GS, MOPSO, NSGA-II, MNMA, and MOMmCAA over 30 independent runs on the Quadratic test suite.

Fn	MOLAPO	GS	MOPSO	NGSGA-II	MNMA	MOMmCAA
Quad-1	0.6240 (0.0944) [5]−	0.7748 (0.1275) [3]−	0.9951 (0.0044) [2]≈	0.6962 (0.3369) [4]−	0.5028 (0.2753) [6]−	**0.9981 (0.0036)** [1]
Quad-2	0.5449 (0.0991) [6]−	0.8027 (0.2635) [3]−	**0.9879 (0.0244)** [1]+	0.5542 (0.2826) [5]−	0.5996 (0.1161) [4]−	0.9779 (0.0194) [2]
Quad-3	0.8620 (0.1774) [5]−	0.7980 (0.1186) [6]−	0.8751 (0.1856) [4]−	**0.9776 (0.0347)** [1]+	0.9108 (0.1406) [3]−	0.9432 (0.0674) [2]
Quad-4	**0.9783 (0.0343)** [1]+	0.9181 (0.1343) [3]−	0.7756 (0.1862) [4]−	0.3984 (0.1191) [6]−	0.4685 (0.1619) [5]−	0.9439 (0.0679) [2]
Quad-5	0.8844 (0.1060) [4]−	0.4184 (0.0458) [6]−	**0.9909 (0.0175)** [1]+	0.9082 (0.0663) [3]−	0.4975 (0.0550) [5]−	0.9383 (0.0701) [2]
Quad-6	0.6550 (0.1195) [6]−	0.7024 (0.1305) [5]−	0.7376 (0.1019) [4]−	0.8806 (0.0519) [3]≈	**0.9658 (0.0430)** [1]+	0.8999 (0.0771) [2]
Quad-7	0.5581 (0.0323) [5]−	0.5366 (0.1128) [6]−	0.9571 (0.1227) [2]−	0.7736 (0.1198) [3]−	0.6356 (0.1532) [4]−	**0.9747 (0.0266)** [1]
Quad-8	0.9234 (0.0884) [3]−	**0.9878 (0.0524)** [1]+	0.9188 (0.0135) [4]−	0.8119 (0.0454) [6]−	0.8986 (0.0560) [5]−	0.9488 (0.0476) [2]
Quad-9	0.6500 (0.1554) [6]−	**0.9671 (0.1128)** [1]+	0.9536 (0.0220) [2]≈	0.7165 (0.1319) [5]−	0.7200 (0.2024) [4]−	0.9484 (0.1139) [3]
Quad-10	0.6713 (0.0765) [4]−	**0.9971 (0.0120)** [1]+	0.9174 (0.0983) [2]+	0.5120 (0.0297) [6]−	0.5753 (0.0895) [5]−	0.7920 (0.0212) [3]
Mean rank	4.50	3.50	2.60	4.20	4.20	2.00
+/−/≈	1/9/0	3/7/0	2/6/2	1/8/1	1/9/0	—

Table 8. Statistics (mean (std. dev.)) of Hypervolume metric values of the final populations obtained by MOLAPO, GS, MOPSO, NSGA-II, MNMA, and MOMmCAA over 30 independent runs on the CEC2020 test suite.

Fn	MOLAPO	GS	MOPSO	NGSGA-II	MNMA	MOMmCAA
MMF-1	0.9955 (0.0038) [2]≈	0.9792 (0.0043) [5]−	0.9952 (0.0036) [3]≈	0.9850 (0.0100) [4]−	0.9533 (0.0468) [6]−	**0.9997 (0.0006)** [1]
MMF-2	0.9451 (0.2137) [3]≈	0.9451 (0.2139) [5]≈	0.9453 (0.2134) [2]≈	0.9446 (0.2154) [6]−	**0.9732 (0.2031)** [1]+	0.9451 (0.2138) [4]
MMF-4	0.9276 (0.3679) [3]−	0.9171 (0.3664) [4]−	0.8989 (0.3692) [5]−	0.8378 (0.3683) [6]−	0.9785 (0.2451) [2]−	**0.9983 (0.3689)** [1]
MMF-5	0.9906 (0.0088) [4]−	0.9851 (0.0072) [6]−	0.9918 (0.0077) [3]−	0.9865 (0.0112) [5]−	0.9962 (0.0048) [2]≈	**0.9976 (0.0042)** [1]
MMF-7	0.9736 (0.0431) [5]−	0.9662 (0.0398) [6]−	0.9755 (0.0419) [2]≈	0.9754 (0.0425) [4]≈	**0.9968 (0.0481)** [1]+	0.9755 (0.0417) [3]
MMF-8	0.9116 (0.0631) [4]−	0.9102 (0.0591) [5]−	0.9555 (0.0519) [2]−	0.9324 (0.0521) [3]−	0.3913 (0.0531) [6]−	**0.9713 (0.0466)** [1]
MMF-10	0.8607 (0.0026) [6]−	0.9331 (0.0489) [4]−	0.9602 (0.0238) [3]−	**0.9793 (0.0029)** [1]≈	0.9055 (0.0967) [5]−	0.9721 (0.0223) [2]
MMF-11	0.8512 (0.0338) [6]−	0.8627 (0.0011) [5]−	**0.9723 (0.0211)** [1]≈	0.9278 (0.0887) [3]−	0.8891 (0.0261) [4]−	0.9715 (0.0018) [2]
MMF-12	0.9618 (0.0038) [5]−	0.9319 (0.0842) [6]−	0.9973 (0.0034) [2]≈	0.9639 (0.0138) [4]−	0.9773 (0.0094) [3]−	**0.9987 (0.0029)** [1]
MMF-13	0.9068 (0.0054) [6]−	0.9192 (0.0026) [5]−	**0.9991 (0.0017)** [1]≈	0.9788 (0.0049) [3]−	0.9689 (0.0247) [4]−	0.9982 (0.0020) [2]
Mean rank	4.4	5.1	2.4	3.9	3.2	1.8
+/−/≈	0/8/2	0/9/1	0/4/6	0/8/2	2/7/1	—

Table 9. Statistics (mean (std. dev.)) of Contribution metric values of the final populations obtained by MOLAPO, GS, MOPSO, NSGA-II, MNMA, and MOMmCAA over 30 independent runs on the CEC2020 test suite.

Fn	MOLAPO	GS	MOPSO	NGSGA-II	MNMA	MOMmCAA
MMF-1	0.2128 (0.0289) [6]−	0.4868 (0.0616) [3]−	0.2676 (0.0319) [5]−	0.3151 (0.0662) [4]−	**0.9763 (0.0461)** [1]+	0.8734 (0.1375) [2]
MMF-2	0.2715 (0.0014) [6]−	0.3240 (0.0122) [4]−	0.3588 (0.0870) [3]−	0.8086 (0.0059) [5]−	**0.9993 (0.1599)** [1]+	0.9369 (0.1297) [2]
MMF-4	0.1328 (0.0583) [6]−	0.5403 (0.2403) [3]−	0.2297 (0.1104) [5]−	0.3263 (0.1457) [4]−	0.8489 (0.2181) [2]−	**0.8717 (0.2841)** [1]
MMF-5	0.1913 (0.0175) [6]−	0.3997 (0.0524) [3]−	0.2309 (0.0374) [5]−	0.3514 (0.0576) [4]−	**0.9977 (0.0074)** [1]+	0.9444 (0.0169) [2]
MMF-7	0.1712 (0.2828) [6]−	0.7725 (0.0934) [3]−	0.3623 (0.0616) [5]−	0.6166 (0.2130) [4]−	**0.9838 (0.1464)** [1]+	0.9029 (0.1444) [2]
MMF-8	0.3752 (0.0038) [5]−	0.8731 (0.0093) [3]−	0.8623 (0.0061) [4]−	**0.9153 (0.0028)** [1]+	0.1718 (0.0014) [6]−	0.9011 (0.0014) [2]
MMF-10	0.8731 (0.0420) [6]−	0.9011 (0.0285) [4]−	0.9302 (0.0131) [3]−	0.9561 (0.0023) [2]≈	0.8925 (0.0367) [5]−	**0.9573 (0.0019)** [1]
MMF-11	0.6796 (0.0416) [6]−	0.7734 (0.0183) [5]−	**0.9407 (0.0223)** [1]+	0.9167 (0.0071) [3]−	0.7992 (0.0331) [4]−	0.9366 (0.0112) [2]
MMF-12	0.5151 (0.0734) [5]−	0.4905 (0.0204) [6]−	0.9854 (0.0601) [2]≈	0.6131 (0.1622) [3]−	0.5951 (0.2830) [4]−	**0.9866 (0.0511)** [1]
MMF-13	0.5081 (0.0882) [6]−	0.6111 (0.0613) [5]−	**0.9618 (0.0883)** [1]+	0.8055 (0.0775) [4]−	0.8305 (0.0798) [3]−	0.9189 (0.0613) [2]
Mean rank	5.8	3.9	3.4	3.4	2.8	1.7
+/−/≈	0/10/0	0/10/0	2/7/1	1/8/1	4/6/0	—

Table 10. Statistics (mean (std. dev.)) of Epsilon Indicator metric values of the final populations obtained by MOLAPO, GS, MOPSO, NSGA-II, MNMA, and MOMmCAA over 30 independent runs on the CEC2020 test suite.

Fn	MOLAPO	GS	MOPSO	NGSGA-II	MNMA	MOMmCAA
MMF-1	0.9787 (0.0493) [3]−	0.9528 (0.0410) [4]−	0.9802 (0.0472) [2]≈	0.9349 (0.0475) [5]−	0.8351 (0.1070) [6]−	**0.9832 (0.0498)** [1]
MMF-2	0.9586 (0.0293) [3]−	0.9328 (0.0209) [4]−	0.9602 (0.0272) [2]≈	0.9149 (0.0275) [5]−	0.7652 (0.1236) [6]−	**0.9632 (0.0298)** [1]
MMF-4	0.6585 (0.0393) [6]−	0.8386 (0.0229) [3]−	0.6602 (0.0282) [5]≈	0.7149 (0.0295) [4]−	0.8402 (0.1036) [2]−	**0.8632 (0.0197)** [1]
MMF-5	0.8870 (0.1226) [4]−	0.8793 (0.1163) [5]−	0.8922 (0.1229) [3]−	0.8787 (0.1204) [6]−	**0.9858 (0.0285)** [1]≈	0.9851 (0.0124) [2]
MMF-7	0.8289 (0.1358) [6]−	0.8314 (0.1339) [4]−	0.8345 (0.1374) [3]−	0.8310 (0.1347) [5]−	**0.9988 (0.0466)** [1]+	0.9333 (0.0137) [2]
MMF-8	0.7739 (0.0043) [5]−	0.7900 (0.0039) [4]−	0.8999 (0.0002) [1]≈	0.8696 (0.0038) [3]−	0.3130 (0.0389) [6]−	0.8994 (0.0010) [2]
MMF-10	0.8611 (0.0041) [6]−	0.8918 (0.0081) [5]−	0.9192 (0.0113) [4]≈	0.9351 (0.0066) [2]−	0.9302 (0.0023) [3]−	**0.9416 (0.0031)** [1]
MMF-11	0.6951 (0.0164) [6]−	0.7959 (0.0217) [5]−	**0.9023 (0.0086)** [1]+	0.8365 (0.0043) [3]−	0.8052 (0.0065) [4]−	0.8974 (0.0019) [2]
MMF-12	0.7487 (0.2003) [6]−	0.9148 (0.0109) [5]−	0.9892 (0.0113) [2]≈	0.9280 (0.0905) [4]−	0.9529 (0.0356) [3]−	**0.9956 (0.0125)** [1]
MMF-13	0.9369 (0.0115) [5]−	0.8624 (0.1193) [6]−	**0.9965 (0.0101)** [1]+	0.9458 (0.0317) [4]−	0.9575 (0.0244) [3]−	0.9754 (0.0301) [2]
Mean rank	4.8	4.5	2.4	4.1	3.5	1.5
+/−/≈	0/10/0	0/10/0	2/6/2	0/10/0	1/8/1	—

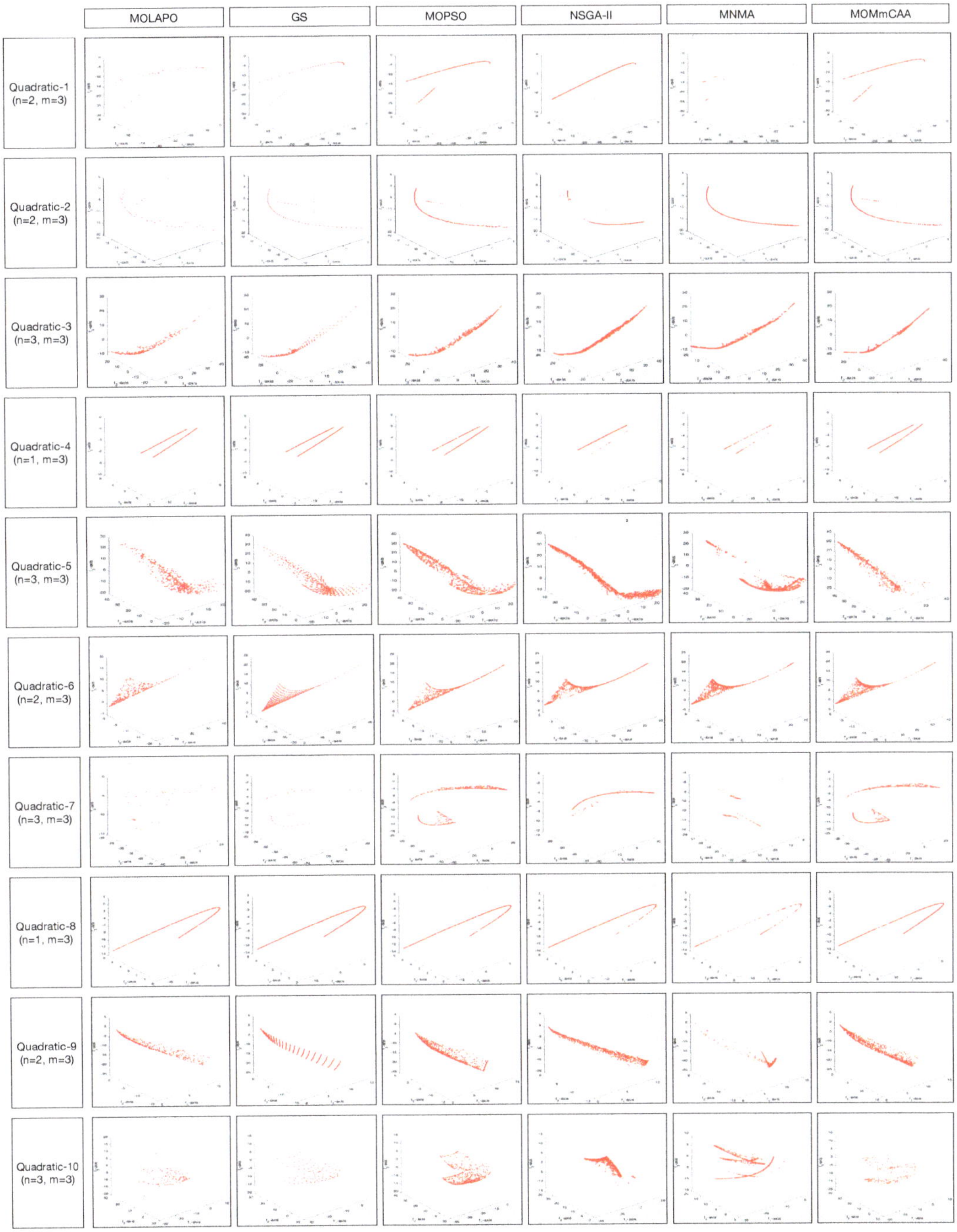

Figure 5. Representative PFs obtained by the six MOEAs on the Quadratic benchmark set.

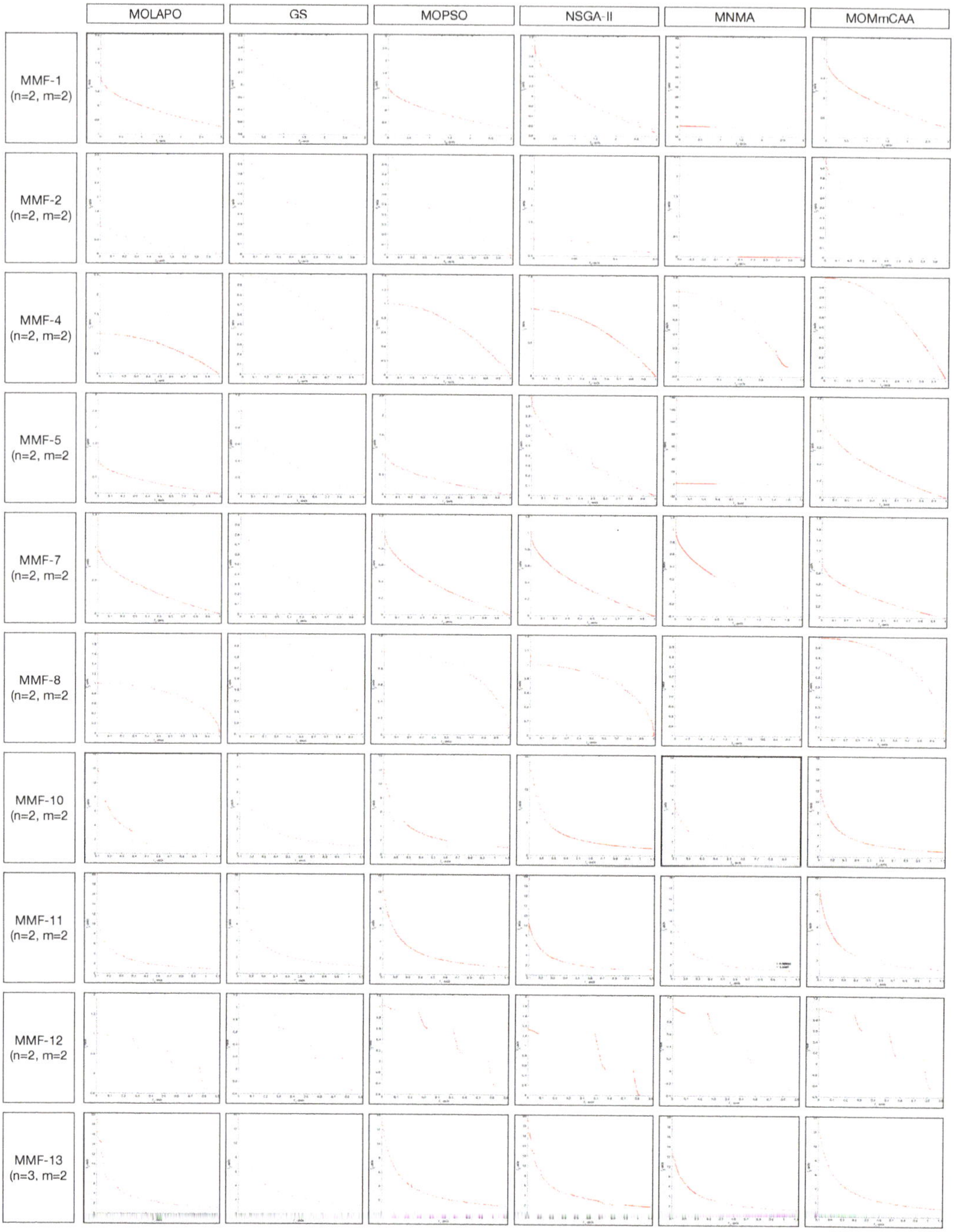

Figure 6. Representative PFs obtained by the six MOEAs on the CEC2020 benchmark set.

4. Engineering Design Problems

In this section, the capability of the MOMmCAA is evaluated in solving two real-world engineering design problems: design of a four-bar truss and design of a disk brake.

4.1. Four-Bar Truss Design Problem

In the four-bar truss design [58], the structural volume h_1 and displacement h_2 have to be minimized. This problem consists of four design variables z_1 to z_4 corresponding to the cross-sectional area of parts 1 to 4, as illustrated in Figure 7 The equations are provided below.

$$\begin{aligned}
&\text{Minimize:}\\
&h_1(z) = 200\left(2z_1 + \sqrt{2z_2} + \sqrt{z_3} + z_4\right)\\
&h_2(z) = 0.01\left(\frac{2}{z_1} + \frac{2\sqrt{2}}{z_2} - \frac{2\sqrt{2}}{z_3} + \frac{2}{z_4}\right)\\
&\text{where:}\\
&1 \leq z_1 \leq 3, 1.4142 \leq z_2 \leq 3\\
&1.4142 \leq z_3 \leq 3, 1 \leq z_4 \leq 3
\end{aligned} \tag{6}$$

This problem involves minimizing two components, h_1 and h_2. One approach is to combine both functions using $\boldsymbol{h}(z) = \lambda_1 h_1(z) + \lambda_2 h_2(z)$ or $\boldsymbol{h}(z) = h_1^{q_1}(z) + h_2^{q_2}(z)$, where λ_i and q_i are the weighting coefficients [59]. However, determining these coefficients relies on experience and a trial-and-error process to achieve the desired results. Therefore, the multi-objective approach considers both functions separately in order to approximate the PF and obtain multiple non-dominated solutions.

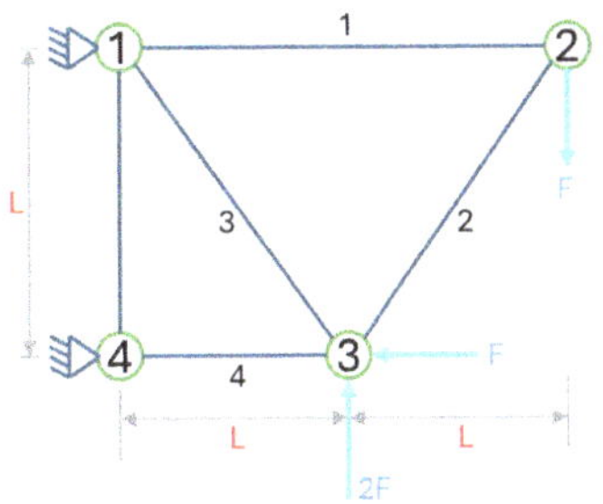

Figure 7. Description of the four-bar truss design problem.

4.2. Disk Brake Design Problem

The multi-objective disc brake design problem proposed in [60] has five constraints and two objectives to be minimized, namely, the stopping time h_1 and brake mass h_2. This problem has four design variables: the inner radius of the disc z_1, the outer radius z_2, the engaging force z_3, and the number of friction surfaces z_4. Figure 8 depicts the system and Equation (7) describes the problem.

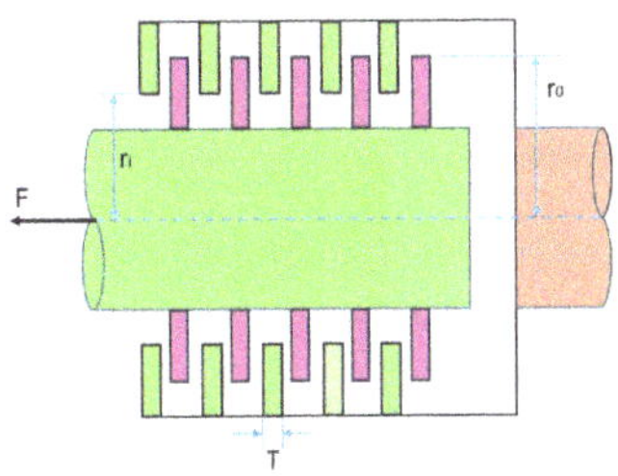

Figure 8. Description of the disk brake design problem.

$$
\begin{aligned}
&\text{Minimize:}\\
&h_1(z) = (4.9 * 10^{-5})(z_2^2 - z_1^2)(z_4 - 1)\\
&h_2(z) = (9.82 * 10^6)\frac{z_2^2 - z_1^2}{(z_2^3 - z_1^3)(z_4 z_3)}\\
&\text{Subject to:}\\
&g_1(z) = z_2 - z_1 - 20\\
&g_2(z) = 30 - (2.5(z_4 - 1))\\
&g_3(z) = 0.4 - \frac{z_3}{3.14 * (z_2^2 - z_1^2)}\\
&g_4(z) = 1 - (2.22 * 10^{-}3)\frac{z_3(z_2^3 - z_1^3)}{(z_2^2 - z_1^2)^2}\\
&g_5(z) = (2.66 * 10^{-2})\frac{z_3 z_4 (z_2^3 - z_1^3)}{(z_2^2 - z_1^2)} - 900\\
&\text{where:}\\
&55 \le z_1 \le 80, 75 \le z_2 \le 110\\
&1000 \le z_3 \le 3000, 2 \le z_4 \le 20
\end{aligned}
\tag{7}
$$

4.3. Design Problem Results

Table 11 presents the statistical results of the MOMmCAA and the other algorithms in dealing with the engineering design problems using the performance metrics of HV, C and EI. It can be seen that the MOMmCAA is one of the two best algorithms for these metrics in both cases, demonstrating the competitiveness of the proposed algorithm. The MOMmCAA is able to calculate better results than MOLAPO, GS, NSGA-II, and MNMA, and its results are very close to those obtained with MOPSO. Figure 9 presents the PFs obtained by the different algorithms for the two engineering design problems.

Table 11. Statistics (mean (std. dev.)) of all metric values of the final populations obtained by MOLAPO, GS, MOPSO, NSGA-II, MNMA, and MOMmCAA over 30 independent runs on the four-bar truss and disk brake design problems.

Four-Bar Truss	**MOLAPO**	**GS**	**MOPSO**	**NGSGA-II**	**MNMA**	**MOMmCAA**
HV	0.8459 (0.0123) [6]	0.8719 (0.0234) [5]	0.9598 (0.0041) [2]	0.9370 (0.0098) [3]	0.9182 (0.0377) [4]	**0.9678 (0.0026)** [1]
C	0.2666 (0.0146) [6]	0.3927 (0.0725) [5]	**0.9790 (0.0704)** [1]	0.7765 (0.3033) [3]	0.7477 (0.0203) [4]	0.9681 (0.0641) [2]
EI	0.5127 (0.0605) [6]	0.5824 (0.0274) [5]	**0.9204 (0.0389)** [1]	0.8913 (0.2059) [3]	0.8603 (0.2389) [4]	0.9184 (0.0270) [2]
Disk Brake	**MOLAPO**	**GS**	**MOPSO**	**NGSGA-II**	**MNMA**	**MOMmCAA**
HV	0.8780 (0.0180) [5]	0.8737 (0.0051) [6]	**0.9947 (0.0024)** [1]	0.9857 (0.0023) [4]	0.9923 (0.0027) [3]	0.9931 (0.0012) [2]
C	0.6498 (0.0882) [5]	0.5346 (0.1352) [6]	0.9128 (0.2202) [2]	0.8296 (0.2406) [3]	0.7847 (0.3748) [4]	**0.9128 (0.2202)** [1]
EI	0.7637 (0.0608) [5]	0.7329 (0.0252) [6]	0.9614 (0.0092) [2]	0.9216 (0.0325) [3]	0.8936 (0.0888) [4]	**0.9971 (0.0080)** [1]

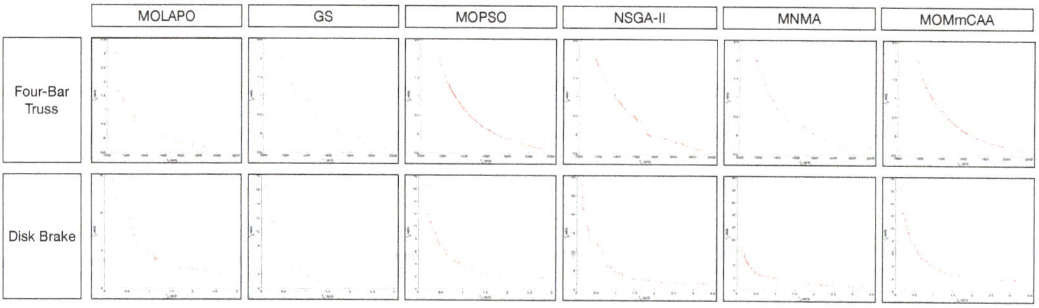

Figure 9. Representative PFs obtained by the six MOEAs on the two engineering design problems.

5. Conclusions and Further Work

This paper presents a new multi-objective optimization algorithm called the MOMmCAA inspired by the neighborhood and local interaction rules of majority and minority cellular automata. The randomness, concurrency, and information exchange generated between the smart-cells by applying different rules produce an appropriate balance between exploration and exploitation actions.

Comparative computational testing was carried out on 27 test functions with various characteristics, including convex, concave, mixed, disconnected, and degenerated PFs. These test functions were used to challenge the MOMmCAA, and its performance was compared against five other algorithms recognized for their efficiency. The experiments showed satisfactory performance on the part of the MOMmCAA.

In addition, two multi-objective engineering problems from the recent literature were used to test the MOMmCAA against the results obtained by the other algorithms. The MOMmCAA again demonstrated its high quality in finding solutions to these problems, proving its competitiveness against other recent metaheuristics.

Compared to classical techniques, the MOMmCAA provides improved flexibility. It can explore large search spaces and adapt to problems with multiple objectives and complex constraints. These features make the MOMmCAA especially useful for solving multi-objective optimization problems, where traditional methods may be inefficient due to assumptions about the problem's nature, the need for derivatives, or the complexity of the objective functions.

As further work, the MOMmCAA has to be proven effective in solving real-world problems such as power grid design, vehicle routing optimization, industrial systems control, and feature selection in bioinformatics. Its ability to balance multiple conflicting criteria makes it suitable for multi-objective situations.

However, the MOMmCAA has limitations in scalability for high-dimensional problems, where managing the repository of non-dominated solutions and correctly selecting the algorithm parameters are critical aspects affecting its performance. Its computational cost can also be high when dealing with complex problems, especially when requiring many iterations or accurate PF estimation.

These limitations provide opportunities for future algorithm refinement, including testing improvements with fewer parameters, dynamic parameter control, or other solution control mechanisms such as niche strategy, clustering, rank dominance, or FP maintenance methods. The richness of cellular automata behaviors also presents new opportunities for proposing new multi-objective optimization algorithms, such as the utilization of periodic, chaotic, universal, complex, or surjective and reversible cellular automata.

Author Contributions: Conceptualization, J.C.S.-T.-M. and J.M.-M.; methodology, J.C.S.-T.-M. and U.H.-H.; validation, J.C.S.-T.-M. and J.M.-M.; formal analysis, J.C.S.-T.-M., L.L.-M. and N.H.-R.; investigation, J.C.S.-T.-M., U.H.-H. and J.M.-M.; resources, N.H.-R. and L.L.-M.; writing—original draft preparation, J.C.S.-T.-M., U.H.-H. and J.M.-M.; writing—review and editing, N.H.-R. and L.L.-M.; visualization, J.C.S.-T.-M. and N.H.-R.; supervision, J.M.-M. and L.L.-M.; funding acquisition, J.C.S.-T.-M. All authors have read and agreed to the published version of the manuscript.

Funding: This study was supported by the Autonomous University of Hidalgo (UAEH) and the National Council for Humanities, Science and Technology (CONAHCYT) with project number F003-320109.

Institutional Review Board Statement: Not applicable.

Informed Consent Statement: Not applicable.

Data Availability Statement: The MOMmCAA source code is available on Github https://github.com/juanseck/MOMmCAA (accessed on 23 August 2024).

Conflicts of Interest: The authors declare that they have no known competing financial interest or personal relationships that could have appeared to influence the work reported in this paper.

References

1. Keeney, R.L.; Raiffa, H. *Decisions with Multiple Objectives: Preferences and Value Tradeoffs*; Wiley: New York, NY, USA, 1976.
2. Ehrgott, M. *Multicriteria Optimization*; Springer: Berlin/Heidelberg, Germany, 2005.
3. Charnes, A.; Cooper, W.W. Programming with Goals. *Eur. J. Oper. Res.* **1961**, *1*, 39–54. [CrossRef]
4. Steuer, R.E. *Multiple Criteria Optimization: Theory, Computation, and Application*; Wiley: New York, NY, USA, 1986.
5. Zhou, A.; Qu, B.Y.; Li, H.; Zhao, S.Z.; Suganthan, P.N.; Zhang, Q. Multiobjective evolutionary algorithms: A survey of the state of the art. *Swarm Evol. Comput.* **2011**, *1*, 32–49. [CrossRef]
6. Holland, J.H. Genetic algorithms. *Sci. Am.* **1992**, *267*, 66–73. [CrossRef]
7. Eberhart, R.; Kennedy, J. Particle swarm optimization. In Proceedings of the IEEE International Conference on Neural Networks, Perth, WA, Australia, 27 November–1 December 1995; Volume 4, pp. 1942–1948.
8. Dorigo, M.; Birattari, M.; Stutzle, T. Ant colony optimization. *IEEE Comput. Intell. Mag.* **2006**, *1*, 28–39. [CrossRef]
9. Van Laarhoven, P.J.; Aarts, E.H.; van Laarhoven, P.J.; Aarts, E.H. *Simulated Annealing*; Springer: Berlin/Heidelberg, Germany, 1987.
10. Deb, K.; Pratap, A.; Agarwal, S.; Meyarivan, T. A fast and elitist multiobjective genetic algorithm: NSGA-II. *IEEE Trans. Evol. Comput.* **2002**, *6*, 182–197. [CrossRef]
11. Zhang, Q.; Li, H. MOEA/D: A Multiobjective Evolutionary Algorithm Based on Decomposition. *IEEE Trans. Evol. Comput.* **2007**, *11*, 712–731. [CrossRef]
12. Zitzler, E.; Laumanns, M.; Thiele, L. SPEA2: Improving the strength Pareto evolutionary algorithm. *TIK Rep.* **2001**, *103*. Available online: https://neo.lcc.uma.es/emoo/zitzler01.ps.gz (accessed on 23 September 2024).
13. Coello, C.C.; Lechuga, M.S. MOPSO: A proposal for multiple objective particle swarm optimization. In Proceedings of the 2002 Congress on Evolutionary Computation, CEC'02 (Cat. No. 02TH8600), Honolulu, HI, USA, 12–17 May 2002; Volume 2, pp. 1051–1056.
14. Smith, K.I.; Everson, R.M.; Fieldsend, J.E. Dominance measures for multi-objective simulated annealing. In Proceedings of the 2004 Congress on Evolutionary Computation (IEEE Cat. No. 04TH8753), Portland, OR, USA, 19–23 June 2004; Volume 1, pp. 23–30.
15. Mirjalili, S.; Jangir, P.; Saremi, S. Multi-objective ant lion optimizer: A multi-objective optimization algorithm for solving engineering problems. *Appl. Intell.* **2017**, *46*, 79–95. [CrossRef]
16. Mirjalili, S.; Jangir, P.; Mirjalili, S.Z.; Saremi, S.; Trivedi, I.N. Optimization of problems with multiple objectives using the multi-verse optimization algorithm. *Knowl.-Based Syst.* **2017**, *134*, 50–71. [CrossRef]
17. Alaya, I.; Solnon, C.; Ghedira, K. Ant colony optimization for multi-objective optimization problems. In Proceedings of the 19th IEEE International Conference on Tools with Artificial Intelligence (ICTAI 2007), Patras, Greece, 29–31 October 2007; Volume 1, pp. 450–457.
18. Chen, Z.G.; Zhan, Z.H.; Lin, Y.; Gong, Y.J.; Gu, T.L.; Zhao, F.; Yuan, H.Q.; Chen, X.; Li, Q.; Zhang, J. Multiobjective cloud workflow scheduling: A multiple populations ant colony system approach. *IEEE Trans. Cybern.* **2018**, *49*, 2912–2926. [CrossRef]
19. Zhou, S.Z.; Zhan, Z.H.; Chen, Z.G.; Kwong, S.; Zhang, J. A multi-objective ant colony system algorithm for airline crew rostering problem with fairness and satisfaction. *IEEE Trans. Intell. Transp. Syst.* **2020**, *22*, 6784–6798. [CrossRef]
20. Zhang, X.; Zhan, Z.H.; Fang, W.; Qian, P.; Zhang, J. Multipopulation ant colony system with knowledge-based local searches for multiobjective supply chain configuration. *IEEE Trans. Evol. Comput.* **2021**, *26*, 512–526. [CrossRef]
21. Mirjalili, S.; Gandomi, A.H.; Mirjalili, S.Z.; Saremi, S.; Faris, H.; Mirjalili, S.M. Salp Swarm Algorithm: A bio-inspired optimizer for engineering design problems. *Adv. Eng. Softw.* **2017**, *114*, 163–191. [CrossRef]
22. Zhan, Z.H.; Li, J.; Cao, J.; Zhang, J.; Chung, H.S.H.; Shi, Y.H. Multiple populations for multiple objectives: A coevolutionary technique for solving multiobjective optimization problems. *IEEE Trans. Cybern.* **2013**, *43*, 445–463. [CrossRef]
23. Liu, S.C.; Chen, Z.G.; Zhan, Z.H.; Jeon, S.W.; Kwong, S.; Zhang, J. Many-objective job-shop scheduling: A multiple populations for multiple objectives-based genetic algorithm approach. *IEEE Trans. Cybern.* **2021**, *53*, 1460–1474. [CrossRef]
24. Tejani, G.G.; Kumar, S.; Gandomi, A.H. Multi-objective heat transfer search algorithm for truss optimization. *Eng. Comput.* **2021**, *37*, 641–662. [CrossRef]
25. Tejani, G.G.; Pholdee, N.; Bureerat, S.; Prayogo, D. Multiobjective adaptive symbiotic organisms search for truss optimization problems. *Knowl.-Based Syst.* **2018**, *161*, 398–414. [CrossRef]
26. Kumar, S.; Tejani, G.G.; Pholdee, N.; Bureerat, S.; Jangir, P. Multi-objective teaching-learning-based optimization for structure optimization. *Smart Sci.* **2022**, *10*, 56–67. [CrossRef]
27. Khodadadi, N.; Talatahari, S.; Dadras Eslamlou, A. MOTEO: A novel multi-objective thermal exchange optimization algorithm for engineering problems. *Soft Comput.* **2022**, *26*, 6659–6684. [CrossRef]
28. Kumar, S.; Jangir, P.; Tejani, G.G.; Premkumar, M.; Alhelou, H.H. MOPGO: A new physics-based multi-objective plasma generation optimizer for solving structural optimization problems. *IEEE Access* **2021**, *9*, 84982–85016. [CrossRef]
29. Khodadadi, N.; Azizi, M.; Talatahari, S.; Sareh, P. Multi-objective crystal structure algorithm (MOCryStAl): Introduction and performance evaluation. *IEEE Access* **2021**, *9*, 117795–117812. [CrossRef]

30. Nouri-Moghaddam, B.; Ghazanfari, M.; Fathian, M. A novel multi-objective forest optimization algorithm for wrapper feature selection. *Expert Syst. Appl.* **2021**, *175*, 114737. [CrossRef]
31. Yüzgeç, U.; Kusoglu, M. Multi-objective harris hawks optimizer for multiobjective optimization problems. *BSEU J. Eng. Res. Technol.* **2020**, *1*, 31–41.
32. Abdel-Basset, M.; Mohamed, R.; Mirjalili, S.; Chakrabortty, R.K.; Ryan, M. An efficient marine predators algorithm for solving multi-objective optimization problems: Analysis and validations. *IEEE Access* **2021**, *9*, 42817–42844. [CrossRef]
33. Tawhid, M.A.; Savsani, V. Multi-objective sine-cosine algorithm (MO-SCA) for multi-objective engineering design problems. *Neural Comput. Appl.* **2019**, *31*, 915–929. [CrossRef]
34. Azizi, M.; Talatahari, S.; Khodadadi, N.; Sareh, P. Multiobjective atomic orbital search (MOAOS) for global and engineering design optimization. *IEEE Access* **2022**, *10*, 67727–67746. [CrossRef]
35. Sabarinath, P.; Thansekhar, M.; Saravanan, R. Multiobjective optimization method based on adaptive parameter harmony search algorithm. *J. Appl. Math.* **2015**, *2015*, 165601. [CrossRef]
36. Got, A.; Moussaoui, A.; Zouache, D. A guided population archive whale optimization algorithm for solving multiobjective optimization problems. *Expert Syst. Appl.* **2020**, *141*, 112972. [CrossRef]
37. Pathak, S.; Mani, A.; Sharma, M.; Chatterjee, A. Decomposition Based Quantum Inspired Salp Swarm Algorithm for Multiobjective Optimization. *IEEE Access* **2022**, *10*, 105421–105436. [CrossRef]
38. Shi, Y.; Liu, H.; Gao, L.; Zhang, G. Cellular particle swarm optimization. *Inf. Sci.* **2011**, *181*, 4460–4493. [CrossRef]
39. Lopes, R.A.; de Freitas, A.R. Island-cellular model differential evolution for large-scale global optimization. In Proceedings of the Genetic and Evolutionary Computation Conference Companion, Berlin, Germany, 15–19 July 2017; pp. 1841–1848.
40. Seck-Tuoh-Mora, J.C.; Hernandez-Romero, N.; Lagos-Eulogio, P.; Medina-Marin, J.; Zuñiga-Peña, N.S. A continuous-state cellular automata algorithm for global optimization. *Expert Syst. Appl.* **2021**, *177*, 114930. [CrossRef]
41. Vafashoar, R.; Morshedlou, H.; Rezvanian, A.; Meybodi, M.R.; Vafashoar, R.; Morshedlou, H.; Rezvanian, A.; Meybodi, M.R. Applications of cellular learning automata and reinforcement learning in global optimization. In *Cellular Learning Automata: Theory and Applications*; Springer: Cham, Switzerland, 2021; pp. 157–224.
42. Seck-Tuoh-Mora, J.C.; Lopez-Arias, O.; Hernandez-Romero, N.; Martínez, G.J.; Volpi-Leon, V. A New Algorithm Inspired on Reversible Elementary Cellular Automata for Global Optimization. *IEEE Access* **2022**, *10*, 112211–112229. [CrossRef]
43. Zhu, G.; Ma, L. Cellular ant algorithm for multi-objective function optimization. *Control Decis.* **2007**, *22*, 1317.
44. Sidiropoulos, E. Multi-objective cellular automata optimization. In Proceedings of the Cellular Automata: 10th International Conference on Cellular Automata for Research and Industry, ACRI 2012, Santorini Island, Greece, 24–27 September 2012; Proceedings 10; Springer: Berlin/Heidelberg, Germany, 2012; pp. 131–140.
45. Zhu, D.; Zhan, T.; Zhang, Y.; Tian, H. Algorithm and application of cellular multi-objective particle swarm optimization. *Trans. Chin. Soc. Agric. Mach.* **2013**, *44*, 280–287.
46. Birashk, A.; Kordestani, J.K.; Meybodi, M.R. Cellular teaching-learning-based optimization approach for dynamic multi-objective problems. *Knowl.-Based Syst.* **2018**, *141*, 148–177. [CrossRef]
47. Seck-Tuoh-Mora, J.C.; Hernandez-Romero, N.; Santander-Baños, F.; Volpi-Leon, V.; Medina-Marin, J.; Lagos-Eulogio, P. A majority–minority cellular automata algorithm for global optimization. *Expert Syst. Appl.* **2022**, *203*, 117379. [CrossRef]
48. Nematollahi, A.F.; Rahiminejad, A.; Vahidi, B. A novel multi-objective optimization algorithm based on Lightning Attachment Procedure Optimization algorithm. *Appl. Soft Comput.* **2019**, *75*, 404–427. [CrossRef]
49. Kokkolaras, M.; Audet, C.; Hare, W. *Derivative-Free and Blackbox Optimization*; Springer Series in Operations Research and Financial Engineering; Springer: Berlin/Heidelberg, Germany, 2017; p. 302. [CrossRef]
50. Nadeau, P.C. Multiobjective Nelder-Mead Algorithm Using a Mesh-Map of Weighted Sums. Ph.D. Thesis, University of British Columbia, Vancouver, BC, USA, 2020.
51. Zenil, H.; Martinez, G.J. Cellular automata. *Scholarpedia* **2024**, *19*, 53227. [CrossRef]
52. Clarke, K.C. Cellular automata and agent-based models. In *Handbook of Regional Science*; Springer: Berlin/Heidelberg, Germany, 2021; pp. 1751–1766.
53. Knowles, J.D.; Corne, D.W. Approximating the nondominated front using the Pareto archived evolution strategy. *Evol. Comput.* **2000**, *8*, 149–172. [CrossRef]
54. Coello, C.A.C.; Pulido, G.T.; Lechuga, M.S. Handling multiple objectives with particle swarm optimization. *IEEE Trans. Evol. Comput.* **2004**, *8*, 256–279. [CrossRef]
55. Ulrich, T.; Bader, J.; Zitzler, E. Integrating decision space diversity into hypervolume-based multiobjective search. In Proceedings of the 12th Annual Conference on Genetic and Evolutionary Computation, Portland, OR, USA, 7–11 July 2010; pp. 455–462.
56. Bandyopadhyay, S.; Pal, S.K.; Aruna, B. Multiobjective GAs, quantitative indices, and pattern classification. *IEEE Trans. Syst. Man Cybern. Part B* **2004**, *34*, 2088–2099. [CrossRef] [PubMed]
57. Zitzler, E.; Thiele, L.; Laumanns, M.; Fonseca, C.M.; Da Fonseca, V.G. Performance assessment of multiobjective optimizers: An analysis and review. *IEEE Trans. Evol. Comput.* **2003**, *7*, 117–132. [CrossRef]
58. Cheng, F.; Li, X. Generalized center method for multiobjective engineering optimization. *Eng. Optim.* **1999**, *31*, 641–661. [CrossRef]

59. Nowak, L.; Knypiński, Ł.; Jedryczka, C.; Kowalski, K. Decomposition of the compromise objective function in the permanent magnet synchronous motor optimization. *COMPEL Int. J. Comput. Math. Electr. Electron. Eng.* **2015**, *34*, 496–504. [CrossRef]
60. Ray, T.; Liew, K.M. Society and civilization: An optimization algorithm based on the simulation of social behavior. *IEEE Trans. Evol. Comput.* **2003**, *7*, 386–396. [CrossRef]

MDPI AG
Grosspeteranlage 5
4052 Basel
Switzerland
Tel.: +41 61 683 77 34

Algorithms Editorial Office
E-mail: algorithms@mdpi.com
www.mdpi.com/journal/algorithms

www.ingramcontent.com/pod-product-compliance
Lightning Source LLC
Chambersburg PA
CBHW040829250926
51850CB00007BA/1299
9783725850754